Advanced Structured Materials

Volume 214

Series Editors

Andreas Öchsner, Faculty of Mechanical and Systems Engineering, Esslingen University of Applied Sciences, Esslingen, Germany

Lucas F. M. da Silva, Department of Mechanical Engineering, Faculty of Engineering, University of Porto, Porto, Portugal

Holm Altenbach, Faculty of Mechanical Engineering, Otto von Guericke University Magdeburg, Magdeburg, Sachsen-Anhalt, Germany

Common engineering materials are reaching their limits in many applications, and new developments are required to meet the increasing demands on engineering materials. The performance of materials can be improved by combining different materials to achieve better properties than with a single constituent, or by shaping the material or constituents into a specific structure. The interaction between material and structure can occur at different length scales, such as the micro, meso, or macro scale, and offers potential applications in very different fields.

This book series addresses the fundamental relationships between materials and their structure on overall properties (e.g., mechanical, thermal, chemical, electrical, or magnetic properties, etc.). Experimental data and procedures are presented, as well as methods for modeling structures and materials using numerical and analytical approaches. In addition, the series shows how these materials engineering and design processes are implemented and how new technologies can be used to optimize materials and processes.

Advanced Structured Materials is indexed in Google Scholar and Scopus.

Alexander D. Pogrebnjak · Yang Bing ·
Martin Sahul
Editors

Nanocomposite and Nanocrystalline Materials and Coatings

Microstructure, Properties and Applications

Springer

Editors
Alexander D. Pogrebnjak
Department of Nanoelectronics
Sumy State University
Sumy, Ukraine

Yang Bing
School of Power and Mechanical Engineering
Wuhan University
Wuhan, Hubei, China

Martin Sahul
Institute of Materials Science
Slovak University of Technology
Trnava, Slovakia

ISSN 1869-8433 ISSN 1869-8441 (electronic)
Advanced Structured Materials
ISBN 978-981-97-2669-1 ISBN 978-981-97-2667-7 (eBook)
https://doi.org/10.1007/978-981-97-2667-7

This Springer imprint is published by the registered company Springer Nature Singapore Pte Ltd.
The registered company address is: 152 Beach Road, #21-01/04 Gateway East, Singapore 189721, Singapore

Paper in this product is recyclable.

Preface

In contemporary industries, the widespread utilization of nanocomposites and other nanomaterials is evident. Notably, these materials have emerged as protective coatings and films, enhancing the performance characteristics of cutting tools, stamps, and various components susceptible to wear. Consequently, this field of solid-state physics and materials science holds considerable appeal for a diverse audience of researchers. Engaging scientists and engineers, it also captivates the interest of Ph.D. and master students in related specialties, such as micro- and nanoelectronics. This book encompasses a collection of chapters that meticulously outline the current state of various research areas. It explores topics such as diborides, their intricate structure, properties, and promising applications, the book delves into the forefront of materials science. Additionally, it delves into the highly sought-after topic of nitrides based on high-entropy alloys, dissecting their properties and structure. Furthermore, the book navigates the cutting-edge advancements in antiferromagnetic spintronic oscillators and sheds light on Terahertz signal detectors based on antiferromagnetic spintronic nanostructures. A series of original chapters dedicated to the deposition and exploration of protective coatings tailored for diverse applications are also included. Through these comprehensive explorations, this book can provide valuable insights for researchers, scholars, and students working in the ever-evolving field of materials science.

Sumy, Ukraine/Trnava, Slovak Republic — Alexander D. Pogrebnjak
Wuhan, China — Yang Bing
Trnava, Slovak Republic — Martin Sahul

Contents

Hard Transition-Metal Diboride Coatings 1
Marián Mikula

AlCrNbSiTiN High-Entropy Nitride Hard Coatings 37
Xiangyu Zhang, Yan Liu, Vasiliy Pelenovich, and Bing Yang

Antiferromagnetic Spintronic Oscillators: Fundamentals and Applications 91
Denys Slobodianiuk, Oleh Shtanko, and Oleksandr Prokopenko

Terahertz Signal Detectors Based on Antiferromagnetic Spintronic Nanostructures 129
Volodymyr Prokopenko and Oleksandr Prokopenko

Layer-by-Layer Chitosan/PCL Electrospun Membrane Loaded with Copper Nanoparticles as Antibacterial Wound Healing Dressing 149
Viktoriia Korniienko, Yuliia Varava, Rafal Banasiuk, Valeriia Korniienko, Kateryna Diedkova, Oksana Petricenko, Disha Arora, Anastasiia Denysenko, Roman Moskalenko, and Maksym Pogorielov

A Comparative Study of Microstructure and Properties of TiZrN/NbN and TiSiN/NbN Nanolaminate Coatings 163
Olga Maksakova, Vyacheslav Beresnev, Serhii Lytovchenko, and Diana Kaynts

The Development of Nanostructuring Method Metal Surfaces by Electrospark Alloying 181
O. Haponova, V. Tarelnyk, S. Marchenko, N. Tarelnyk, and I. Konoplianchenko

Phase Composition and Structure of Nanocrystalline Films of NiCu, NiCo, and NiFe Alloys 201
V. B. Loboda, S. M. Khursenko, and V. O. Kravchenko

Hard Transition-Metal Diboride Coatings

Marián Mikula

Abstract Transition-metal diborides (TMB_2, TM = Ti, Zr, Hf, V, Nb, Ta, Co, Mo, and W) form extremely temperature-stable single-phase compositions belonging to ultra-high temperature ceramics. The strong ionic and covalent bonds between atoms of transition metal and boron atoms in the hexagonal a-AlB_2 or w-W_2B_5 type of TMB_2 structure led to the high hardness, excellent wear resistance, and chemical inertness of these materials. Physical deposition methods enhance TMB_2's intrinsic character by forming hard coatings. Hard TMB_2 coatings typically have a nanocomposite structure consisting of nanocolumns with a diameter of a few nanometers and an amorphous, so-called B-tissue phase. These coatings are very hard, often more than 40 GPa. However, TMB_2 coatings still do not fulfill their application potential in demanding thermal–mechanical conditions due to their low oxidation resistance compared to the bulk equivalent, inherent brittleness, and poor crack propagation resistance. This chapter demonstrates a variety of approaches to tuning the mechanical properties and improving the weaknesses of these promising materials using both deposition methods (changing deposition parameters and ionizing sputtered species) and material design (changing stoichiometry, alloying, and multilayer architecture).

Keywords Transition-metal diborides · Hard coatings · Magnetron sputtering · Nanostructure · Mechanical properties · Alloying · Multilayers

---

M. Mikula (✉)
Fakulty of Mathematics, Physics, and Informatics, Comenius University in Bratislava, Mlynská Dolina F2, Bratislava, Slovak Republic
e-mail: marian.mikula@fmph.uniba.sk

Institute of Materials and Machine Mechanics, Slovak Academy of Sciences, Dúbravská Cesta 9/6319, Bratislava, Slovak Republic

A. D. Pogrebnjak et al. (eds.), *Nanocomposite and Nanocrystalline Materials and Coatings*, Advanced Structured Materials 214,
https://doi.org/10.1007/978-981-97-2667-7_1

1 Introduction

In general, hard coatings represent a group of materials that are attractive due to their mechanical properties, especially their high hardness and good resistance to abrasive and erosive wear, but also their low coefficient of friction under certain environmental conditions. These properties are also accompanied by excellent oxidation and corrosion resistance at elevated temperatures. Hard coatings, usually a few micrometers thick, have found their application on cutting plates of machining tools, on forming and casting molds, or on components (engine blocks, nozzles, turbine blades, etc.) in the automotive and aviation industries. In addition, hard coatings also serve as diffusion barriers in microelectronics or as decorative coatings.

Hard coatings based on transition-metal nitrides (TMN) have dominated the deposition industry for decades, where physical deposition techniques (PVD), especially ARC evaporation or magnetron sputtering, are used in their preparation.

The structure of TMN coatings is most often face-centered cubic single-phase (TiN, CrN) [1–4] or formed by a metastable multi-element solid solution (TiAlN, CrAlN, TaAlN, TiAlTaN) [5–14], which tends to spinodally decompose at high temperatures (700–1000 °C) into a stable multiphase structure. These decomposition processes are often accompanied by an increase in hardness—the so-called age hardening.

Another group of materials to which this chapter is devoted are transition-metal diborides (TMB_2), representing a promising alternative to hard TMN coatings in demanding high-temperature applications. In general, the combination of boron and metals represents a large family of materials with different crystallographic structures and physical properties, from hard orthorhombic monoborides (TiB, CrB, FeB) [15–18], through superconducting hexagonal MgB_2 [19], hard and superconducting tetragonal Mo_2B [20], tetragonal magnetic ReB_4 [21], hexagonal LaB_6 used in hot cathodes [22], to very hard and low-friction orthorhombic ternary so-called BAM systems with $AlMgB_{14}$ composition [23].

The hard coatings described here are based on diborides of IVB–VIB transition metals belonging to the group of so-called ultra-high temperature ceramics (UHTC). These materials crystallize in the hexagonal a-AlB_2 or w-$W_2B_{5\text{-}x}$-type of structure, where the atoms are bonded by strong covalent and ionic bonds, which leads to high-temperature chemical stability and excellent mechanical properties [24]. Unlike TM nitride coatings, TM-diboride coatings prepared by PVD methods usually have a nanocomposite character formed by TMB_2 nanocolumns surrounded by amorphous boron, the so-called B-tissue phase, and such a structure shows an extremely high hardness >40 GPa [25]. Unfortunately, the presence of the B-tissue phase causes significantly lower oxidation resistance of TMB_2 coatings at elevated temperatures compared to their bulk equivalents [26, 27]. In addition, TMB_2 coatings have a typical brittle ceramic character and low resistance to crack propagation [28]. These weaknesses significantly limit their application potential, and, although TMB_2 coatings achieve higher hardness than TMN coatings, they are currently unable to effectively alternate them.

In the following section, transition-metal diborides are described from the point of view of their basic properties, chemical bonds, band structure, and electronic structure. The influence of deposition effects on the formation of the coating nanostructure, stoichiometry, and their mechanical properties is discussed in the next section. The following is a discussion of the B-tissue phase and methods of its reduction through progressive physical deposition methods. Another section is devoted to the formation of ternary diboride systems with increased temperature stability and improved oxidation resistance. The last section is focused on the fracture behavior of TMB_2 coatings and promising ways to improve the fracture toughness by alloying with ductile metal or through the concept of superlattices and multilayer systems.

2 Transition-Metal Diborides

Diborides of transition metals from groups IVB to VIB of the periodic table are classified as ultra-high-temperature ceramics. In the transition-metal-boron equilibrium binary diagrams, TMB_2 are formed as single-phase compositions in a very narrow concentration ratio, which are characterized by high melting temperature values in the range of ~2200 °C (CrB_2) to 3400 °C (HfB_2). However, some TM diborides from the IIIB (YB_2) and VIIB (ReB_2) groups are also interesting in demanding thermal and mechanical conditions.

In general, TMB_2 crystallize in two different hexagonal structures. So-called early TMB_2 (e.g., TiB_2, ZrB_2, HfB_2, and NbB_2) prefer the α-AlB_2 structure type (space group 191-P6/mmm) where α-structure consists of a hexagonal-shaped unit cell with an alternating stacking of covalently bonded boron hexagons and metal layers. So-called late TMB_2 (e.g., MoB_2, WB_2) form the ω-W_2B_{5-x} (space group 194-P63/mmc) consisting of hexagonal-shaped unit cells with alternating flat and puckered boron hexagons between the metal layers [24], as can be seen in Fig. 1. However, as will be shown in the next chapters, thermodynamically non-equilibrium TMB_2 synthesis processes, e.g., into coatings, led to the formation of the α-AlB_2 structure across the diborides of all transition metals, even in the case of lower chemical stability of the a-structure compared to the w-structure. For this reason, we will demonstrate the electronic nature of bonds and their influence on stability, mechanical properties, and lattice parameters only on the α-AlB_2 type of structure.

A closer description of the α-AlB_2 structure (Fig. 1a) shows that this structure is quite tightly packed. Boron atoms lie on the corners of hexagons, with the three nearest neighboring boron atoms in each plane. The TM atoms lie directly in the center of each boron hexagon but midway between adjacent boron layers; each TM atom has 12 nearest neighboring B atoms and six nearest neighboring atoms in the TM plane. The in-plane distance between adjacent TM–TM atoms is equal to the lattice parameter *a*. The lattice parameter *c* represents the distance between the nearest planes occupied by TM atoms (Fig. 1a).

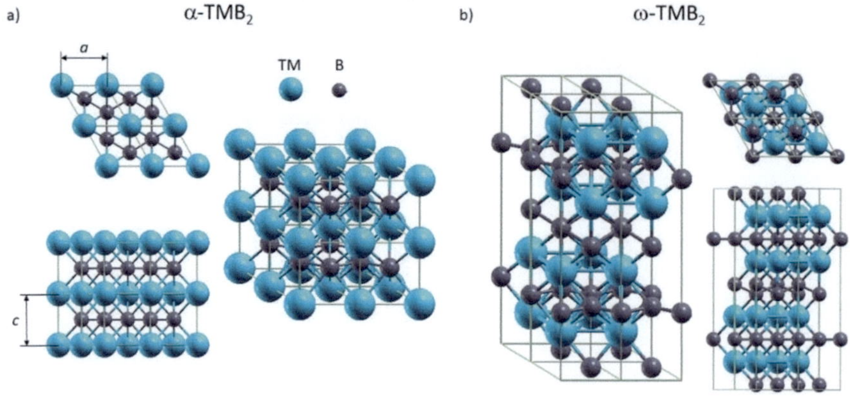

Fig. 1 Illustration of the supercell of **a** the a and **b** the w phases of transition-metal diborides

2.1 Density of States

The total electron density of states (TDOS) histogram provides a qualitative view of the chemical bonds between boron atoms and transition metals. Figure 2 shows the projected densities of states of selected systems representing diborides of elements from the IIIB to VIB groups. The vertical line indicates the Fermi level (E_F). The TDOS histograms of TMB_2 consist of several parts [29]: (a) a region present in the low-energy part of the curve (~ -14 to -8 eV), which is mainly formed by tightly bonded B-2s electrons; these electrons are localized and their effect on bonding is small; (b) a region with a maximum around -5 eV showing, at lower energies, bonding states that are mainly formed by overlapping B sp^2–B sp^2 and B-2p–B-2p orbitals. At higher energies approaching the Fermi level, this region ends with a sharp valley—the so-called pseudogap. To the left of the valley, the density of states is formed by mutual interactions of B-2p–TM-d orbitals as well as interactions of TM-d–TM-d and B-2p–B-2p orbitals [30]; (c) to the right of the valley, there is a peak showing high-energy bonding states on the TDOS curve, which are dominantly formed by the contribution from antibonding TM-d-states. This peak shifts toward lower energies due to an increase in the number of valence electrons from transition metals with three valence electrons (Y), through four valence electrons (Zr, Ti) to the group with five (Nb, Ta), and six valence electrons (Mo, W), respectively. This also changes the nature of interactions between atoms of transition metals and boron when the increasing number of valence electrons strengthens TM–B interactions and weakens TM-d–TM-d interactions.

This is an important fact that significantly determines not only the chemical stability of TM diborides but also their electrical conductivity and mechanical properties. As can be seen in YB_2 (Fig. 2a), the Fermi level is located below the peak of the bonding states, which indicates that the valence bands are partially filled and therefore the chemical stability of YB_2 is lower. In addition, the final states at the

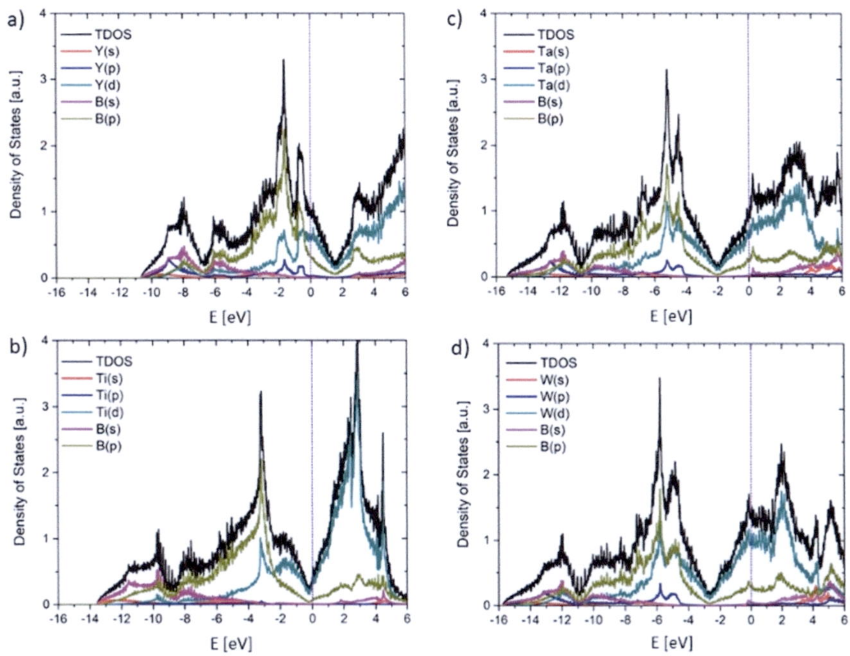

Fig. 2 Total electron density of states for **a** *3d* YB_2; **b** *4d* TiB_2; **c** *5d* TaB2; **d** *6d* WB_2

Fermi level formed by contributions from the Y-4d and B-2p states indicate metallic electrical conductivity in this ceramic material [30, 31].

A higher number of electrons—four in TiB_2 (Fig. 2b)—primarily shifts the TDOS to lower energies. At the same time, the Fermi level lies on the pseudogap, which means that the valence bands are fully filled and the chemical stability of TiB_2 is significantly higher compared to YB_2 (T_{mTiB2} ~ 3220 °C > T_{mYB2} ~ 2100 °C); on the contrary, the electrical conductivity is lower due to the lower number of states near the E_F. This situation changes in TaB2 and WB_2 with a higher number of valence electrons, where, according to Mott's law [32], the electrical conductivity increases due to the increasing number of states near E_F. On the other hand, as can be seen in Fig. 2c and d, if the number of valence electrons is 5 or 6, the Fermi level is above the bottom of the antibonding states that are occupied by the excess electrons. This leads to lower chemical stability and lower melting temperatures of these TMB_2 compounds (T_{mTaB2} ~ 3000 °C, T_{mWB2} ~ 2400 °C) compared to TM diborides from the IVB group.

The projected density of states also explains the mechanical behavior of TMB_2 from the perspective of chemical bonds. The presence of B2p–TM-d bonds and especially TM-d–TM-d bonds determines the bulk modulus (B) as well as the shear modulus (G) of these ceramic materials. Increasing values of B correlate with an increasing number of valence electrons. The highest bulk moduli have diborides VB and VIB of transition metals, where, unlike diborides IIIB and IVB, the presence of both B2p–TM-d bonds and TM-d–TM-d bonds contributes to increasing B values [33].

In the case of shear moduli, B2p–TM-d states play an important role, where their gradual filling with an increasing number of valence electrons from three to four leads to an increase in G. However, their subsequent saturation and occupation of antibonding d-states (in the transition metals with five and six valence electrons, respectively) lead to a decrease in shear modulus values [30].

The G/B ratio defined as Pugh's criterion for distinguishing ductile from brittle materials indicates that while TM (IIIB-VB)-B_2 exhibits brittle behavior, TM(VIB)-B_2 behaves ductile during mechanical loading [30].

In addition, the values of B and G determine Young's modulus E of TMB_2 and can also be used in the empirical model of Chen et al. [34] to estimate hardness. Young's modulus follows a similar trend to shear modulus, where TM (IVB-VB)-diborides (TiB_2, ZrB_2, HfB_2, NbB_2, or VB_2) have the highest E values exceeding 500 GPa, leading to poor thermal shock resistance. Conversely, TM (IIIB and VIB)-diborides exhibiting lower E values (YB_2 ~ 342 GPa, MoB_2 ~ 452 GPa, and WB_2 ~ 423 GPa) are more thermally resistant [30]. According to [34] TM diborides such as TiB_2 or ZrB_2 with high E values are extremely hard, with hardness exceeding 40 GPa. The decreasing trend of hardness is correlated with the decrease of Young's modulus, where the hardness of, e.g., TaB2 and WB_2 is 25 GPa and 15 GPa, respectively [30]. It is important to note that the hardness values calculated in this empirical model may not accurately reflect the true hardness of TMB_2. However, as will be shown in the following sections, this prediction of hardness, e.g., in the case of ZrB_2 or TiB_2, is in good agreement with superhard ZrB_2 [35] and TiB_2 [25, 36] coatings; in the case of CrB_2 [37] or WB_2 [38] the predicted hardness is rather close to their sintered bulk forms.

2.2 *Charge Density Maps*

Charge density maps show charge redistributions due to the formation of chemical bonds between atoms in transition-metal diborides. This can be illustrated for ZrB_2 [39]. Here, the electron charge density in the 0001 plane across the Zr atoms (Fig. 3a) shows the metallic Zr–Zr bonds, as demonstrated by the uniform distribution of the charge density between neighboring Zr atoms. On the contrary, strong electron localization is visible between neighboring B atoms in the graphic-like 0002 plane, and B–B bonds have an s-type covalent character formed by overlapping hybridized B sp^2–B sp^2 orbitals (Fig. 3b). However, strong electron localization is also found between neighboring B atoms in the (11–20) plane, where B–B bonds are p-type covalent in nature and are formed by overlapping $B2p_z$–$B2p_z$ orbitals. The Zr–B bonds shown on the (11–20) plane are ionic-covalent (Fig. 3c); here, the ionic character is given by charge transfer from Zr to B atoms, and lobes formed by weak interactions between Zr $4d(t_{2g})$ and B $2p_z$ orbitals indicate Zr–B covalent bonds.

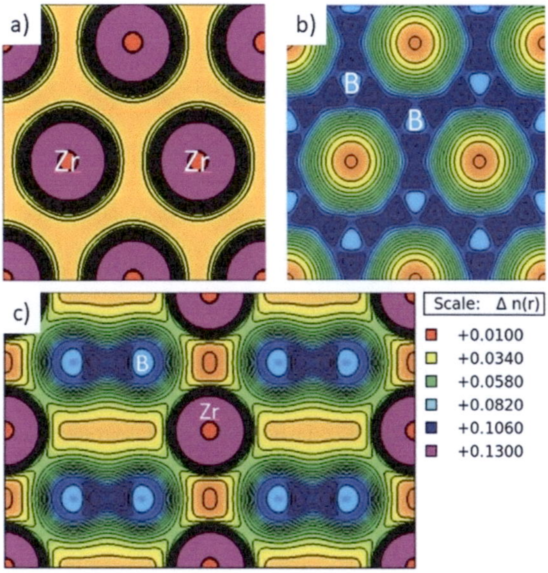

Fig. 3 Charge density maps for ZrB_2 in **a** the (0001) plane where the Zr–Zr bonding is present, **b** the (0002) plane where the B–B bonding is present, and **c** the (10–10) plane where the Zr–Zr, B–B, and Zr–B bondings are present

Changes in the lattice parameters are also related to the nature of the chemical bonds between atoms of transition metals and boron. In the hexagonal α-AlB_2-type structure, the lattice constant *c* is larger than *a* due to the chemical bond anisotropy since the B–B covalent s-type as well as the p-type bonds are stronger than the ionic-covalent TM–B bonds. In addition, both lattice constants *a* and *c*, respectively, decrease with increasing numbers of valence electrons in transition metals. This decrease is caused by the strengthening of the bonds by gradually filling the valence bands from partially filled (YB_2) to fully filled (ZrB_2). In the case of TMB_2 with a high number of valence electrons (CrB_2, WB_2), where the excess electrons fill the antibonding d-states, the decrease in the lattice constant *c* is due to the enhancement of TM–B bonds by TM d(t_{2g})–B $2p_z$ and TM d(t_{2g})–B sp^2.

3 Deposition of Transition-Metal Diboride Coatings

Looking at the history of the twentieth century, many technological approaches, based on chemical (CVD) or physical (PVD) vapor deposition, have been developed to prepare TMB_2 coatings [40]. CVD methods use toxic borane-based precursors that condense at very high temperatures (>900 °C) for the growth of TiB_2 [41, 42] and ZrB_2 coatings [43–45]. However, these extreme conditions preclude the use of technical materials, e.g., high-speed steels and alloys, and represent a burden on the environment. From the PVD methods, E-beam vaporization [46], pulsed laser deposition [47, 48], and reactive sputter process [49] proved to be suitable for the growth of very hard, crystalline, highly oriented, or epitaxial TiB_2 or ZrB_2 films on monocrystalline substrates such as sapphire, SiC, and silicon. However, again, these depositions are accompanied by high temperatures of 600–900 °C, and, for that reason, these PVD methods are more suitable for basic research on TMB_2 coatings than for producing coatings on an industrial scale.

Logically, one could expect that cathodic arc deposition, dominantly used for the industrial production of hard coatings, could also be suitable for the deposition of TMB_2 coatings. But early research on TiB_2 arc coatings showed problems: stabilizing the arc discharge and generating large amounts of macroparticles due to the high melting temperature of TiB_2 targets [50]. Further modifications of the cathode processes, from alloying the target with a small amount (a few wt.%) of aluminum or carbon [50] through arc pulsing [51] to controlling the movement of the arc-spot by an external magnetic field [52], led to higher stability of the processes and the suppression of macroparticle generation and subsequent growth of hard TiB_2 coatings on externally unheated steel substrates. Despite these positive results, the optimization processes have not yet been completed, and studies of the behavior of other arc TM-diboride cathodes are also lacking. For these reasons, the commercial production of TMB_2 coatings using cathodic arc deposition has not yet started.

Undoubtedly, the most widely used method for preparing TM-diboride coatings in lab-scale or industry-scale equipment, respectively, is direct current (DC) magnetron sputtering. Here, a sintered stoichiometric diboride compound target is sputtered using an inert (argon) glow discharge, and high-quality TMB_2 coatings can be prepared already at low deposition temperatures (on an externally unheated substrate).

However, there are several important aspects that affect the resulting chemical composition, structure, and physical properties of the coatings. The first aspect is the quality of the ceramic TMB_2 target. Commercial targets sintered at high temperatures approaching 2000 °C often reach only around 80% of the theoretical density [17]. Therefore, the targets contain hidden porosity filled with contaminants (water vapor, oxygen, carbon, etc.). Since these elements can significantly influence the nucleation and growth of TMB_2 coatings, it is necessary to first pre-clean the target before deposition and then apply sufficiently high target power to reach a high proportion of sputtered species compared to impurities. Tengdelius et al. [53] reported an oxygen content of $3 \div 26$ at.% in highly oriented and epitaxial ZrB_2 coatings at target powers of 400 W and 100 W, respectively, which was manifested by an increase in electrical resistivity. However, it must be said that epitaxial growth requires specific conditions, such as high temperatures and slow film growth. In general, the presence of impurities on the growth and structure of TMB_2 coatings is not so dramatic and usually does not exceed 3 at. % if the chamber is sufficiently evacuated ($<10^{-3}$ Pa) and high target power densities are used [25, 36, 54].

The second aspect is the different angular distribution of sputtered atoms from the compound target. According to sputtering theory, the angular distribution of atoms should follow the cosine law. However, the sputtering of TMB_2 targets leads to inhomogeneous emission in terms of the spatial and energy distribution of sputtered atoms. Subsequent transport through the plasma further boosts these effects, resulting in discrepancies between the stoichiometry of the target and the TMB_2

coating. Significant deviations in the composition of coatings are usually observed in combinations of constituents with significant weight differences. Experimental results by Olson et al. [55] indicate that boron atoms are more easily emitted from the target after only a few collisions (primary knock-in mode). Due to the energy obtained from the directly incident ions, they scatter along the normal of the target. Conversely, heavier metal atoms are emitted by a classical collision cascade (secondary knock-in mode) and therefore scattered in a wide angular distribution [55]. This is in good agreement with, e.g., TiB_2 and ZrB_2 coatings, respectively, where the B/Ti (Zr) ratio often exceeds the value of 2 and the coatings are overstoichiometric. However, for TM diborides such as NbB_2 or TaB2, the B/Nb (Ta) ratio is usually much lower than 2, and the coatings are understoichiometric [54, 56, 57]. This is due to the third aspect—the resputtering of light boron from the growing coating. Resputtering is mainly the result of backscattered neutrals (neutralized gaseous Ar ions) from the target. Ar neutrals backscattered from heavy targets (TaB2) retain very high kinetic energy, which they transfer to knocking out large amounts of boron atoms from the coating. Resputtering is also a part of the deposition of lighter TM diborides such as TiB_2. But in this case, the kinetic energy of the reflected Ar neutrals is significantly lower, and the coatings still retain their overstoichiometric character [56].

All the mentioned aspects can be influenced by a suitable combination of deposition parameters (bias, substrate temperature, substrate-target distance, ion flux, target thickness, and others). This leads to the growth of TMB_2 coatings with the desired chemical composition and structure, as will be demonstrated in the following sections.

3.1 Compositional and Structural Evolution in Early Transition-Metal Diboride Coatings

The TiB_2 coating is an ideal representative for demonstrating differences in chemical composition due to the different angular distribution of sputtered species from a stoichiometric diboride compound target. The first important factor for observing differences in B/Ti is the location of the substrates relative to the normal axis of the magnetron. In cases where the substrate is located close to the intersection with the normal axis, TiB_2 are characterized by a large excess of boron and are overstoichiometric TiB_{2+z}, where $+z$ represents the excess of B ($TiB_{2.4}$ [25], $TiB_{2.7}$ [36], superstoichiometric $TiB_{4.42}$ [28], $TiB_{4.5}$ [58]). The more the surface of the substrate moves horizontally away from the normal axis, the more the amount of boron in TiB_2 coatings decreases, and the ratio approaches the stoichiometric ratio of 2 (Fuger et al. [28] reported the interval 2.07 $\div$ 3.19).

In general, the structure of the sputtered TiB_2 coatings is unique, having a nanocomposite character (see Fig. 4), where crystalline hexagonal a-TiB_2 nanocolumns (nanofilaments) with a diameter in the range of typically 5–20 nm nanometers are surrounded by an amorphous so-called B-tissue phase [25]. The

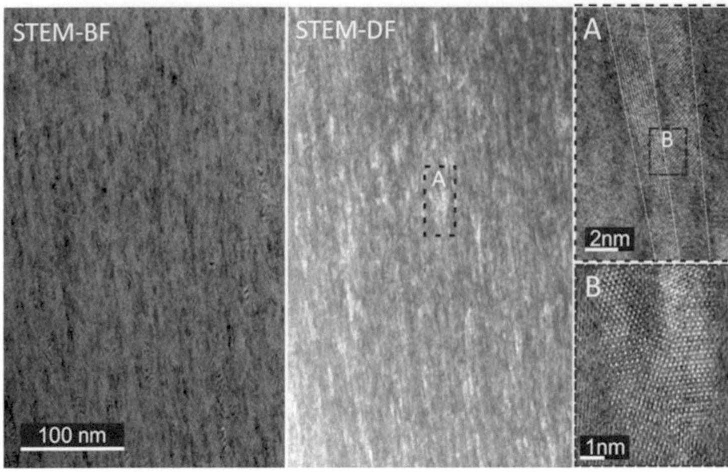

Fig. 4 Scanning transmission electron microscopy **a** bright-field (STEM-BF) and **b** dark-field (STEM-DF) cross-section image of as-deposited $TiB_{3.1}$ coating showing nanocolumnar character of the structure. **c** STEM-DF detail from region A showing interface between nanocolumns. **d** STEM-DF from region B showing detail of a very fine nanocolumns embedded in the disordered B-excess region

volume fraction of the tissue phase is given by the boron excess in the coating, and the increasing amount of the B-tissue phase suppresses the growth of crystalline TiB_2 nanocolumns (the diameter of the nanocolumns is close to ~3 ÷ 4 nm). However, the B/Ti ratio plays another important role from the point of view of structure, which is the preferential growth of crystalline planes of nanocolumns—texture. The preferential {0001} orientation is typical for slightly overstoichiometric TiB_2 coatings, but a change occurs at high boron contents when the {0001} texture is lost and the crystalline phase grows predominantly in the 10–11 direction [28, 36].

Significant differences in the stoichiometry of TiB_{2+z} coatings are reflected in the mechanical properties, especially in the hardness values. If the B/Ti ratio is in the range of about 2 to 3 and the {0001} texture dominates the coatings, the hardness of the coatings usually exceeds 40 GPa [25, 28, 36]. Here, the hardness is mainly achieved by the "ideal" size of the nanocolumns from the point of view of the Hall–Petch relation and the high cohesive strength between the very thin (thickness of a few atomic monolayers) B-tissue phase and the nanocolumns. The second reason for the superhardness is the intrinsic nature of the crystalline phase, given by strong covalent bonds. Finally, the high atomic density in the {0001} planes parallel to the substrate surface, acting very resistant to the diamond tip of the nanoindenter, also contributes significantly to the increase in hardness. In the case of superstoichiometric ratios B/Ti > 3, the hardness decreases significantly (even by tens of % [28]). This behavior is mainly due to excessive grain refinement (inverse Hall–Petch effect) and the loss of the 0001 preferential orientation. A large proportion of the soft B-tissue phase and its decreasing cohesive strength also have a non-negligible effect on the decrease in hardness.

Magnetron sputtering is a highly variable technique, and the composition of TiB_2 coatings can be influenced by tuning of the tuneable deposition parameters. However, several of them are interconnected, and a change in one parameter can have a different effect on the growth of the TiB_2 coating when another parameter is simultaneously changed. The first deposition parameter is the pressure of the working (Ar) atmosphere—plasma. An increase in the pressure from the order of tenths of Pa to 1–2 Pa leads to a decrease in the B/Ti ratio from usually highly overstoichiometric toward ~2 [28, 59]. Increasing the pressure increases the plasma density, and light boron atoms are more scattered due to more frequent collisions with heavy Ar ions. On the other hand, for a given constant pressure, B/Ti increases with increasing distance (larger deposition angle) between target and substrate. In this case, the plasma density decreases with increasing distance from the target, and the collision-induced effects on the scattering of B atoms are weaker. It should be noted that the ionization energies of Ti and B atoms are lower compared to those of Ar atoms, meaning that their ionization in the plasma is easier. However, the number of Ar^+ ions found in the magnetron glow discharge is significantly higher than Ti^+ and B^+, respectively, due to their low sputtering yield. Therefore, the influence of ionized target species on the composition of TiB_2 coatings is negligible. The change of both deposition parameters, pressure and distance, affects the structure of the coatings, when the growth of the TiB_2 coating at higher pressures is accompanied by the formation of the {0001} texture and these coatings are harder. The 0001 highly oriented structure in TiB_2 coatings also dominates during depositions, when the target-substrate distance or the deposition angle, respectively, corresponds to the achievement of the B/Ti ratio <~3 [28].

Applying a negative substrate bias proved to be very important for the chemical and structural changes of TiB_2 coatings. In general, the bombardment of a growing film essentially causes atomic "peening", where the sputtered atoms localized on the surface are hit by heavy Ar ions and forced into voids and interstitial lattice spaces. This process leads to the densification of the material, reduction of grain size, and the generation of compressive macrostresses in the coating [60].

In the published results of Mikula et al., it was shown that increasing the bias voltage increases the kinetic energy of the incident Ar ions on the growing coating. This leads to a decrease in the B/Ti ratio due to the resputtering of boron atoms. But the implantation of a small amount of Ar ions (1–3 at.%) into the lattice was also observed. The resulting structure of such TiB_2 is typically nanocomposite, where the proportion of the 0001-oriented TiB_2 phase increases with the increasing energy of Ar ions. In addition, the coatings contain a number of point defects, which are manifested by high values of compressive macrostresses (σ ~ several GPa). The {0001} texture in combination with large macrostrains leads to an increase in hardness well exceeding ~40 GPa (~55 GPa [25], ~69 GPa [61] and ~73 GPa in the extremely strained TiB_2 coating [62]).

It is possible to suppress the negative effect of bias—the incorporation of Ar ions—by increasing the temperature of the substrates during deposition. TiB_2 coatings are highly temperature-stable, and their chemical composition and nanostructure practically do not change even at substrate temperatures of 700 °C [61, 62]. Argon begins to be released from the growing coating after exceeding the deposition temperature of 300 °C. The crystal lattice essentially behaves as relaxed, and the compressive macrostress values σ do not exceed 1 GPa. However, almost stress-free TiB_2 coatings still show extremely high hardness values exceeding 40 GPa due to their nanocomposite nature [61, 62].

The above results pointed out one important fact: magnetron-sputtered TiB_2 coatings are always overstoichiometric (B/Ti > 2) regardless of the combination of deposition parameters, and the nanostructure of TiB_2 coatings always contains a B-tissue phase. As will be shown in Sect. 4, the presence of the B-tissue phase plays a key role in the oxidation resistance or fracture toughness of diboride coatings, and its absence is often desirable. Petrov et al. [63] reported a way to achieve B/Ti $\approx$ 2 even in magnetron-sputtered TiB_2 coatings by amplifying the axial magnetic field using Helmholtz coils. In this way, the Ar plasma density increased, and the ionization of Ti atoms also improved (the number of Ti^+ ions increased). Subsequently, energetic Ti^+ ions were incorporated into the growing film and compensated for the quantitative difference between boron and titanium atoms [63].

ZrB_2 coatings also belong to the group of early transition-metal diboride coatings, which have received more attention. Tengdelius et al. reported several summary results in which they demonstrated the growth and structure of sputtered ZrB_2 coatings under different temperature conditions, such as no heating substrates up to a deposition temperature of 550 °C [64] or epitaxial growth on highly oriented substrates at temperatures reaching 900 °C [35, 53]. In the first case, the chemical composition of ZrB_2 coatings grown on Si (001) substrates in an industry-scaled device is close to stoichiometry B/Zr $\approx$ 2, especially at lower deposition temperatures [64]. The tendency toward a slight understoichiometry of B/Zr ~ 1.84 is associated with an increase in temperature to 550 °C. Despite worse vacuum conditions (high vacuum, HV) with a basic pressure of 10^{-3} Pa, typical for an industrial system, the amount of C and O_2 contaminants does not exceed 2 at.%. The microstructure of ZrB_2 coatings has a typical nanocolumnar character. Small differences in stoichiometry affected the preferential orientation of the coatings; while stoichiometric ZrB_2 grows preferentially in the 0001 direction, Zr-rich ZrB_{2-z} coatings are randomly oriented [64]. Ultra-high vacuum (UHV) conditions accompanied by a high deposition temperature (900 °C) together with the use of single-crystal 4H-SiC (0001) and Si (111) substrates enable the epitaxial growth of ZrB_2 coatings in the 0001 direction [65]. Decreasing the deposition temperature to 550 °C and sputtering power, respectively, leads to a change in the preferential orientation of nanocolumnar grains from 0001 to 10–11, which is mainly caused by the increasing concentration of impurities [53]. A detailed STEM study of epitaxially grown stoichiometric ZrB_2 coatings on Al_2O_3 (0001) at 900 °C shows a nanocomposite structure formed by ZrB_2 nanocolumns with a diameter of ~20 nm separated by a semi-coherent interface with a small B-excess [35]. In this context, the structures of slightly overstoichiometric

TiB_{2+z} and stoichiometric ZrB_2 coatings are the same. The ultrathin B-tissue phase increases the cohesive strength in the boundary region, which usually has the lowest resistance to distortion during nanoindentation. The hardness of 0001 epitaxially grown ZrB_2 is about 45 GPa, similar to TiB_{2+z} coatings [35].

3.2 Compositional and Structural Evolution in Late-Transition-Metal Diboride Coatings

Late-transition-metal diboride coatings well demonstrate the resputtering effect of the growing films. This becomes dominant during the reflection of Ar neutrals from the heavy compound target, resulting in the growth of understoichiometric TMB_{2-z} coatings, where *-z* means boron deficiency. As will be shown, the amount and localization of boron vacancies significantly affect the formation of the nanostructure and its mechanical properties. From the group of TMs (VB and VIB) diborides, TaB2 coatings represent a very promising alternative to TiB_2 and ZrB_2 coatings since tantalum itself is very interesting due to its temperature stability, oxidation resistance, and toughness. Grančič et al. [58] reported TaB2 coatings prepared by magnetron sputtering under standard deposition conditions that were highly substoichiometric B/Ta ~ 1.2, and transmission electron microscopy (TEM) investigation revealed the amorphous nature of their structure. Only the maximum reduction of the target voltage (~370 V), when the glow discharge was still working, reduced the kinetic energy of Ar neutrals reflected from the TaB2 target, which resulted in the growth of very fine nanocrystalline TaB1.7, but without the typical non-columnar character.

An extensive view of the compositional and structural changes of TaB2 coatings due to the change in the kinetic energy of Ar neutrals was reported in 2020 in the work of Šroba et al. [56]. Here, the experiments were performed using High Target Utilisation Sputtering (HiTUS) technology, where it is possible to independently change the kinetic energy of argon ions accelerated toward the target (target voltage control) while maintaining the same amount (constant target current). In this way, it was possible to vary the target voltage values from an extremely low 157 V to a very high 840 V. As can be seen from the elemental analysis results (Fig. 5a), TaB2 coatings grow in a wide composition ratio B/Ta from highly overstoichiometric TaB3.84 (157 V) to highly understoichiometric TaB1.36 (840 V).

To the explanation of resputtering effect, the involvement of energetic reflected Ar neutrals in sputtering from TaB2 target using incident Ar^+ ions was analyzed by simulation stopping and range of ions in matter (SRIM) software. The simulations assumed that the energy of incident Ar^+ ions on the target (in electron volts) was equal to the target potential (in volts) [56]. The output of this simulation is the energy distribution of reflected Ar neutrals at the substrate position (Fig. 5b). The percentage of reflected Ar neutrals above the target is approximately 14% with respect to the number of impinging Ar^+ ions. It was clearly seen that the number of reflected Ar

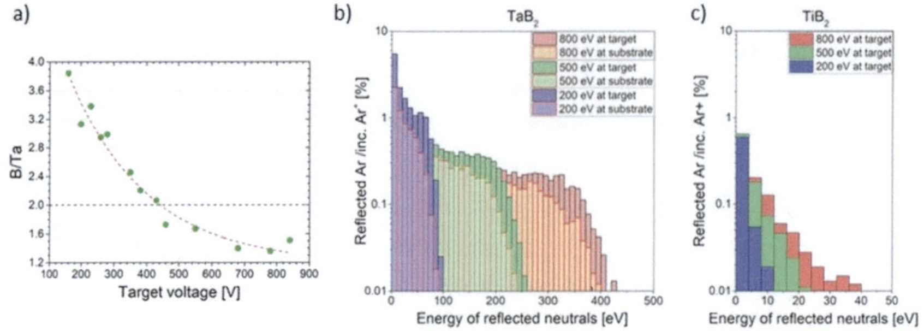

Fig. 5 **a** Boron-to-tantalum (B/Ta) ratio of films plotted as a function of target voltage. **b** Energy distribution of reflected Ar neutrals above TaB2 target position and at the substrate position (lighter color bars) for different energies of incident Ar^{+}. **c** Energy distribution of reflected Ar neutrals above the TiB_2 target [56]

neutrals at the substrate position is reduced compared to the target position (8%, 6%, and 4% for 200, 500, and 800 V, respectively); however, the shapes of energy distributions remain constant, and maximum values of energies are only slightly reduced, meaning that a significant number of highly energetic reflected Ar neutrals at substrate position can strongly affect the chemical composition of growing $TaB2_{\pm z}$ films, as was observed on magnetron sputtered $TaB2_{-z}$ films [58]. In comparison, the substantially smaller numbers, and values of energies of reflected neutrals were calculated for the well-known TiB_2 target (Fig. 5c). Here, the resputtering of boron is negligible typically leading to the formation of overstoichiometric TiB_{2+z} films.

The wide compositional range is also reflected in the character of the structure of the growing TaB2 films. Essentially, all TaB2 coatings prepared by HiTUS are crystalline, consisting of a hexagonal α-TaB2 phase that forms a {0001} texture. However, this disappears at high boron contents, and the coatings have a rather randomly oriented polycrystalline character. A significant difference occurs when looking at the nanostructure of the coatings, as seen in Fig. 6a. The TaB3.84 coatings are nanocomposite with a very fine nanocolumnar TaB2 phase embedded in the B matrix (Fig. 6a). As the amount of boron in the almost stoichiometric TaB2.06 decreases, the in-plane size of the nanocolumns increases, and the volume fraction of the B-tissue phase decreases significantly (Fig. 6b). In highly substoichiometric TaB1.36 coatings, the B-tissue phase is absent, and the nanostructure is formed by large hexagonal crystalline areas (Fig. 6c). These regions contain numerous boron vacancies and Ta-rich stacking faults; however, interestingly, they are still able to retain their crystalline form. A certain explanation of the structural stability of TaB2 coatings with a high boron deficit is offered by the density functional theory (DFT) calculation of formation energies and density of states (DOS). As mentioned in Sect. 2, in TaB2, the Fermi level is not lying on the pseudogap but above the bottom of antibonding states, indicating that the number of valence electrons five also leads to filling the antibonding states and to a weakening of the chemical stability of TaB2. The introduction of boron vacancies moves E_F toward the minimum of the

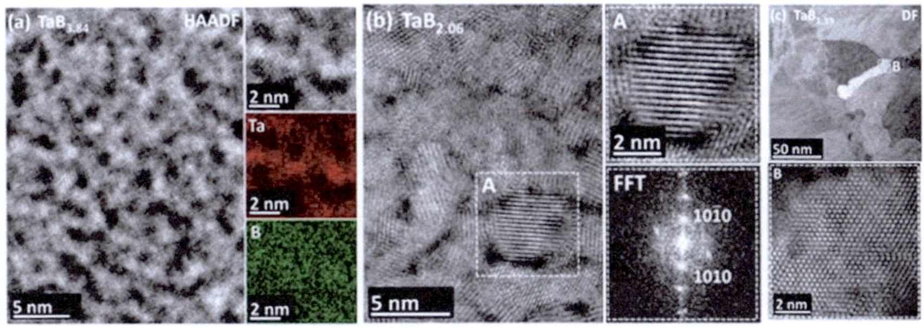

Fig. 6 **a** STEM-HAADF plane-view image of as-deposited highly overstoichiometric TaB3.84 film. The inset in (**a**) shows STEM-HAADF detail of a very fine-grained nanostructure with corresponding STEM-EDS maps of elemental distributions. **b** STEM-HAADF plane-view image of as-deposited TaB2.06 film showing the circle like outline of the nanocolumns (in the detail from region A is an individual nanocolumn with the size of a few nanometers) with the corresponding FFT pattern. **c** STEM-DF plane-view image of the as-deposited highly understoichiometric TaB1.39 film showing large nanocolumns formed by the α-TaB2 phase (atomic resolution image obtained from selected region B) [56]

pseudogap, which is an indication that TaB2$_{-z}$ is stabilized upon the formation of vacancies ($z > 1.625$) [56].

TaB2$_{\pm z}$ coatings are generally very hard, but hardness values vary depending on stoichiometry. First, the decisive factor determining the hardness is the volume fraction of the amorphous matrix, which is higher the lower the hardness value. In the case of TaB3.84 the hardness is 27 GPa. The highest hardness of 42 GPa achieved in TaB2.98 is the result of the optimal thickness of the B-tissue phase, where the cohesive force at the interface between the crystalline and amorphous phases is the highest. However, superhardness >40 GPa is preserved even in substoichiometric TaB2$_{-z}$ coatings. In these cases, the hardness is attributed to so-called vacancy-induced hardening. Namely, the presence of boron vacancies in the sublattice causes the strengthening of ion-covalent bonds between the neighboring boron and tantalum atoms, which may have a positive effect on hardness values [56].

Although HiTUS technology offers great possibilities in terms of the compositional and structural design of TaB2 coatings, it is not as widely used as conventional DC magnetron sputtering. Viskupová et al. [57] found the possibility of varying the B/Ta stoichiometry in a wider concentration range from 1.3 to 2, even during deposition by magnetron sputtering. Here, the decrease of target thickness from 6 to 4 mm and accompanied increase of permanent magnetic field above the target surface influences the plasma density and the target voltage (decrease from ~460 to ~260 V) corresponding to specific target current. This way the energy distribution of Ar neutrals shifts to smaller energies, which ultimately leads to the reduced resputtering of light boron from the coating and thus increased the B/Ta ratio toward 2.

An additional method that affects the B/Ta ratio is the application of an external magnetic field using Helmholtz coils. This causes an increase in the plasma density near the substrates and leads to the preferential ionization of Ta atoms and consequently to the reduction of the B/Ta ratio. Stoichiometric TaB2 was crystalline, containing the 0001-oriented α-TaB2 phase. The reduction of B/Ta from 2 to 1.8 was accompanied by a change in the preferred orientation from 0001 to 10–11. A further decrease of B/Ta toward 1.3 led to the formation of an almost X-ray amorphous structure. The hardness of the TaB2 coatings agreed with the structure, when the highest hardness of 48 GPa was shown by the almost stoichiometric, highly oriented TaB1.91 coatings. On the contrary, the hardness of highly substoichiometric, almost amorphous coatings reached about 31 GPa [57].

An important fact emerges from this study: since the energy of Ar neutrals is influenced by the thickness of the target, its gradual erosion changes the B/Ta ratio. This can affect the reproducibility of the coating properties during long-term use of the target. Therefore, the lifetime of the target must be estimated by regular inspection of test samples.

Several works also pay attention to other transition-metal (VB and VIB) diboride coatings, such as NbB_2, MoB_2, and WB_2, where their growth is accompanied by a significant resputtering effect [54, 66, 67]. As discussed in Sect. 2, the chemical stability of the α-type structure of late TM diborides is reduced due to the occupation of antibonding states by excess valence electrons. However, the formation of vacancies in the boron sublattice reduces the number of states near the Fermi level and strengthens the TM–B interactions, leading to a chemical stabilization of the α-type structure [68]. This fact is in good agreement with the following experimental results, where a slight deficit of boron in the coatings does not lead to the collapse of the hexagonal lattice.

Nedfors et al. [54] reported the structural evolution of understoichiometric $NbB_{1.8}$ coatings sputtered from a stoichiometric NbB_2 target. Despite the slight boron deficiency, the coatings were crystalline, formed by the α-NbB_2 phase, with a weak {0001} texture. A deeper look into the nanostructure revealed the presence of a self-organized nanocomposite formed by vacancies containing $NbB_{2\text{-}z}$ nanocolumns surrounded by a very thin B-tissue phase. As in the $TaB2_{-z}$ coatings, the $NbB_{2\text{-}z}$ coatings are superhard due to vacancy-induced hardening, when the hardness of the $NbB_{1.8}$ coating reaches 42 GPa.

Similar behavior was also observed in the case of $MoB_{2\text{-}z}$ coatings obtained by sputtering a stoichiometric MoB_2 target [66]. The understoichiometric $MoB_{1.6}$ coatings had a dense columnar character consisting of $MoB_{2\text{-}z}$ nanocolumns with a diameter of several nanometers surrounded by a B-tissue phase. Nanoindentation measurements revealed a relatively high hardness of 29 GPa.

The α-AlB_2-type structure stabilized by boron vacancies was also observed in sputtered $WB_{2\text{-}z}$ coatings [67]. The hardness of $WB_{2\text{-}z}$ coatings was significantly influenced by the preferential orientation of the nanocolumns. Purely 0001-oriented $WB_{1.87}$ coatings achieved a superhardness of 40 GPa; however, texture suppression and concurrent growth of nanocolumns in the 0001 as well as 10–11 directions occurred during the deposition of $WB_{1.55}$ coatings led to a significant decrease in hardness to 25 GPa.

The group of TM (VIB)-diborides also includes CrB_2, although in terms of atomic weight it is significantly lighter than MoB_2 and WB_2, respectively, and its atomic weight is closer to TiB_2. Thus, one would not expect a significant effect of resputtering but rather a similar growth as in the case of overstoichiometric TiB_{2+z} coatings. Dahlquist et al., in a theoretical view of the chemical stability of CrB_2, does not indicate the need for stabilization of the α-type structure by B-vacancies [68]. Experimental works devoted to magnetron-sputtered CrB_2 coatings report certain differences in the achieved stoichiometry [69, 70]. Zhang et al. [69] synthesized almost stoichiometric CrB_2 coatings (B/Cr in the interval from 1.88 to 2.06 measured by XPS). The decrease in stoichiometry was achieved by changing the temperature of the substrate from 100 to 400 °C. A higher deposition temperature led to the growth of a 0001-oriented nanocolumnar structure exhibiting a superhardness of up to 51 GPa. Conversely, randomly 0001 and 10–11 oriented structures were formed during low-temperature deposition, and the hardness of the coatings decreased to 30 GPa. Nedfors et al. [70] prepared understoichiometric $CrB_{1.5}$ coatings (XPS measurement of chemical composition) using DC magnetron sputtering. The differences from the work of Zhang et al. [69] were mainly in the larger target-substrate distance and significantly higher magnetron power. $CrB_{1.5}$ coatings consist of B-vacancy-containing nanocolumns embedded in a boron matrix. The formation of the {10–11} texture during the growth of the films resulted in a relatively low hardness of 25 GPa.

4 Amorphous Boron-Tissue Phase

In the previous Sect. 3, the evolution of the structure of TMB_2 coatings prepared mainly by conventional magnetron sputtering was described. The different angular distribution of the sputtered species from the compound target causes the B/TM ratio in the coatings to be overstoichiometric (>2) in many cases. On the other hand, due to the resputtering effect, the amount of boron is lower in some cases, which leads to the growth of substoichiometric coatings (B/TM < 2). But in all cases, the resulting structure is a nanocomposite formed by fine crystalline nanocolumns surrounded by a thin amorphous B-tissue phase and exhibits excellent mechanical properties.

However, the B-tissue phase also represents the "Achilles heel" of the TMB_2 coating. Namely, its presence determines the oxidation resistance of TMB_2 coatings (generally much lower compared to bulk equivalents) and the resistance of TMB_2 coatings to the formation and propagation of micro-cracks that form in this grain boundary region during mechanical loading.

For this reason, new possibilities for preparing TMB_2 coatings without forming the B-tissue phase have been intensively sought in recent years. One of the advanced approaches is the use of progressive physical deposition methods (PVD), which use highly efficient ionization of source materials (HiPPMS, high-power pulse magnetron sputtering). In general, HiPPMS is characterized by short (tens of microseconds) high-energy pulses with an energy density on the target of up to 10 kW cm^{-2} [71, 72]. This causes an increase in the plasma density to ~10^{19} ions per cm^3, when up to 80% ionization of the sputtered material occurs without thermal damage to the target (with conventional magnetron sputtering, the ionization of the sputtered atoms is less than 5%). This process is also accompanied by a strong gas rarefaction in front of the target, leading to a temporary separation of the gas-ion and target-ion flows, which allows additional control of the coating composition. Furthermore, the positively charged ions of the target species can be accelerated by a sufficiently high negative DC or pulsed bias voltage on the substrate. All this makes it possible to strongly influence the growth of the coating, densify its structure, introduce or eliminate the formation of defects, and thus control the magnitude of internal stresses.

Hellgren et al. [73] reported the growth of TiB_2 coatings by HiPPMS as a function of Ar pressure and substrate temperature. Based on the obtained results, they demonstrated that significantly understoichiometric $TiB_{2\text{-}z}$ coatings (B/Ti = 1.4–1.5) grow at low deposition pressures of 0.67 Pa in a wide temperature range from room temperature to 900 °C. They found an explanation in the comparison of the fluxes of ionized Ti and B atoms. In the HiPPMS discharge, Ti is ionized at a higher rate than B because its ionization potential (6.8 eV) is lower compared to B (8.3 eV), and titanium also has a higher ionization cross section. The metal-ion-rich plasma is further directed toward the substrate by steered magnetic field lines from the unbalanced magnetron, which increases the Ti deposition flux compared to the B flux. This effect is also present during dcMS but is significantly more pronounced for HiPPMS due to the higher degree of ionization [73]. Polyakov et al. [74] used a cylindrical rotating magnetron operating in HiPPMS mode to deposit understoichiometric $TiB_{2\text{-}z}$ coatings. The strongly {0001} textured $TiB_{1.62}$ coatings had a dense columnar structure without the presence of pores and voids and exhibited a superhardness of ~45 GPa. However, the authors did not perform a detailed analysis of the nanostructure to confirm or exclude the presence of the B-tissue phase. This was reported by Thörnberg et al. [75] on understoichiometric $TiB_{1.43}$ coatings synthesized using HiPPMS. Non-trivial plasma-deposition HiPPMS processes cause the formation of an understoichiometric structure that does not contain a B-tissue phase. Here, the structure is formed by hexagonal, tightly packed TiB_2 nanocolumns containing Ti–rich stacking faults, which serve to compensate for boron atom deficiency. In addition to the defective B-tissue phase-free structure, the $TiB_{1.43}$ coatings exhibited a strong {0001} texture, leading to a hardness of up to ~44 GPa. Palisaitis et al. [76] performed a detailed analysis of the nanostructure of the understoichiometric, epitaxially grown $TiB_{1.9}$ coating using scanning transmission electron microscopy (STEM) in combination with the electron energy-loss spectroscopy (EELS) technique. They found that unpaired Ti is accommodated in the structure through the formation of Ti stacking faults at the level of several atomic levels, which are localized in the prismatic {1–100} atomic planes of the TiB_2 crystal structure. The mechanism of formation of these defects is controlled by boron deficiency and the nature of the TiB_2 line compound. The structure around the defects is indeed stoichiometric TiB_2, while the locally increased concentration of Ti in planar defects results in an average stoichiometry of $TiB_{1.9}$ [76].

So far, less attention has been paid to the preparation of ZrB_2 coatings sputtered from a stoichiometric ZrB_2 compound target using HiPPMS technology. Samuelsson et al. [77] were the first to report the growth of slightly understoichiometric $ZrB_{2\text{-}z}$ coatings with a B/Zr ratio in the range of 1.82 to 1.98 as a function of pulse repetition frequency (300–900 Hz). All coatings are strongly {0001} textured, containing residual macrostresses. By tuning the HiPPMS process in terms of the degree of ionization in the deposition flow, it was possible to adjust the residual stress in the coatings in the range from tensile to compressive with increasing degrees of ionization [77]. However, the study did not include a deeper analysis of the nanostructure using transmission electron microscopy or measurements of mechanical properties. Mikula and Šroba in a more extensive unpublished study described the influence of steering HiPPMS deposition parameters (pulse width, frequency, and synchronized substrate bias) on the composition, structure, and mechanical properties of ZrB_2 coatings.

As-deposited ZrB_2 coatings were understoichiometric, with a wide B/Zr ratio in the range of 1.38–1.94. X-ray diffraction analysis confirmed the presence of a crystalline, 0001 preferentially oriented α-ZrB_2 phase in all coatings. A more detailed STEM analysis of the significantly understoichiometric $ZrB_{1.57}$ coating (Fig. 7a) revealed the presence of nanocolumnar, tightly packed ZrB_2 grains, which, however, are not segregated by the amorphous B-tissue phase but directly form mutual boundaries. In addition, even in this case, the deficit of boron atoms in the grains is compensated by the formation of stacking faults in the Zr sublattice (Fig. 7b). Increasing boron content in $ZrB_{1.9}$ coatings already causes the formation of the B-tissue phase, as seen in Fig. 7c. $ZrB_{2\text{-}z}$ coatings exhibit excellent mechanical properties with hardness values above ~34 GPa for understoichiometric $ZrB_{1.38}$, with an increasing tendency to a maximum value of ~46 GPa with an increasing B/Zr ratio to 1.94.

5 Alloying of Transition-Metal Diboride Coatings

As previously shown, DC magnetron sputtering from stoichiometric compound targets leads to the growth of TMB_2 coatings over a wide B/TM concentration range and the formation of a self-organized nanocomposite structure containing the B-tissue phase. Logically, the next step to modify the nanostructure, and suppress weaknesses is alloying TMB_2 coatings with other elements, especially metals. While the first part shows the effect of alloying on thermally induced decomposition processes accompanied by a change in mechanical properties, the next part describes the effect of alloying on the oxidation behavior of alloyed diboride coatings.

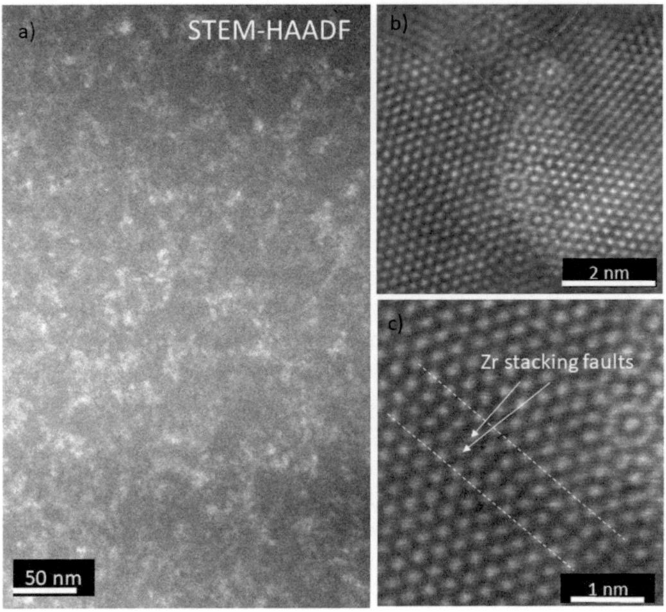

Fig. 7 **a** Plan-view STEM image acquired from the $ZrB_{1.57}$ coating grown with HiPPMS technology. **b** Atomic resolution STEM-DF image showing long-ordered lattice arrangement in a-ZrB_2 nanocolumns. The orange line indicates the boundaries between the nanograins without the presence of the B-tissue phase. **c** Detail acquired from the STEM-HAADF image reveals Zr-enriched stacking planar faults within nanograins

5.1 The Influence of Alloying on the Mechanical Properties of Transition-Metal Diboride Coatings

In this context, Alling et al. [78], in a pioneering theoretical study, proposed the concept of alloying binary diborides with another metal in order to calculate their temperature stability. They chose the volume misfit dV and the differences in the bulk moduli B of the MB_2 binaries as input parameters. The tendency to form stable solid solutions with the structure type hex-AlB_2 at 0 K and 1273 K (1000 °C) was predicted for several ternaries from 45 considered types of stoichiometric $M^1{}_{1-x}M^2{}_{1-x}B_2$ (M^i = Mg, Al, Sc, Y, Ti, Zr, Hf, V, Nb, Ta), e.g., $V_{0.5}Y_{0.5}B_2$, $Nb_{0.5}Y_{0.5}B_2$, $Ti_{0.5}Ta_{0.5}B_2$, $Zr_{0.5}Nb_{0.5}B_2$, and $Ta_{0.5}V_{0.5}B_2$. In other cases, the authors indicated the tendency of solid solutions to isostructural spinodal decomposition, which could be accompanied by the so-called age hardening. This group of diborides includes $Ti_{0.5}Al_{0.5}B_2$, which decomposes spinodally in the entire concentration range into binary TiB_2 and AlB_2 constituents at both considered temperatures of 0 K and 1273 K, respectively. In addition, similar behavior is also shown by other diborides alloyed with aluminum, e.g., $Zr_{0.5}Al_{0.5}B_2$, $V_{0.5}Al_{0.5}B_2$, $Hf_{0.5}Al_{0.5}B_2$, and yttrium $Ti_{0.5}Y_{0.5}B_2$, but also combinations of $V_{0.5}Nb_{0.5}B_2$ and $Ti_{0.5}Zr_{0.5}B_2$ [78]. In any case, this work has opened a new horizon for experimental approaches to confirm these first-principle predictions. However, it is important to note that crystalline DFT models are very simplified because they do not reflect the nanocomposite nature of real coatings, especially the presence of an amorphous tissue phase, but considering these aspects when creating structural models would lead to extreme computing time consumption.

Mockute et al. [79] presented the first experimental study of ternary $Ti_{1-x}Al_xB_{2+D}$ coatings prepared by DC magnetron co-deposition from tilted TiB_2 and AlB_2 targets. The overstoichiometric $Ti_{0.71}Al_{0.29}B_{3.08}$ coating contained 7 at.% aluminum, and its structure was formed by crystalline 0001-oriented a-$(TiAl)B_2$ nanocolumns embedded in an amorphous B matrix. This means that aluminum atoms substituted titanium atoms in the metal sublattice and formed a solid solution in the coating, contrary to ab initio predictions. But as can be seen in Fig. 8, phase separation occurs during annealing at a temperature of 1000 °C for 1 h in a vacuum when Ti-deficient and Al-enriched nanoregions begin to form in the interior of $(TiAl)B_2$ nanocolumns. Extending the annealing time to 12 h does not lead to further phase separation and in-plane enlargement of $(TiAl)B_2$ nanocolumns, which were prevented by excess boron segregation along the grain boundaries. Here, the thermally induced decomposition process was accompanied by an increase in hardness of 15% from 32 to 37 GPa, in agreement with Alling's calculations [79].

A different approach to the growth and tuning of the chemical composition of $TiAlB_2$ coatings was chosen by Nedfors et al. [80] by using a segmented TiB_2/AlB_2 target. The B stoichiometry of the $Ti_xAl_{1-x}B_{2+y}$ coatings varies from understoichiometric ($y = 1.70$) for the Al-rich coatings to overstoichiometric ($y = 2.92$) for the Ti–rich coatings. The aluminum content determined the character of the structure of the coatings. In Al-rich ($0.25 \leq x \leq 0.45$) coatings, a nanocomposite structure grows, which is formed by $(TiAl)B_2$ nanocolumns surrounded by excess aluminum.

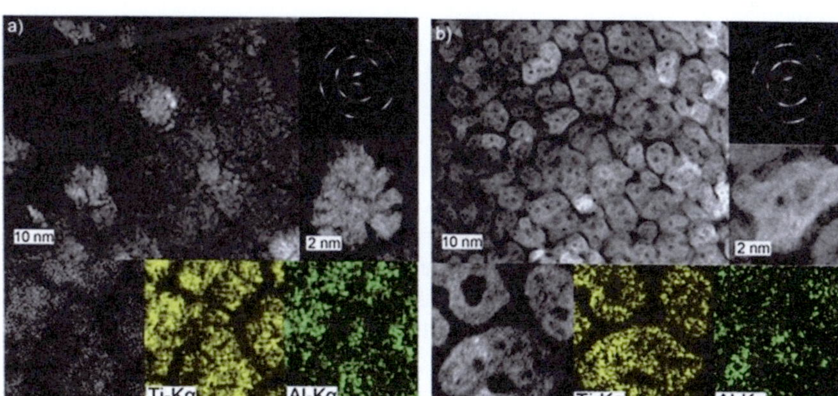

Fig. 8 Plan-view STEM-HAADF images and elemental maps with corresponding SAED patterns of **a** as-deposited $Ti_{0.71}Al_{0.29}B_{3.08}$ and **b** annealed coatings clearly demonstrate the different microstructures of the two states [79]

Excessive aluminum content ($x < 0.15$) led to the formation of an amorphous structure. In the case of Ti–rich ($x > 0.45$) coatings, the tissue phase is formed by boron atoms and some aluminum, and $(TiAl)B_2$ nanocolumns grow preferentially in the 0001 direction. Increasing the aluminum content has a negative effect on the hardness, which decreased from 39 GPa for the $Ti_{0.79}Al_{0.21}B_{2.70}$ coating to about 16 GPa for the amorphous coating. However, Nedfors does not report the high-temperature structural evolution of the coatings in the study, so the question of possible age hardening remains unanswered.

Bakhit et al. [81] used a combination of DC (TiB_2 target) and HiPPMS (Al target) magnetron power supplies for the deposition of $TiAlB_2$ coatings. The structure of understoichiometric 0001-textured $Ti_{0.68}Al_{0.32}B_{1.35}$ coatings is formed by hex. Ti–rich nanocolumns (solid solution) embedded in an Al-rich $Ti_{1-x}Al_xB_y$ tissue phase, which is B-deficient. The authors do not mention hardness values or thermally induced decomposition in the article.

Another method of preparation, but in an industrial facility, of $TiAlB_2$ coatings was published by Stüber et al. [82], which rotated the substrates between oppositely placed TiB_2 and Al targets. By controlling the power supplies, they varied the aluminum content in the coatings between 10 and 16 at.% and observed a decreasing boron content with increasing aluminum content. The structure was composed of a hex. $(Ti,Al)B_2$ solid solution in the form of nanocolumns that were separated by the B-tissue phase at lower aluminum contents ($Ti_{0.90}Al_{0.10}B_{2.22}$). The B-tissue boundary phase disappeared at higher aluminum contents ($Ti_{0.84}Al_{0.16}B_{2.00}$). The incorporation of aluminum into the growing coatings suppressed the preferential growth of nanocolumns in the 0001 direction. The hardness of $Ti_{0.90}Al_{0.10}B_{2.22}$ coatings shows approximately 34 GPa, but it slightly decreased at higher aluminum content to 31 GPa for $Ti_{0.84}Al_{0.16}B_{2.00}$ [82].

Fiantok et al. [39] presented the effect of Al alloying on structural evolution and mechanical properties in ternary $ZrAlB_2$ coatings. A $Zr_{0.72}Al_{0.28}B_{2.64}$ coating containing 8 at.% aluminum was grown using magnetron co-deposition. Unlike a similar system $Ti_{0.71}Al_{0.29}B_{3.08}$ published by Mockute et al. [79], this coating does not contain a solid solution, but already in the as-deposited state it has a nanocomposite character, exhibiting hexagonal α-ZrB_2 nanocolumns surrounded by an amorphous Al-rich tissue phase. This fact is in good agreement with the calculations indicating a strong tendency for phase separation of $Zr_{1-x}Al_xB_2$ solid solution into binary constituents over the entire concentration range, x [39]. Thermally induced processes lead to the coarsening of ZrB_2 nanocolumns and the coalescence of the Al-rich tissue phase into larger regions, which still retain their amorphous character even at high annealing temperatures. The hardness of the alloyed coatings is lower (23 GPa) compared to the reference $ZrB_{2.2}$ (39 GPa). The expected age hardening during annealing was not observed, although a slight increase in hardness to 26.5 GPa occurs at a temperature of 800 °C, and this value does not change up to a temperature of 1000 °C. In this case, it is the result of an increased concentration of point defects at grain boundaries and not phase decomposition. In addition, the authors revealed that aluminum should diffuse from the coating during high-temperature exposure when its reduced content of 3 at.% was observed in the Al-depleted porous structure after annealing at a temperature of 1100 °C.

A compositionally different Al-containing ternary $TaAlB_2$ system was investigated by Šroba et al. [83] combining ab initio calculations and experiments. Ab initio calculations predict a strong tendency toward the spinodal phase decomposition of the α-$TaAlB_2$ solid solution into isostructural binaries in the entire concentration range at $T = 0$ K. However, an increase in temperature ($T > 773$ K) leads to the stabilization of the α-$Ta_{1-x}Al_xB_2$ solid solution at $x < 0.5$ [83]. The structure of magnetron co-sputtered (TaB2 and AlB_2 targets) $Ta_{0.75}Al_{0.25}B_{2.14}$ and $Ta_{0.69}Al_{0.31}B_{2.16}$ coatings with aluminum concentrations of 8 and 12.5 at. %, respectively, has a nanocomposite character formed by a-TaB2 nanocolumns surrounded by an Al-rich tissue phase. This confirms the calculated prediction of the insolubility of aluminum in TaB2. High-temperature exposure in vacuum confirmed structural stability up to 1200 °C, but Al started to diffuse from the tissue regions at 1000 °C. The hardness of both alloy coatings is lower compared to the reference binary of 34, 28, and 27 GPa, respectively, and these values are maintained up to a temperature of 1200°, when an approximate 10% decrease occurs.

Mikula et al. [31] theoretically and experimentally investigated the thermally induced structural evolution in the yttrium alloyed ternary $Ti_{1-x}Y_xB_{2+D}$ system. DFT calculations indicated a strong driving force for spinodal phase separation into the basic binary constituents TiB_2 and YB_2 over the entire concentration range x. Magnetron sputtered $Ti_{0.78}Y_{0.22}B_{2.2}$ coating with yttrium content of 7 at. % has a nanocolumnar structure formed by a-$TiYB_2$ filaments embedded in a boron matrix with a negligible yttrium content, stable up to a temperature >1100 °C. During this exposure, the spinodal decomposition of the $TiYB_2$ solid solution occurs, and the simultaneous interdiffusion processes between the crystalline phase and the B-tissue phase led to the formation of a dual nanocrystalline TiB_2-YB_4 structure. However, age hardening is not observed; on the contrary, the formation of the soft YB_4 phase causes the hardness to decrease from 31 GPa in the as-deposited state to 28 GPa at 1200 °C.

Other works devoted to the preparation of ternary diboride coatings have also been added, e.g., Zr-Ta-B_2, Zr-Ta-B_2, W-Ta-B_2, and V-Mo-B_2. Bakhit et al. [84] grew a ternary $Zr_{1-x}Ta_xB_y$ ($0 \leq x \leq 0.3$) coating using hybrid Ta-HiPPMS/ZrB_2-DC magnetron sputtering with a substrate bias synchronized to the metal-rich portion of each HiPPMS pulse. Alloying $ZrB_{2.4}$ with tantalum leads to resputtering of boron from the growing film, which results in the growth of understoichiometric coatings ($0.1 < y < 0.3$). All coatings were formed by the hexagonal $Zr_{1-x}Ta_xB_y$ phase, where the addition of tantalum during the deposition promotes the preferential growth of the crystalline phase in the 0001 direction. STEM-EDS analysis revealed that the boundaries of the nanocolumns remain B-rich in the case of $Zr_{0.9}Ta_{0.1}B_{2.1}$, but if the content of tantalum increases, the column boundary tissue phase changes from B-rich to Ta-rich. Hardness increases slightly from 35 GPa for reference $ZrB_{2.4}$ to ~37 GPa for $Zr_{0.9}Ta_{0.1}B_{2.1}$, which the authors attribute primarily to solid-solution hardening. $Zr_{0.8}Ta_{0.2}B_{1.8}$ and $Zr_{0.7}Ta_{0.3}B_{1.5}$ have even higher hardness values, ~42

GPa, due to increased solid-solution hardening and decreasing width of nanocolumns in agreement with Hall–Petch behavior [84].

A similar concept was chosen when alloying the ZrB_2 coating with chromium using hybrid CrB_2-HiPPMS/ZrB_2 DC magnetron sputtering [85]. In $Zr_{1-x}Cr_xB_y$ coatings, the Cr/(Zr + Cr) ratio, x, increases from 0.13 to 0.9, while the B/(Zr + Cr) ratio, y, decreases from 2.92 to 1.81. The nanocolumn structure $Zr_{1-x}Cr_xB_y$ grows dominantly in the 0001 direction. The reference binaries $ZrB_{2.18}$ and $CrB_{1.81}$ have a porous columnar nanostructure. The formation of ternary systems leads to the growth of a dense nanostructure with point-defect clusters induced due to ion irradiation during film growth. However, the authors did not perform a detailed chemical analysis and distribution of elements in the nanostructure. The hardness of the coatings depends on the density of the nanostructure. In the case of underdense coatings ($x < 0.32$), the hardness varies between 27 and 29 GPa. In the case of dense coatings ($0.32 \leq x \leq 0.9$), the hardness is higher when the maximum value of ~32 GPa is shown by the $Zr_{0.68}Cr_{0.32}B_{2.80}$ coating.

Moraes et al. [86] reported the structure and temperature stability of the ternary W-Ta-B_2 system depending on the amount of tantalum. $W_{1-x}Ta_xB_{2-z}$ coatings crystallize in a metastable a-structure and form a solid solution up to a tantalum content of 26 at.%. The authors experimentally (with the support of DFT calculations) showed that point defects such as boron vacancies increase the stability of the metastable a-structure (a-WB_{2-z}) compared to the preferred stable ω-W_2B_{5-z} structure. However, it was further shown that tantalum also significantly contributes to the stabilization of the a-structure when the temperature of decomposition and phase transformation from a-type to stable w-type is shifted from 800 to 1000 °C (for WB_{2-z}) to 1200–1400 °C ($W_{0.74}Ta_{0.26}B_{2-z}$).

It has also been shown that thermally induced processes, which, according to predictions, should lead to the decomposition of ternary solid diboride solutions into basic constituents, can be accompanied by the formation of various planar defects. Palisaitis et al. [76, 87] were the first to report that slight understoichiometry in binary TM diborides can lead to the development of planar defects, such as stacking faults. It was shown (Fig. 9) on slightly understoichiometric $TiB_{1.9}$ that the unpaired Ti is included in the structure by the formation of B-deficient planar defects residing on {1–100} planes of the otherwise stoichiometric TiB_2 crystal structure [76]. Similar behavior was observed for $CrB_{2.1}$ and $CrB_{1.90}$, where planar defects were classified as anti-phase boundaries without (APB-1 type) and with (APB-2 type) local changes in stoichiometry [87]. Viskupová et al. [88] investigated in detail the high-temperature behavior and mechanical properties of sputtered ternary $V_{1-x}Mo_xB_{2-\Delta}$ coatings. Here, the authors showed, in the case of slightly understoichiometric $V_{0.58}Mo_{0.42}B_{1.9}$, the formation of APB-2 boundary defects accompanied by the chemical separation of V and Mo. The increased concentration of molybdenum at the anti-phase boundaries leads to the local formation of MoB regions in the hexagonal VB_2 structure. Annealing above 1000 °C of highly understoichiometric $V_{0.37}Mo_{0.63}B_{1.2}$ coatings causes the formation of a crystalline orthorhombic MoB phase with a large number of planar defects, which results in lamellar twinning. While, in the first case, the structural changes in the annealed $V_{0.58}Mo_{0.42}B_{1.9}$ coating did not affect the mechanical

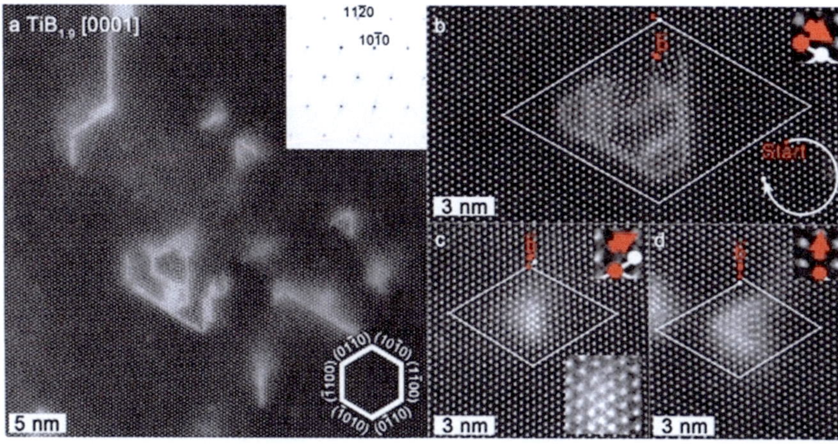

Fig. 9 Plan-view HAADF-HRSTEM images acquired from the $TiB_{1.9}$ thin film along the [0001] zone axis together with corresponding FFT pattern in the inset **a**. **b–d** the right-hand Burgers circuits drawn around the defects, Burgers vector indicated by an arrow [76]

properties, the formation of a crystalline phase with a high density of twin lamellae due to annealing in the $V_{0.37}Mo_{0.63}B_{1.2}$ coating was accompanied by an increase in hardness from 26.6 to 35.2 GPa [88].

5.2 *The Influence of Alloying on the Oxidation Behavior of Transition-Metal Diboride Coatings*

Diborides of transition metals belonging to ultra-high temperature ceramics declare high oxidation resistance in demanding thermo-mechanical applications. However, this fact can so far be attributed only to their bulk form. The oxidation behavior of TMB_2 in the form of coatings is determined by their nanocomposite nature, especially the presence of the B-tissue phase, which very easily interacts with the nearby atmosphere. The excess boron surrounding the crystalline grains reacts with the in-diffusing oxygen and, together with atmospheric moisture, creates volatile products in the form of B_2O_3 and H_3BO_3. These oxidation processes start at very low temperatures (TiB_2 coating at ~450 °C [27]) and lead to gradual oxidation of the coating and its destruction. This fact significantly limits the application potential of TMB_2 coatings, and they fall behind in competition with TM nitride coatings.

In this context, the development of TMB_2 coatings is focused on increasing their oxidation resistance in several ways: (i) suppressing the formation of the B-tissue phase; and (ii) alloying with oxide-forming elements, leading to the formation of protective surface oxide scales with very slow oxidation kinetics.

Dorri et al. [26] reported in a detailed study the oxidation behavior of overstoichiometric TiB_{2+z} coatings containing the B-tissue phase. Isothermally annealed $TiB_{2.5}$ coatings in humid air (humidity 40%) show linear non-protective oxidation kinetics in the temperature range of 300–700 °C, except for the temperature of 600 °C, where the formation of oxide scale and the slowing down of the oxidation rate indicate an oxidation-protective behavior. During oxidation up to a temperature of 500 °C, a B-deficient TiO_2 skeleton is formed in $TiB_{2.5}$, while boron reacts with a humid oxidizing atmosphere, forming B_2O_3. The latter subsequently hydrates into volatile H_mBO_n acid and evaporates from the coating. In the early stages of oxidation (lowest temperature, 300 °C), dense scales with a uniform distribution of Ti, B, and O are formed on the surface of the coating, which indicates the formation of a protective barrier. After exceeding 400 °C, B-rich grain boundaries begin to open as a result of the formation of volatile products. The resulting inter-columnar cracks create an easy path for oxygen to penetrate deeper into the non-oxidized structure. After increasing the temperature above 500 °C, crystallization of amorphous oxide scales occurs. This process is accompanied by a rapid increase in the oxidation rate due to the formation of large voids and cracks in the oxide scales. The subsequent significant reduction of the oxidation rate at a temperature of 600 °C is attributed to the formation of a highly dense TiO_2 sublayer near the oxide/film interface with a top amorphous thin layer. This protective oxide system can maintain the oxidation resistance of $TiB_{2.5}$ coatings up to a temperature of approximately 500 °C. After annealing at a temperature of 700 °C, the result of oxidation is an oxide layer with a rutile-TiO_2 structure, divided into two different regions. While the top layer consists of submicrometric crystallites similar to those observed at 600 °C, the inner layer is highly porous and formed by elongated grains. In addition, wide cracks also propagate through both layers.

Thörnberg et al. [89] compared the oxidation behavior of several sputtered $TiB_{2\pm z}$ coatings at a temperature of 400 °C, where the B-tissue phase was present or absent. In the case of overstoichiometric DC magnetron-sputtered TiB_{2+z} coatings, the B-tissue phase surrounding the crystalline TiB_2 nanofilaments accelerated the oxidation processes and created an easy path for the in-diffusion of oxygen and the formation of volatile boric acid. The decrease in tissue phase volume and lower B/Ti ratio in understoichiometric TiB_{2-z} coatings prepared by HiPPMS lead to a denser TiO_x oxide scale. The average oxidation rate in terms of consumed film thickness for $TiB_{1.43}$, $TiB_{2.20}$, and $TiB_{2.70}$ was determined to be 2.9 ± 1.5, 7.1 ± 1.0, and 20.0 ± 5.0 nm/h [89]. However, the authors do not describe the oxidation behavior of $TiB_{2\pm z}$ coatings at temperatures higher than 400 °C.

The oxidation resistance of a hard, understoichiometric $CrB_{1.5}$ coating was described by Glechner et al. [90]. XRD structural analysis of the coating revealed a typical hexagonal a-type structure; however, a deeper TEM view of the nanostructure that would reveal the presence of the B-tissue phase is not included. In any case, the oxidation resistance of the $CrB_{1.5}$ coating is higher compared to $TiB_{2\pm z}$, when the thermogravimetric analysis recorded an oxidation onset temperature of approximately 610 °C. The $CrB_{1.5}$ coating oxidizes in several steps, forming a protective Cr_2O_3 oxide scale, but after exceeding 1100 °C, volatile CrO_2 and B_2O_3 oxides are formed. Glechner et al. also performed thermogravimetric analyses of other binary

TMB_2 coatings crystallizing in the AlB_2-type structure, where the onset temperature of oxidation was $TiB_{2.72}$ (490 °C) and $WB_{1.88}$ (500 °C), followed by TaB1.53 (620 °C), and highest for $HfB_{2.36}$ (800 °C). Above this onset temperature, the mass increases until the coating is fully oxidized and then subsequently decreases, which is assumed to be due to volatilization of B_2O_3 (typically above 1100 °C) and WO_3 in the case of $WB_{1.88}$ coatings [90].

A very promising way to increase the oxidation resistance of TMB_2 coatings is to alloy them with silicon and form TM-Si-B_2 ternaries. Grančič et al. [27] in 2014 were the first to report the influence of silicon addition on the structure, mechanical properties, and oxidation behavior of Ti-Si-B_2 coatings prepared by DC magnetron co-deposition from TiB_2 and Si targets. The content of silicon varied from 5 to 20 at.%, and with its increasing amount, the amount of boron decreased toward understoichiometric coatings. In addition, the crystalline a-type structure changed to X-ray amorphous at high Si contents, which has a negative effect on the hardness, which decreased from 24 GPa (Si content 5 at.%) to 14 GPa (Si content 20 at.%). As can be seen in Fig. 10a, thermogravimetric analysis observed the onset of oxidation at 480 °C at all silicon concentrations. However, with increasing Si content, the oxidation rate decreases when the slope of the weight gain becomes smaller. TEM analysis of $TiSiB_2$ coatings with a Si content of 10 at.% and 20 at.%, respectively, annealed in humid air at a temperature of 500 °C for one hour revealed the formation of a top nanocrystalline oxide scale formed by homogeneously distributed Ti, Si, and O elements. The oxide scale thickness was significantly reduced in the case of higher Si content, and XRD analysis confirmed the presence of the anatase-TiO_2 phase. The high-temperature behavior changed after annealing at 800 °C for 1 h, when the coating containing 10 at.% was fully oxidized and, as can be seen in Fig. 10b, modulation of chemically and structurally different oxide layers occurs. The surface thin film is formed by a protective amorphous SiO_x scale, under which there is a coarse-grained, columnar rutile-TiO_2 layer, followed by an amorphous boronsilicate-based layer, and again a rutile-TiO_2 layer with a finer columnar morphology. A higher Si content (20 at.%) reduced the oxidation rate at a temperature of 800 °C, when the structural character of the oxide layers remains the same but approximately 30% of the coating thickness is not oxidized.

Glechner et al. [90] have grown ternary $TiSiB_{2+z}$ coatings by DC magnetron sputtering from TiB_2 on which Si wafers were placed. The coatings contained 0 to 40 at.% Si, and with increasing Si content, the boron content decreases. The beginning of oxidation in synthetic air of Si-containing coatings was similar to unalloyed $TiB_{2.72}$; however, TGA analysis revealed for each higher Si content a plateau of a steady mass signal (above 1000 to 1100 °C), suggesting a strongly delayed oxidation process due to a formation of protective scale on top with a remaining unaffected coating beneath [90]. The authors investigated the oxidation behavior of a selected $Ti_{0.13}Si_{0.41}B_{0.46}$ coating isothermally annealed at 900 °C and 1100 °C for 1 h. After isothermal oxidation at 900 °C for 60 min, $Ti_{0.13}Si_{0.41}B_{0.46}$ exhibits an oxide scale that contains rutile-TiO_2, and the remaining coating recrystallized from an X-ray amorphous structure in the as-deposited state to a mixture of TiB_2 and Si grains. After annealing at 900 °C, the oxide scale has a vitreous, non-porous fracture morphology.

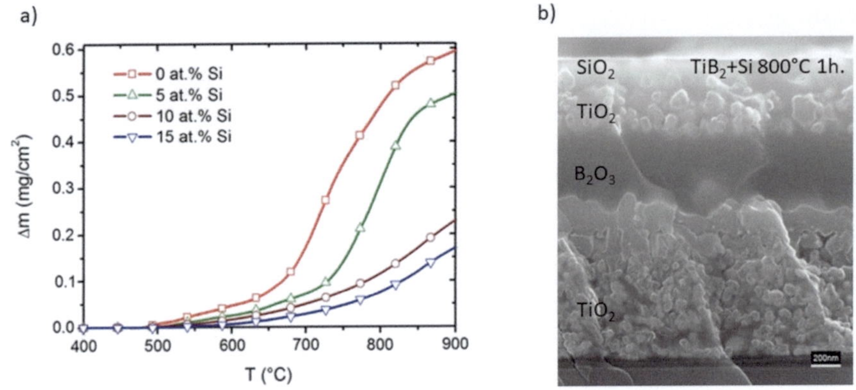

Fig. 10 **a** High-resolution thermogravimetry curves of the coatings with several Si concentrations; **b** cross-sectional SEM images of the 10 at.% annealing at 800 °C in ambient air (10 °C/min temperature ramping) [27]

The same nature of fracture is maintained by the oxide scale even after annealing the coating at 1100 °C, which is even significantly thinner. Structural analysis confirmed the presence of crystalline TiB_2 and Si phases, with a very small amount of rutile-TiO_2 in this oxide scale. At this temperature, a very protective, amorphous Si-rich oxide scale is formed on top. In contrast to the scale formed at 900 °C, the formation of a crystalline metal oxide is mainly suppressed.

A significant increase in the oxidation resistance of TMB_2 due to the addition of Si was also observed on the systems $Hf_{0.21}Si_{0.18}B_{0.61}$, $Ta_{0.26}Si_{0.12}B_{0.62}$, and $W_{0.28}Si_{0.16}B_{0.56}$ [90]. Onset oxidation temperatures start at 500 °C for W-Si-B_2, 600 °C for Ta-Si-B_2, and approximately 900 °C for Hf-Si-B_2 coatings. While the increase in mass gain is gradual in the case of Hf- and W-Si-B_2 coatings, in the case of Ta-Si-B_2 coatings, the mass gain and thus oxidation are very rapid. The result of high-temperature exposure at 900 °C for one hour in synthetic air is the formation of featureless oxide scales on the surface of the coatings, and their structure is formed by HfO_2, Ta_2O_5, and $W_{18}0_{49}$ phases containing also diboride phases HfB_2, TaB2, and WB_2 according to the individual composition of the coatings. Increasing the temperature to 1100 °C activates the formation of Si-rich oxide scale on the surface, which prevents the formation of metal oxides. The result is significantly thinner and denser oxide scales on the surface of the coatings. However, this behavior was not observed in $Ta_{0.26}Si_{0.12}B_{0.62}$ coatings, which exhibit a different oxidation mechanism and do not prefer the formation of a Si-rich oxide scale [90].

Alloying with aluminum seems to be a promising way to improve the oxidation behavior of TMB_2 coatings. Bakhit et al. [81] deposited $Ti_{1-x}Al_xB_y$ coatings using hybrid Al-HiPPMS/TiB_2-DCMS co-sputtering. When the Al/(Ti + Al) ratio, x, decreases from x = 0.35 to 0.29, the B/(Ti + Al) ratio, *y*, increases from 1.30 to 1.54. All as-deposited thin films show columnar structure. The column boundaries of $TiB_{2.4}$ films are B-rich, while the $Ti_{0.68}Al_{0.32}B_{1.35}$ alloys have Ti–rich columns surrounded by an Al-rich $Ti_{1-x}Al_xB_y$ tissue phase, which is B-deficient. A significant improvement in oxidation resistance was observed after rapid high-temperature annealing at 800 °C for 30 min, when the oxide scale thickness is 1900 nm on the surface of $TiB_{2.4}$ and 470 nm on the surface of $Ti_{0.68}Al_{0.32}B_{1.35}$ coating. The enhanced oxidation resistance is mainly attributed to the formation of a dense, protective Al-containing oxide scale, which decreases the oxygen diffusion rate by suppressing the oxide-crystallite coarsening [81].

6 Transition-Metal diborides—concept of Multilayer/superlattice Systems

The concept of multilayers, especially in the form of periodically alternating lattice-coherent nanolayers called superlattices (SL), is known to be a very effective way to increase hardness and improve fracture toughness.

From the point of view of mechanical properties, the optimization of the bilayer thickness (bi-layer period, L) turned out to be crucial in controlling the number of dislocations and their mobility inside individual SL nanolayers. The driving force for sliding dislocations across interfaces is reduced by coherent stresses that arise from lattice mismatch (Lattice mismatch, Da) and differences between the shear moduli of individual nanolayer materials (DG). These strengthening parameters effectively lead to an increase in hardness, which was reported primarily on TM nitride coatings TiN/VN [91] and TiN/NbN [92] with a large lattice mismatch. Optimizing the bi-layer period in the range of 5 to 9 nm led to an extreme increase in hardness of up to 200% compared to monolithic constituents. Similar SL nitride systems, TiN/WN [93] or TiN/ZrAlN [94], where the lattice mismatch was essentially zero, showed a significant increase in hardness due to large differences in shear moduli. Current studies of TiN/CrN [95] and TaN/MoN [96] superlattice coatings demonstrate that coherency strains, shear modulus mismatch, as well as chemical and structural modulations of individual nanolayers, improve fracture toughness. The authors of these studies, performing micromechanical bending tests of coating cantilevers, report an increase in the K_{IC} value by up to tens of percent.

The research on multilayer/superlattice coatings based on transition-metal diborides is still an underexplored area, and attention has been focused on it only in recent years. Dorri et al. [97] experimentally grew epitaxial CrB_x/TiB_y superlattice coatings on an Al_2O_3 (0001) substrate heated to 600 °C using DC magnetron sputtering. The authors prepared a series of superlattices with a constant bi-layer period L = 6 nm and a thickness ratio G = 0.5–0.8 ($G = D_{TiBy}/L$) of individual nanolayers. The second series maintained a constant ratio G between the thicknesses of individual nanolayers and varied the thickness of bi-layer periods L in the interval from 1 to 8.2 nm (Fig. 11). The stoichiometry of B/Cr = 1.72 and B/Ti = 3.3, respectively, of the basic binary CrB_x and TiB_y constituents was typical for these types of diboride coatings. This study indicated the increased difficulty of preparing superlattices based on nanocomposite TMB_2 containing the B-tissue phase compared to stoichiometric cubic TMNs. It was shown that overstoichiometric TiB_y nanolayers containing excess boron-tissue phase significantly affect the structural quality of superlattices. SLs containing a thinner TiB_y nanolayer than the CrB_x nanolayer (G = 0.2–0.4) show the highest structural quality with an almost stoichiometric B/TM ratio. Increasing the thickness of overstoichiometric TiB_y leads to columnar SL growth with distorted B-rich boundaries. On the other hand, increasing the thickness of the understoichiometric CrB_y nanolayer prevents the formation of B-rich boundaries, and the SL quality is high. From the point of view of mechanical properties, it was shown that the modulation of the bi-layer period L affects the hardness of SL coatings. The lowest hardness (at optimal growth G = 0.3) of 29 GPa was measured for L = 1 nm as a consequence of the nanocolumnar structure with locally interrupted interfaces of 34 GPa; on the contrary, the highest hardness of 34 GPa was shown by SLs with a bi-layer period of L = 7 nm.

In 2023, Fiantok et al. [98] published a pioneering theoretical work that, considering selected criteria for lattice mismatch and shear moduli, identified potential candidates for superlattices based on transition-metal diborides with improved ductility and fracture toughness. The authors studied 264 possible combinations of α/α, α/ω, or ω/ω MB_2 (where M = Al or group 3–6 transition metal) diboride superlattices. Considering the energetic stability and restrictions Da < 4%, DG > 40 GPa, 33 superlattice systems were selected, where they investigated the mechanical stability and elastic constants C_{ij}. The superlattice's ability to resist brittle cleavage along interfaces was estimated by Griffith's formula for fracture toughness (K_{IC}). According to calculations, the α/α-type TiB_2/MB_2 (M = Mo, W), HfB_2/WB_2, VB_2/MB_2 (M = Cr, Mo), NbB_2/MB_2 (M = Mo, W), and α/ω-type AlB_2/MB_2 (M = Nb, Ta, Mo, W) were suggested as the most promising candidates, providing an atomic-scale basis for enhanced toughness and resistance to crack growth. It's important to note that the supercells used to model superlattices are crystalline, stoichiometric, and lack a B-tissue phase. In any case, this study can very effectively help in the material design of superlattice coatings.

Hahn et al. [99] investigated in the only published study so far, the mechanical properties and fracture toughness of sputtered TMB_2/TMB_2 superlattice coatings, where TM = Ti, Zr, and W. In the case of the TiB_2/ZrB_2 system, characterized by a large lattice mismatch and a small difference in shear modulus (ΔG = 27 GPa and Δa = 0.14 Å), the superlattice effect was observed in a wide range of bi-layer period L = 2–15 nm. Conversely, in the TiB_2/WB_2 system with a small lattice mismatch and a large difference in shear modulus (ΔG = 112 GPa and Δa = 0.01 Å), coherent growth was observed in a narrower range of bi-layer period L = 2–9 nm. The results of the mechanical properties measurements show that the hardness increase is associated with the DG differences in the TiB_2/WB_2 system, where the highest hardness of 45 GPa (an increase of 12% compared to binary coatings) was observed for L = 6 nm. A large lattice mismatch in the TiB_2/ZrB_2 coating does not have a significant effect on hardening. The situation was the opposite in the fracture-mechanical behavior of SL coatings. Here, it is shown that the optimal thickness of the bi-layer period and the

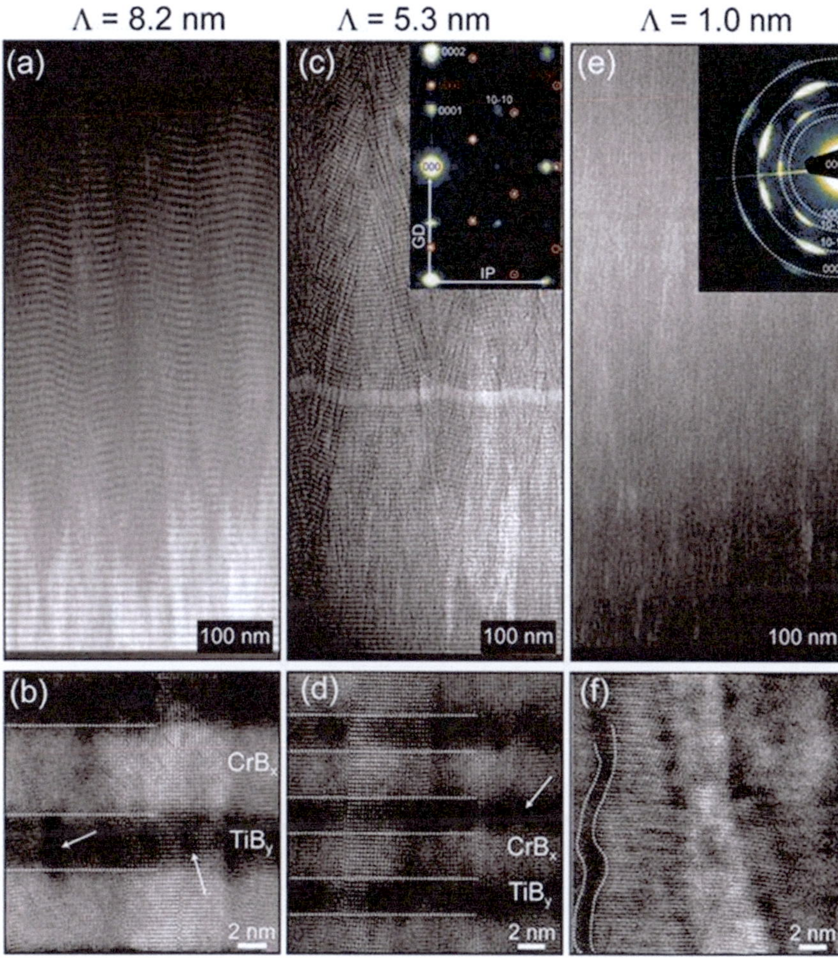

Fig. 11 Low-magnification cross-sectional STEM-HAADF micrographs of CrB_x/TiB_y superlattices with modulation periods **a** L = 8.2 nm, **c** 5.3 nm, and **e** 1 nm. Lattice-resolved cross-sectional HAADF-STEM micrographs of superlattices with modulation periods **b** L = 8.2 nm, **d** 5.3 nm, and **f** 1 nm. In the inset in (**c**), IP denotes in-plane and GD denotes growth direction and the diffraction pattern of the Al_2O_3 substrate is shown by red hexagons. The selected area electron diffraction pattern in the inset (**e**) was acquired from the middle of the film. Note: extra wide single layer in (**c**) is due to unintentional drop out of TiB_y deposition during three periods following which the SL growth mode is recovered [97]

large lattice mismatch between the individual nanolayers in the TiB_2/ZrB_2 system have a positive effect on the increase of the K_{IC} value, when the highest value of K_{IC} = 3.7 MPa m$^{-1/2}$ was measured for L = 6 nm, which is almost 20% higher than the K_{IC} values for TiB_2 and ZrB_2, respectively. In the case of TiB_2/WB_2, the differences in shear moduli proved to be less effective in increasing the fracture toughness, and the coatings show an average K_{IC} value of about 2.8 MPa $m^{-1/2}$ over the entire L range.

7 Outlook

In this chapter, the state of the art in the field of hard transition-metal diboride (TMB_2) coatings is described in detail. As shown, strong ionic and covalent bonds between metals and boron atoms in the hexagonal TMB_2 structure led to high-temperature stability and excellent mechanical properties. Despite this, the application potential of TMB_2 coatings in demanding thermal–mechanical conditions is still not fulfilled.

Magnetron sputtering of a stoichiometric ceramic compound target represents a typical method of deposition of TMB_2 coatings. However, this process is accompanied by several important aspects, especially the different angular distribution of the sputtered atoms and the resputtering of light boron from the growing layer. The resulting chemical composition of the coatings is different from the target composition, and the coatings show an excess of boron or its deficiency. This results in the formation of an overstoichiometric TMB_{2+z} nanocomposite structure formed by TMB_2 nanocolumns surrounded by an amorphous B-rich tissue phase. In the second case, understoichiometric TMB_{2-z} coatings consist of TMB_2 nanocolumns containing boron vacancies, and amorphous B-rich tissue may also be present or absent. All these coatings show excellent mechanical properties, in particular a significantly higher hardness of ~ 40 GPa compared to the bulk equivalent (~20–24 GPa). This superhardness is determined by the intrinsic hardness + of the TMB_2 nanocolumns and the cohesive strength of the B-tissue phase, which serves as the "glue" that strongly holds the nanocolumns together. Boron vacancies have a certain contribution to increasing hardness, especially in the case of understoichiometric coatings, and it is a so-called boron-vacancy-induced hardening.

As seen, the B-tissue phase also represents the "Achilles' heel" of the nanocomposite structure, as it creates an easy path for oxygen penetration into the TMB_2 coatings and their oxidation and destruction at significantly lower temperatures compared to TMB_2 bulks. In addition, the B-tissue phase is prone to the formation of micro-cracks and their propagation during mechanical loading, so TMB_2 coatings have low fracture toughness.

In the chapter, several successful concepts were presented to suppress the negative effects of the B-tissue phase. The first of them is the use of progressive HiPPMS deposition technology with a high degree of ionization of sputtered species, which leads to the formation of understoichiometric B-tissue phase-free TMB_{2-z} coatings containing nanocolumns with B vacancies in the boron sublattice and stacking faults in the metal sublattice. Such coatings are very hard and show improved oxidation resistance and increased resistance to the formation of micro-cracks.

Alloying TMB_2 with other metals or silicon also seems to be a promising way to improve thermo-mechanical properties and oxidation resistance. Several ternary

$TM^1TM^2B_2$ coatings form metastable solid solutions ($TlAlB_2$), whose decomposition at high temperatures is accompanied by age hardening. In another case, alloying ZrB_2 with tantalum improves its ductility. The addition of Si to TMB_2 coatings effectively improved oxidation resistance as it increased the onset temperature of oxidation and promoted the formation of protective top oxide scales with slow parabolic oxidation kinetics.

Multilayer architecture in the form of TMB_2/TMB_2 superlattices also represents a very promising way to improve the mechanical properties and fracture toughness of these ceramic coatings. As it turned out, the differences in the lattice parameters and the differences in the shear moduli of individual nanolayers can effectively prevent not only the movement of dislocations, which leads to hardening, but also deflect the sliding of micro-cracks at the interfaces and slow down their movement, thus improving the fracture toughness.

The achieved results presented in this chapter open new possibilities in the material design of TMB_2 coatings. HiPPMS technologies and their "tuning" of the structure during the growth of TMB_2 coatings are very promising. Also, the new method of micromechanical bending testing of coatings significantly expands the knowledge of the fracture-mechanical behavior of TMB_2 coatings and gives feedback on their structure. It can be expected that in the next few years, it will be possible to prepare TMB_2 coatings that will, in many respects, surpass the dominant coatings based on TM nitrides.

Acknowledgements This work was supported by the Slovak Research and Development Agency (Grant No. APVV-21-0042), Scientific Grant Agency (Grant No. VEGA 1/0296/22), and the Operational Program Integrated Infrastructure for the project: Advancing University Capacity and Competence in Research, Development, and Innovation (ACCORD), co-financed by the European Regional Development Fund. Special thanks to Dr. Tomáš Fiantok for the illustrations on Figs. 1 and 2.

References

1. Sundgren, J.-E.: Thin Solid Films **128**, 21 (1985)
2. Hultman, L.: Vacuum **57**, 1–30 (2000)
3. Vetter, J.: Surf. Coat. Technol. **76–77**, 719–724 (1993)
4. Münz, W.-D., Lewis, D.B., Hovsepian, P.E., Schönjahn, C., Ehiasarian, A., Smith, I.J.: Surf. Eng. **17**, 15 (2001)
5. Mayrhofer, P.H., Mitterer, C., Hultman, L., Clemens, H.: Prog. Mater. Sci. **51**, 1032 (2006)
6. Mayrhofer, P.H., Hörling, A., Karlsson, L., Sjölén, J., Larsson, T., Mitterer, C., Hultman, L.: Appl. Phys. Lett. **83**, 2049 (2003)
7. Mayrhofer, P.H., Willmann, H., Reiter, A.E.: Surf. Coat. Technol. **202**, 4935 (2008)
8. Mikula, M., Plašienka, D., Sangiovanni, D.G., Sahul, M., Roch, T., Truchlý, M., Gregor, M., Čaplovič, L., Plecenik, A., Kúš, P.: Acta Mater. **121**, 59 (2016)
9. Mikula, M., Truchlý, M., Sangiovanni, D.G., Plašienka, D., Roch, T., Gregor, M., Ďurina, P., Janík, M., Kúš, P.: J. Vacuum Sci. Technol. A: Vac. Surf. Films **35**, 060602 (2017)
10. Mikula, M., Sangiovanni, D.G., Plašienka, D., Roch, T., Čaplovičová, M., Truchlý, M., Satrapinskyy, L., Bystrický, R., Tonhauzerová, D., Vlčková, D., Kúš, P.: J. Appl. Phys. **121**, 155304 (2017)

11. Rachbauer, R., Blutmager, A., Holec, D., Mayrhofer, P.H.: Surf. Coat. Technol. **206**, 2667 (2012)
12. Rachbauer, R., Holec, D., Mayrhofer, P.H.: Surf. Coat. Technol. **211**, 98 (2012)
13. Xu, Y.X., Chen, L., Pei, F., Du, Y.: Surf. Coat. Technol. **304**, 512 (2016)
14. Xu, Y.X., Riedl, H., Holec, D., Chen, L., Du, Y., Mayrhofer, P.H.: Surf. Coat. Technol. **324**, 48 (2017)
15. Madtha, S., Lee, C., Ravi Chandran, K.S.: J. Am. Ceramic Soc. **91**, 1319 (2008)
16. Mitterer, C.: J. Solid State Chem. **133**, 279 (1997)
17. Mitterer, C., Komenda-Stallmaier, J., Losbichler, P., Schmölz, P., Werner, W.S.M., Störi, H.: Vacuum **46**, 1281 (1995)
18. Gidikova, N., Simeonova, J., Petkova, D.: Mater. Sci. Eng. A **184**, L1 (1994)
19. Buzea, C., Yamashita, T.: Supercond. Sci. Technol. **14**, R115 (2001)
20. Ge, Y., Bao, K., Ma, T., Zhang, J., Zhou, C., Ma, S., Tao, Q., Zhu, P., Cui, T.: ACS Omega **6**, 21436 (2021)
21. Mat'aš, S., Siemensmeyer, K., Wheeler, E., Wulf, E., Beyer, R., Hermannsdörfer, T., Ignatchik, O., Uhlarz, M., Flachbart, K., Gabáni, S., Priputen, P., Efdokimova, A., Shitsevalova, N.: J. Phys.: Conf. Ser. **200**, 032041 (2010)
22. Cahill, J.T., Graeve, O.A.: J. Market. Res. **8**, 6321 (2019)
23. Cook, B.A., Harringa, J.L., Lewis, T.L., Russell, A.M.: Scripta Mater. **42**, 597 (2000)
24. Moraes, V., Riedl, H., Fuger, C., Polcik, P., Bolvardi, H., Holec, D., Mayrhofer, P.H.: Sci. Rep. **8**, 9288 (2018)
25. Mayrhofer, P.H., Mitterer, C., Wen, J.G., Greene, J.E., Petrov, I.: Appl. Phys. Lett. **86**, 131909 (2005)
26. Dorri, S., Palisaitis, J., Greczynski, G., Petrov, I., Birch, J., Hultman, L., Bakhit, B.: Corros. Sci. **206**, 110493 (2022)
27. Grančič, B., Mikula, M., Roch, T., Zeman, P., Satrapinskyy, L., Gregor, M., Plecenik, T., Dobročka, E., Hájovská, Z., Mičušík, M., Šatka, A., Zahoran, M., Plecenik, A., Kúš, P.: Surf. Coat. Technol. **240**, 48 (2014)
28. Fuger, C., Hahn, R., Hirle, A., Wojcik, T., Kutrowatz, P., Bohrn, F., Hunold, O., Polcik, P., Riedl, H.: Mater. Res. Lett. **11**, 613 (2023)
29. Vajeeston, P., Ravindran, P., Ravi, C., Asokamani, R.: Phys. Rev. B **63**, 045115 (2001)
30. Zhou, Y., Xiang, H., Feng, Z., Li, Z.: J. Mater. Sci. Technol. **31**, 285 (2015)
31. Mikula, M., Grančič, B., Švec, P., Roch, T., Truchlý, M., Kohuľák, O., Satrapinskyy, L., Fiantok, T., Izaii, V., Haršáni, M., Orovčík, Ľ, Kúš, P.: Scripta Mater. **164**, 91 (2019)
32. Mott, N.F.: Phil. Mag. **19**, 835 (1969)
33. Shein, I.R., Ivanovskii, A.L.: J. Phys. Condens. Matter **20**, 415218 (2008)
34. Chen, X.-Q., Niu, H., Li, D., Li, Y.: Intermetallics **19**, 1275 (2011)
35. Tengdelius, L., Broitman, E., Lu, J., Eriksson, F., Birch, J., Nyberg, T., Hultman, L., Högberg, H.: Acta Mater. **111**, 166 (2016)
36. Mikula, M., Grančič, B., Roch, T., Plecenik, T., Vávra, I., Dobročka, E., Šatka, A., Buršíková, V., Držík, M., Zahoran, M., Plecenik, A., Kúš, P.: Vacuum **85**, 866 (2011)
37. Sonber, J.K., Murthy, T.S.R.C., Subramanian, C., Kumar, S., Fotedar, R.K., Suri, A.K.: Int. J. Refract Metal Hard Mater. **27**, 912 (2009)
38. Okada, S., Kudou, K., Lundström, T.: Jpn. J. Appl. Phys. **34**, 226 (1995)
39. Fiantok, T., Šroba, V., Koutná, N., Izai, V., Roch, T., Truchlý, M., Vidiš, M., Satrapinskyy, L., Nagy, Š, Grančič, B., Kúš, P., Mikula, M.: J. Vac. Sci. Technol. A **40**, 033414 (2022)
40. Magnuson, M., Hultman, L., Högberg, H.: Vacuum **196**, 110567 (2022)
41. Takahashi, T., Kamiya, H.: J. Cryst. Growth **26**, 203 (1974)
42. Pierson, H.O., Mullendore, A.W.: Thin Solid Films **72**, 511 (1980)
43. Sung, J., Goedde, D.M., Girolami, G.S., Abelson, J.R.: J. Appl. Phys. **91**, 3904 (2002)
44. Tolle, J., Roucka, R., Tsong, I.S.T., Ritter, C., Crozier, P.A., Chizmeshya, A.V.G., Kouvetakis, J.: Appl. Phys. Lett. **82**, 2398 (2003)
45. Bera, S., Sumiyoshi, Y., Yamada-Takamura, Y.: J. Appl. Phys. **106**, 063531 (2009)

46. Bunshah, R.F., Nimmagadda, R., Dunford, W., Movchan, B.A., Demchishin, A.V., Chursanov, N.A.: Thin Solid Films **54**, 85 (1978)
47. Zhai, H.Y., Christen, H.M., Cantoni, C., Goyal, A., Lowndes, D.H.: Appl. Phys. Lett. **80**, 1963 (2002)
48. Ferrando, V., Marré, D., Manfrinetti, P., Pallecchi, I., Tarantini, C., Ferdeghini, C.: Thin Solid Films **444**, 91 (2003)
49. Padmanabhan, K.R., Sørensen, G.: Thin Solid Films **81**, 13 (1981)
50. Knotek, O., Löffler, F., Böhmer, M., Breidenbach, R., Stößssel, C.: Metall. Coat. Thin Films 263–267 (1991)
51. Treglio, J.R., Trujillo, S., Perry, A.J.: Surf. Coat. Technol. **61**, 315 (1993)
52. Zhirkov, I., Petruhins, A., Naslund, L.-A., Kolozsvári, S., Polcik, P., Rosen, J.: Appl. Phys. Lett. **107**, 184103 (2015)
53. Tengdelius, L., Greczynski, G., Chubarov, M., Lu, J., Forsberg, U., Hultman, L., Janzén, E., Högberg, H.: J. Cryst. Growth **430**, 55 (2015)
54. Nedfors, N., Tengstrand, O., Lu, J., Eklund, P., Persson, P.O.Å., Hultman, L., Jansson, U.: Surf. Coat. Technol. **257**, 295 (2014)
55. Olson, R.R., King, M.E., Wehner, G.K.: J. Appl. Phys. **50**, 3677 (1979)
56. Šroba, V., Fiantok, T., Truchlý, M., Roch, T., Zahoran, M., Grančič, B., Švec, P., Nagy, Š, Izai, V., Kúš, P., Mikula, M.: J. Vac. Sci. Technol. A **38**, 033408 (2020)
57. Viskupová, K., Grančič, B., Roch, T., Satrapinskyy, L., Truchlý, M., Mikula, M., Šroba, V., Ďurina, P., Kúš, P.: Surf. Coat. Technol. **421**, 127463 (2021)
58. Grančič, B., Pleva, M., Mikula, M., Čaplovičová, M., Satrapinskyy, L., Roch, T., Truchlý, M., Sahul, M., Gregor, M., Švec, P., Zahoran, M., Kúš, P.: Surf. Coat. Technol. **367**, 341 (2019)
59. Bakhit, B., Petrov, I., Greene, J.E., Hultman, L., Rosén, J., Greczynski, G.: J. Vac. Sci. Technol. A: Vac. Surf. Films **36**, 030604 (2018)
60. Daniel, R., Musil, J.: Novel Nanocomposite Coatings: Advances and Industrial Applications. Pan Stanford Publ, Singapore (2015)
61. Kunc, F., Musil, J., Mayrhofer, P.H., Mitterer, C.: Surf. Coat. Technol. **174–175**, 744 (2003)
62. Mikula, M., Grančič, B., Buršíková, V., Csuba, A., Držík, M., Kavecký, Š, Plecenik, A., Kúš, P.: Vacuum **82**, 278 (2007)
63. Petrov, I., Hall, A., Mei, A.B., Nedfors, N., Zhirkov, I., Rosen, J., Reed, A., Howe, B., Greczynski, G., Birch, J., Hultman, L., Greene, J.E.: J. Vac. Sci. Technol. A: Vac. Surf. Films **35**, 050601 (2017)
64. Tengdelius, L., Samuelsson, M., Jensen, J., Lu, J., Hultman, L., Forsberg, U., Janzén, E., Högberg, H.: Thin Solid Films **550**, 285 (2014)
65. Tengdelius, L., Birch, J., Lu, J., Hultman, L., Forsberg, U., Janzén, E., Högberg, H.: Phys. Status Solidi A **211**, 636 (2014)
66. Malinovskis, P., Palisaitis, J., Persson, P.O.Å., Lewin, E., Jansson, U.: J. Vac. Sci. Technol. A: Vac. Surf. Films **34**, 031511 (2016)
67. Fuger, C., Hahn, R., Zauner, L., Wojcik, T., Weiss, M., Limbeck, A., Hunold, O., Polcik, P., Riedl, H.: Mater. Res. Lett. **10**, 70 (2022)
68. Dahlqvist, M., Jansson, U., Rosen, J.: J. Phys. Condens. Matter **27**, 435702 (2015)
69. Zhang, S., Wang, Z., Guo, P., Ke, P., Odén, M., Wang, A.: Surf. Coat. Technol. **322**, 134 (2017)
70. Nedfors, N., Primetzhofer, D., Wang, L., Lu, J., Hultman, L., Jansson, U.: Surf. Coat. Technol. **266**, 167 (2015)
71. Lundin, D., Brenning, N., Jädernäs, D., Larsson, P., Wallin, E., Lattemann, M., Raadu, M.A., Helmersson, U.: Plasma Sourc. Sci. Technol. **18**, 045008 (2009)
72. Sarakinos, K., Alami, J., Konstantinidis, S.: Surf. Coat. Technol. **204**, 1661 (2010)
73. Hellgren, N., Thörnberg, J., Zhirkov, I., Sortica, M.A., Petrov, I., Greene, J.E., Hultman, L., Rosen, J.: Vacuum **169**, 108884 (2019)
74. Polyakov, M.N., Morstein, M., Maeder, X., Nelis, T., Lundin, D., Wehrs, J., Best, J.P., Edwards, T.E.J., Döbeli, M., Michler, J.: Surf. Coat. Technol. **368**, 88 (2019)
75. Thörnberg, J., Palisaitis, J., Hellgren, N., Klimashin, F.F., Ghafoor, N., Zhirkov, I., Azina, C., Battaglia, J.-L., Kusiak, A., Sortica, M.A., Greene, J.E., Hultman, L., Petrov, I., Persson, P.O.Å., Rosen, J.: Surf. Coat. Technol. **404**, 126537 (2020)

76. Palisaitis, J., Dahlqvist, M., Hall, A.J., Thörnberg, J., Persson, I., Nedfors, N., Hultman, L., Greene, J.E., Petrov, I., Rosen, J., Persson, P.O.Å.: Acta Mater. **204**, 116510 (2021)
77. Samuelsson, M., Jensen, J., Helmersson, U., Hultman, L., Högberg, H.: Thin Solid Films **526**, 163 (2012)
78. Alling, B., Högberg, H., Armiento, R., Rosen, J., Hultman, L.: Sci. Rep. **5**, 9888 (2015)
79. Mockute, A., Palisaitis, J., Alling, B., Berastegui, P., Broitman, E., Näslund, L.-Å., Nedfors, N., Lu, J., Jensen, J., Hultman, L., Patscheider, J., Jansson, U., Persson, P.O.Å., Rosen, J.: Scripta Mater. **127**, 122 (2017)
80. Nedfors, N., Mráz, S., Palisaitis, J., Persson, P.O.Å., Lind, H., Kolozsvari, S., Schneider, J.M., Rosen, J.: Surf. Coat. Technol. **364**, 89 (2019)
81. Bakhit, B., Palisaitis, J., Thörnberg, J., Rosen, J., Persson, P.O.Å., Hultman, L., Petrov, I., Greene, J.E., Greczynski, G.: Acta Mater. **196**, 677 (2020)
82. Stüber, M., Riedl, H., Wojcik, T., Ulrich, S., Leiste, H., Mayrhofer, P.H.: Thin Solid Films **688**, 137361 (2019)
83. Šroba, V., Fiantok, T., Truchlý, M., Roch, T., Grančič, B., Viskupová, K., Satrapinskyy, L., Švec, P., Nagy, Š., Izai, V., Kúš, P., Mikula, M.: J. Vac. Sci. Technol. A **41**, 023410 (2023)
84. Bakhit, B., Engberg, D.L.J., Lu, J., Rosen, J., Högberg, H., Hultman, L., Petrov, I., Greene, J.E., Greczynski, G.: J. Vac. Sci. Technol. A **37**, 031506 (2019)
85. Bakhit, B., Dorri, S., Kosari, A., Mol, A., Petrov, I., Birch, J., Hultman, L., Greczynski, G.: Appl. Surf. Sci. **591**, 153164 (2022)
86. Moraes, V., Fuger, C., Paneta, V., Primetzhofer, D., Polcik, P., Bolvardi, H., Arndt, M., Riedl, H., Mayrhofer, P.H.: Scripta Mater. **155**, 5 (2018)
87. Palisaitis, J., Dahlqvist, M., Hultman, L., Petrov, I., Rosen, J., Persson, P.O.Å.: Materialia **24**, 101478 (2022)
88. Viskupová, K., Grančič, B., Roch, T., Nagy, Š, Satrapinskyy, L., Šroba, V., Truchlý, M., Šilha, J., Kúš, P., Mikula, M.: Scripta Mater. **229**, 115365 (2023)
89. Thörnberg, J., Bakhit, B., Palisaitis, J., Hellgren, N., Hultman, L., Greczynski, G., Persson, P.O.Å., Petrov, I., Rosen, J.: Surf. Coat. Technol. **420**, 127353 (2021)
90. Glechner, T., Oemer, H.G., Wojcik, T., Weiss, M., Limbeck, A., Ramm, J., Polcik, P., Riedl, H.: Surf. Coat. Technol. **434**, 128178 (2022)
91. Mirkarimi, P.B., Hultman, L., Barnett, S.A.: Appl. Phys. Lett. **57**, 2654 (1990)
92. Helmersson, U., Todorova, S., Barnett, S.A., Sundgren, J.-E., Markert, L.C., Greene, J.E.: J. Appl. Phys. **62**, 481 (1987)
93. Buchinger, J., Koutná, N., Chen, Z., Zhang, Z., Mayrhofer, P.H., Holec, D., Bartosik, M.: Acta Mater. **172**, 18 (2019)
94. Yalamanchili, K., Wang, F., Aboulfadl, H., Barrirero, J., Rogström, L., Jiménez-Pique, E., Mücklich, F., Tasnadi, F., Odén, M., Ghafoor, N.: Acta Mater. **121**, 396 (2016)
95. Hahn, R., Bartosik, M., Soler, R., Kirchlechner, C., Dehm, G., Mayrhofer, P.H.: Scripta Mater. **124**, 67 (2016)
96. Hahn, R., Koutná, N., Wójcik, T., Davydok, A., Kolozsvári, S., Krywka, C., Holec, D., Bartosik, M., Mayrhofer, P.H.: Commun Mater **1**, 62 (2020)
97. Dorri, S., Birch, J., Eriksson, F., Palisaitis, J., Persson, P.O.Å., Bakhit, B., Hultman, L., Ghafoor, N.: Mater. Des. **228**, 111842 (2023)
98. Fiantok, T., Koutná, N., Sangiovanni, D.G., Mikula, M.: Sci. Rep. **13**, 12835 (2023)
99. Hahn, R., Tymoszuk, A.A., Wojcik, T., Ntemou, E., Hunold, O., Polcik, P., Kolozsvári, S., Primetzhofer, D., Mayrhofer, P.H., Riedl, H.: Scripta Mater. **235**, 115599 (2023)

AlCrNbSiTiN High-Entropy Nitride Hard Coatings

Xiangyu Zhang, Yan Liu, Vasiliy Pelenovich, and Bing Yang

Abstract High entropy alloys which are composed of at least five primary elements exhibit better performance than the conventional alloys in terms of high hardness, high toughness and high-temperature resistance, thus showing great potential for application in the field of protective coatings. However, the research on high-entropy alloy coatings is still in its initial stage, and there is a lack of in-depth research on the structure and performance of the coatings. In this chapter, the nanocomposite high entropy alloy nitride coatings based on AlCrNbSiTi system were fabricated using reactive RF magnetron sputtering and arc ion plating techniques. The effects of negative bias voltage applied to the sample on the phase composition, grain size, orientation and microstructures of the coatings were characterized, and the mechanical properties, friction behaviour, and temperature resistance of coatings were investigated systematically. The wear friction behaviour and the elemental diffusion and the oxidation mechanism of the coatings at high temperatures were studied.

Keywords High-entropy alloy nitride · Hard coating · RF magnetron sputtering · Nanocomposite structure · Mechanical properties · Thermal stability

X. Zhang · V. Pelenovich · B. Yang
School of Power and Mechanical Engineering, Wuhan University, Wuhan 430072, China
e-mail: toyangbing@whu.edu.cn

Y. Liu
Hubei JiuFengShan Future Science and Technology Laboratory, Wuhan 420000, China

V. Pelenovich (✉)
Institute of Technological Sciences of Wuhan University, Wuhan 430072, China
e-mail: v.pelenovich@whu.edu.cn

Hubei Key Laboratory of Electronic Manufacturing and Packaging Integration, Wuhan University, Wuhan 430072, China

A. D. Pogrebnjak et al. (eds.), *Nanocomposite and Nanocrystalline Materials and Coatings*, Advanced Structured Materials 214,
https://doi.org/10.1007/978-981-97-2667-7_2

1 High-Entropy Alloys and High-Entropy Nitrides

High-entropy alloy (HEA), a novel alloy system developed in recent years [1, 2], has transcended the limitations of the traditional alloy design concept. The concept of HEA is considered one of the three major breakthroughs in alloying theory over recent decades, alongside bulk metallic glasses and rubbery metals. HEAs, characterized by their multi-principal composition, exhibit the high-entropy effect, lattice distortion effect in crystal structure, sluggish diffusion effect in dynamics, and cocktail effect in properties. As a result, they possess remarkable attributes such as high strength [3], excellent wear resistance [4], and outstanding corrosion resistance [5]. These properties have attracted significant international attention. Consequently, the concept of HEAs has been extended to coatings through various preparation techniques. In contrast to bulk materials, HEAs coatings effectively reduce production costs and hinder the formation of multiphase structures, thanks to their small-scale effect and non-equilibrium phase transition. This enables the creation of fine, uniform structures with enhanced properties. HEAs coatings, when applied to the surfaces of desired substrates, not only maintain the excellent properties of the substrates, but also imbue them with new characteristics such as exceptional hardness, toughness, temperature and wear resistance, and self-lubrication. This development significantly enhances the adaptability and service life of components and holds great promise in various industrial applications [6].

1.1 High-Entropy Alloys

Conventional alloys have historically been designed with one or two primary elements, supplemented by small additions of other elements to enhance alloy properties. Under this traditional design, approximately 30 alloy systems have been developed and commercialized, including Fe-based [7], Cu-based [8], Al-based [9], Mg-based [10], Ti-based [11], and Ni-based alloys [12]. As industrial society has advanced, the performance requirements for alloys have increasingly grown demanding. In response, alloy design has evolved toward greater compositional complexity, as illustrated in Fig. 1. However, according to traditional design principles, the addition of more elements to alloys can lead to material embrittlement and the precipitation of intermetallic compounds, ultimately degrading alloy performance. This has resulted in a bottleneck in optimizing alloy performance, making it challenging to meet the growing demand for higher-performance materials across various industries.

As early as 1788, the renowned metallurgist Franz proposed the concept of multi-component alloys. However, this concept was not emphasized at the time due to the lack of experimental evidence. In 2004, Cantor et al. achieved a significant breakthrough by successfully preparing an equiatomic CoCrFeMnNi alloy with a single face-centered cubic structure [14]. This marked the first experimental confirmation

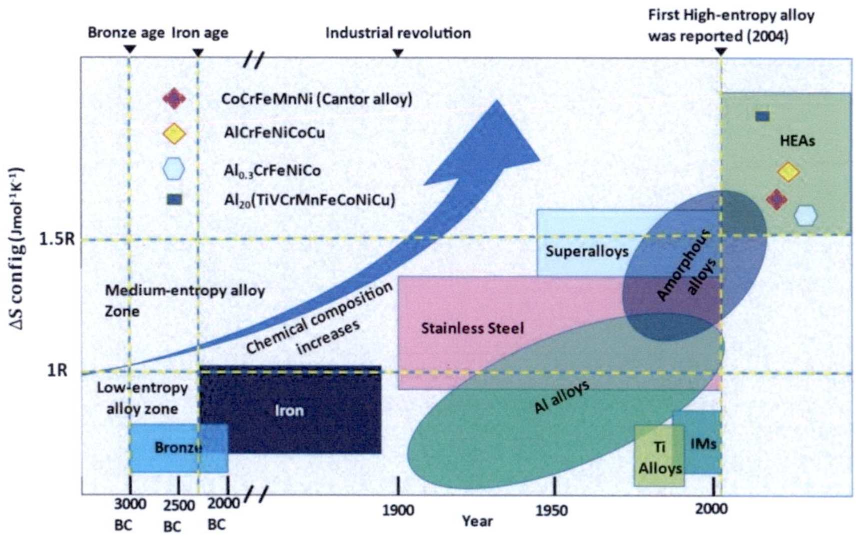

Fig. 1 The process of development of engineering materials [13]. Reprinted from Journal of Materials Research and Technology, 17, Kumar, A. et al., Mechanically alloyed high-entropy alloys: existing challenges and opportunities, 2431–2456, Copyright (2022), with permission from Elsevier

that a five-element alloy could exist entirely as a simple solid solution, without the formation of intermetallic compounds. This alloy is now known as Cantor's alloy. In the same year, Yeh [1] introduced the concept of HEAs and provided the first definition of these alloys, which consist of at least five elements, each present in a range of 5–35 atomic percent. The Boltzmann equation, Eq. (1), suggests that as the number of elements in an alloy increases, so does the configurational entropy:

$$\Delta Sconf = -k \ln w = R \ln n \tag{1}$$

where k is the Boltzmann constant, w is the elemental disorder, n is the number of elements, and R is the ideal gas constant with a value of 8.314 JK^{-1} mol^{-1}.

The relationship between the configurational entropy of the alloy and the number of elements is illustrated in Fig. 2a. The configurational entropy of an alloy composed of five equiatomic elements equals 1.61R. Consequently, HEAs are defined as alloys with a configurational entropy exceeding 1.61R. The high configurational entropy associated with HEAs results in alloys with lower free energy and enhanced phase stability. As a result, HEAs tend to form single-phase solid solutions rather than complex phases or intermetallic compounds [14].

With a deeper understanding of HEAs, the definition is no longer limited to alloys composed of at least five elements with equal or nearly equal atomic ratios, and their structures are no longer limited to single-phase solid solutions. The characteristics of conventional and HEAs are shown in Table 1. It can be seen that the design of traditional alloys is in the edge region of the phase diagram, which only accounts

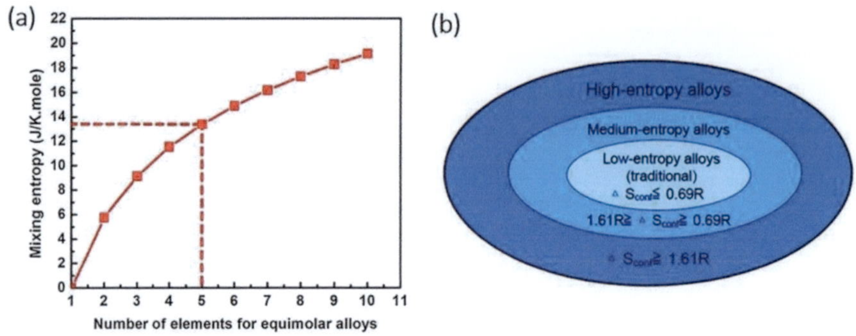

Fig. 2 **a** Effect of the number of equimolar elements on the configurational entropy of alloys. **b** Classification of alloys according to configurational entropy [15]. Reprinted from Progress in Materials Science, 102, Li, Z. et al., Mechanical properties of high-entropy alloys with emphasis on face-centered cubic alloys, 296–345, Copyright (2019), with permission from Elsevier

for a very small part of the design space, while the emergence of HEAs transfers the design of alloys to the center region of the phase diagram, especially the emergence of non-equimolar HEAs, which greatly broadens the space of the alloy design, and opens up a broad new field for the development of alloys.

With the growing body of research on HEAs, four major effects, namely, the high-entropy effect, lattice distortion effect, sluggish diffusion effect, and the cocktail effect, have emerged as core factors contributing to their outstanding performance [16]. The high-entropy effect [17] posits that higher mixing entropy reduces Gibbs free energy, thereby enhancing mutual solubility among constituent elements

Table 1 Characteristics of conventional alloys and HEAs [13]

Alloy	Elements range	Entropy	Lattice	Examples
High entropy alloys	$5 \leq n \leq 13$	$\Delta S_{config} > 1.5\ R$	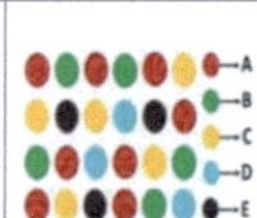	AlCoCrFeMnNi, CoCrFeMnNi, AlCrCuFeNi, MoNbTaW
Medium entropy alloys	$3 \leq n < 5$	$1R \leq \Delta S_{config} \leq 1.5R$	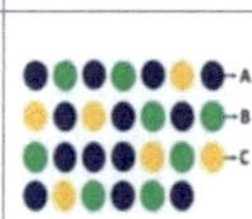	CoCrNi, CoCrFeNi
Low entropy alloys	$1 \leq n \leq 3$	$\Delta S_{config} < 1R$	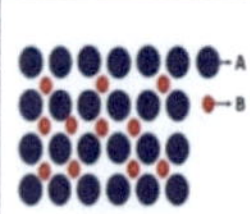	AlMg, AlCu, NiCo

Reprinted from Journal of Materials Research and Technology, 17, Kumar, A. et al., Mechanically alloyed high-entropy alloys: existing challenges and opportunities, 2431–2456, Copyright (2022), with permission from Elsevier and promoting the formation of solid solutions. This tendency favors the development of stable, simple solid solution structures over intermetallic compounds. In HEAs, where each atom is surrounded by different types of atoms, variations in atomic sizes among constituent elements, differences in their bonding energies, and distinct crystal structure trends lead to pronounced lattice distortions. The lattice distortion effect profoundly impacts the structure and properties of HEAs. Severe lattice distortion can impede dislocation motion, consequently enhancing material hardness. However, excessive atomic size disparities may elevate lattice distortion energy to a point where it leads to lattice collapse, resulting in an amorphous structure [18]. The sluggish diffusion effect suggests that phase transitions in HEAs depend on the concerted diffusion of elements to establish an equilibrium distribution between phases. Lattice distortions in HEAs obstruct atomic mobility, causing the diffusion rate of atoms in these alloys to be slower than in conventional alloys. This phenomenon contributes to greater structural stability [19]. The cocktail effect, initially proposed by Ranganathan [20], underscores the collective influence of distinct elements characteristics and interactions between them on HEA performance. This effect augments the diversity and complexity of alloy properties. HEAs exhibit a range of properties, as illustrated in Fig. 3, owing to these unique effects, setting them apart from traditional alloys and making them highly versatile materials with unmatched capabilities.

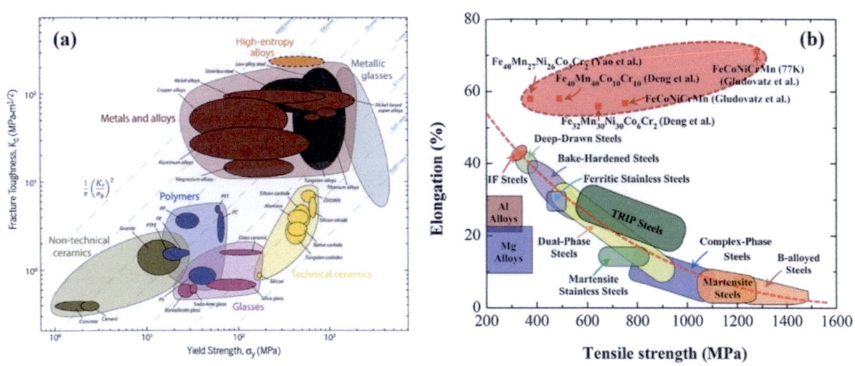

Fig. 3 **a** Fracture toughness as a function of yield strength for CoCrFeMnNi alloy and other materials [15]. Reprinted from Progress in Materials Science, 102, Li, Z. et al., Mechanical properties of high-entropy alloys with emphasis on face-centered cubic alloys, 296–345, Copyright (2019), with permission from Elsevier. **b** Strength versus ductility of HEAs and other conventional alloys [21]. Reprinted from Materials Today, 19(6), Ye, Y. F. et al., High-entropy alloy: challenges and prospects, 349–362, Copyright (2016), with permission from Elsevier

1.2 Synthesis of HEA Coatings

HEA materials can be categorized into bulk materials and coatings. Coatings applied to surfaces can enhance the performance of base materials while preserving their inherent characteristics, leading to cost savings and extended lifespans for structural components. Additionally, the composition and structure of these coatings can be adjusted with flexibility. By introducing reactive gases during the deposition process, a wider design space becomes available, enabling the creation of various high-entropy materials, such as high-entropy nitrides (HENs), high-entropy oxides (HEOs), high-entropy carbides (HECs), etc. Similar to bulk materials, HEA coatings also exhibit the unique characteristics of the four major effects mentioned earlier. As a result, they outperform traditional protective coatings. This achievement has garnered global attention and experienced rapid development over the past decade. It has become a significant focus in the field of HEAs research [6].

Laser Deposition

Laser deposition (LD) is a process that involves a scanning high-energy–density laser beam over the substrate surface, causing the cladding material to melt and rapidly solidify, forming a coating. The technical principles of this process are illustrated in Fig. 4. LD offers several advantages, including high- energy density, rapid deposition rates, and the ability to create coatings with thicknesses on the millimeter scale. These coatings exhibit fine and uniform microstructures and feature a robust metallurgical bond between the coating and the substrate. Consequently, LD has experienced significant development in recent years and has emerged as a primary manufacturing method for HEA coatings [22]. LD can be further categorized into two techniques: laser cladding (LC) and laser surface alloying (LSA). While these methods share similarities, the primary distinction lies in the degree of dilution between the coating and the substrate. LSA coatings represent a mixture of the coating material and the substrate, resulting in a higher degree of dilution compared to LC. In LC, a new layer of plating is applied over the surface of the melted substrate, resulting in lower dilution levels.

The HEA coatings produced using LD primarily consist of Cantor-based HEAs, such as AlCrFeCoNi, AlTiCrFeCoNi, AlSiCrFeCoNi, and CrFeCoNi (Mn/Nb/Mo), among others. These coatings possess lower melting points and cost advantages. In comparison to conventional alloys or metals, these coatings exhibit significantly higher levels of hardness and wear resistance. Additionally, they demonstrate exceptional corrosion resistance owing to their dense, homogenous microstructure and minimal elemental segregation, characteristics achieved through LD processes. For instance, Wang et al. [24] utilized the LD technique to prepare the $MoFe_{1.5}CrTiWAlNb_x$ coating, which boasts an impressive hardness of 910 $HV_{0.2}$. This coating exhibits markedly enhanced wear resistance compared to M2 tool steel. Similarly, Guo et al. [25] employed LD to produce the MoFeCrTiWAlNb coating on

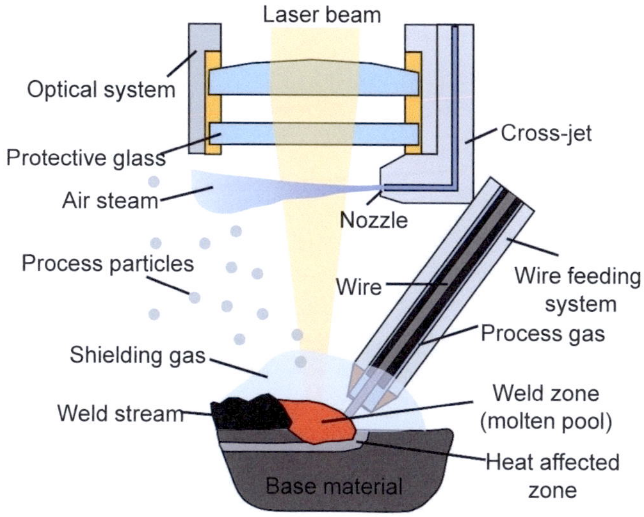

Fig. 4 Schematic diagram of the laser cladding deposition process [23]. Reprinted from Physics Procedia, 39, Klocke, F. et al., Scanner-based Laser Cladding, 346–353, Copyright (2012), with permission from Elsevier

an M2 steel substrate. This coating shows a low coefficient of friction (0.51) and a minimal wear volume (0.0787 mm^3). Its outstanding tribological performance can be attributed to the precipitation of the (Nb,Ti)C phase and the HCP-Fe_2Nb hard phase, along with the self-lubricating effect provided by the Al_2O_3 oxide layer within the coating [26]. Moreover, Wan et al. [27] applied the LD technique to prepare the AlCoCuNiSiTi coating on X70 steel. This coating exhibited excellent corrosion resistance when exposed to a corrosive solution containing NaCl + $Na_2S_2O_3$. The corrosion current density of the coating was approximately 0.5 times lower than that of the substrate, with the coating's superior corrosion resistance primarily attributed to the formation of a dense Al_2O_3 layer on its surface.

However, it is important to note that laser technique comes with inherent challenges, and high residual stresses, cracks, and porosity remain common issues encountered during laser surface treatment [28]. To mitigate these concerns, adjustments to laser power and heat treatment procedures have been explored. Tong et al. [29], for instance, prepared MnFeCoCrNi coatings using the LD technique under varying laser power conditions and subsequently subjected the coatings to high-temperature heat treatment. The impact of laser power and heat treatment on the residual stresses in the coatings is illustrated in Fig. 5. The results indicate that heat treatment, coupled with a moderate increase in laser power, can alleviate stresses within the coatings. However, excessive laser power can exacerbate residual stress levels. Besides residual stresses, challenges related to dilution are also prevalent in HEA coatings produced using LD. This is especially true when coatings are applied to substrates with low-melting points, such as pure Al [30] and pure Mg

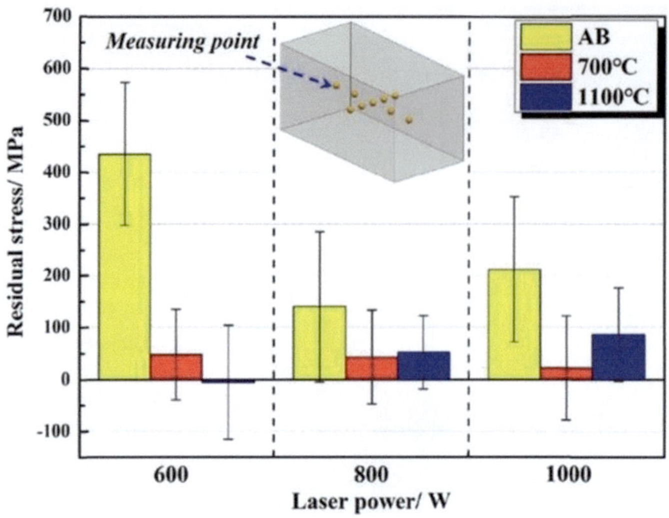

Fig. 5 Variation of residual stresses in the coatings before and after heat treatment of MnFeCoCrNi coatings prepared using different laser powers based on the LD technique, where AB stands for coatings without heat treatment [29]. Reprinted from Journal of Alloys and Compounds, 785, Tong, Z. et al., Laser additive manufacturing of FeCrCoMnNi high-entropy alloy: Effect of heat treatment on microstructure, residual stress, and mechanical property, 1144–1159, Copyright (2019), with permission from Elsevier

[31], leading to extensive mixing of the substrate and coating materials. Strategies such as increasing scanning speed, elevating energy input, and modifying deposition methods [32] have been explored to effectively control the dilution rate. Chao et al. [32] successfully fabricated uniform, nearly defect-free $CoCrFeAl_xNi$ coatings by optimizing parameters such as laser power, scanning speed, laser beam size, powder feed rate, and incubation distance. These optimizations resulted in minimal dilution and a coating-to-substrate interdiffusion thickness of only 30 μm.

The LD technique, owing to its high-energy density, rapid deposition rate, and robust bonding between coatings and substrates, demonstrates significant potential for the fabrication of HEA coatings with low-melting points. Materials produced using this technique exhibit remarkable mechanical properties, wear resistance, and corrosion resistance, among other desirable qualities. However, it is worth noting that coatings prepared by this method are susceptible to issues like heat-affected zones, elevated residual stresses, and dilution of the coating material. Recent research efforts in the field of LD for HEA coatings have primarily focused on reducing residual stresses, mitigating coating defects, and overall improving coating quality through the optimization of process parameters and post-treatment methods.

Thermal Spraying

The thermal spraying (TS) technique involves the process of using a heat source to elevate raw materials to a molten or semi-molten state, propelling them through a flame, and then applying them onto a substrate to create a coating. This principle is visually illustrated in Fig. 6. Depending on the heat source employed, TS can be further subdivided into atmospheric plasma spraying (APS), high-velocity oxy-fuel spraying (HVOF), and cold spraying (CS).

APS employs a hot plasma generated by a DC arc or RF discharge as the heat source for the deposition process. This method has minimal restrictions on the materials that can be coated. The high flame temperature of the heat source concentrates thermal energy, allowing for a rapid deposition rate, the creation of millimeter-scale coatings, and an overall cost-effective and straightforward process. Consequently, it holds great potential for industrial applications and is among the most commonly employed TS techniques [36]. However, one drawback of APS is that powder particles tend to oxidize during flight, resulting in the formation of oxide inclusion defects within the coating and reduced coating density. HVOF spraying, on the other hand, relies on the heat generated from the combustion of fuel and oxygen for the deposition process [37]. This method boasts a high particle spraying rate and operates at relatively lower processing temperatures. The increased particle velocity minimizes oxidation of particles during flight and reduces coating porosity [38]. CS technique involves accelerating fine powder particles in a high-speed air stream to achieve exceptionally high velocities. When the powder particles reach a certain speed, they undergo plastic deformation and adhere to the substrate surface to create a coating. This process operates at low temperatures, boasts high impact speeds, features a fast

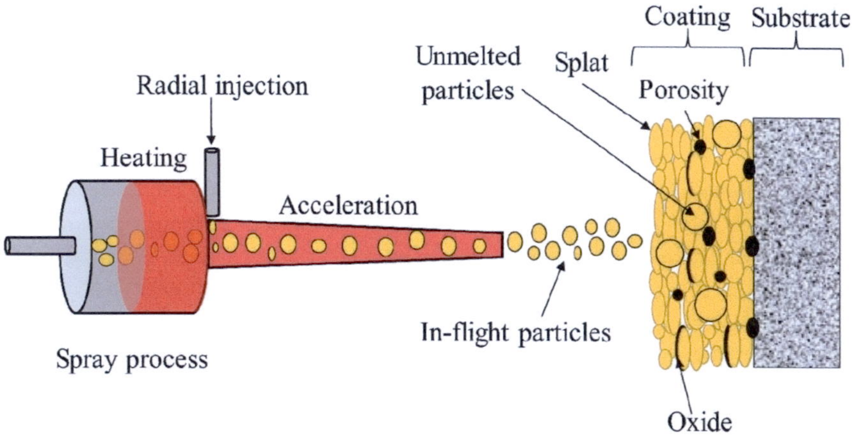

Fig. 6 Schematic diagram of thermal spraying technique [33]. Reprinted from Materials Today: Proceedings, 50, Satyavathi Yedida, V.V. et al., A review on the development of thermal barrier coatings by using thermal spray techniques, 1458–1464, Copyright (2022), with permission from Elsevier

deposition rate, virtually eliminates material oxidation during spraying, and minimizes thermal impact on the substrate. Nevertheless, the method consumes a significant amount of gas for the deposition process, which can be costly, and is typically limited to low-strength and high-plasticity materials due to its inherent mechanism [39].

Similar to the LD technique, HEA coatings prepared using TS mainly consist of low-melting point elements. However, there are significant differences in the structure and morphology of coatings produced by these two techniques [40]. Typically, TS-prepared coatings exhibit a layered structure with randomly distributed defects, including pores, oxide phases, and interlayer cracks. These TS-produced coatings possess outstanding mechanical properties and temperature resistance, making them excellent candidates for replacing traditional thermal barrier coatings. For instance, the AlCoCrFeNi HEA coating, prepared by Ang et al. using APS, boasts a remarkable hardness of 4.32 GPa, surpassing that of conventional APS-sprayed $NiCrAlCoY_2O_3$ thermal barrier coatings. This HEA coating demonstrates excellent mechanical properties [35]. Additionally, the AlSiTiCrFeCoNiMo coatings prepared by Huang et al. through APS spraying exhibit impressive oxidation resistance at high temperatures of 1000 °C with precipitation hardening observed at temperatures up to 1100 °C [41]. In another study, Bhattacharya et al. utilized HVOF-sprayed $Al_{30}Si_2Cr_{23}Co_{22}Ni_{23}$ coating. After subjecting the coating to 100 thermal cycles at 1050 °C, it displayed minimal oxidized weight gain, measuring only 0.2 mg/cm^2. This result contrasts significantly with the data for MCrAlX thermal barrier coatings, which registered an oxidized weight gain of 1 mg/cm^2. These findings highlight the excellent oxidation resistance of the coating, aligning it with the stringent requirements of aerospace applications [42].

The main drawbacks of high-entropy alloy coatings prepared by the TS technique are defects such as stress, laminar structure, and oxide inclusions. Such defects can be effectively reduced or eliminated by post-treatment. Heat treatment and laser remelting of the coating surface are two commonly used post-treatments. Annealing heat treatment can reduce the strain generated during rapid solidification and promote the precipitation of saturated solid solution, thus improving the hardness and wear resistance of the coating, but it is difficult to eliminate the laminar structure and oxide inclusions in the coating. Wang et al. [43] prepared AlCoCrFeNi coatings using the APS technique, and after annealing at 600 °C a σ-phase was generated in the coatings, which led to an increase in the hardness of the coatings from 468 to 575 HV; however, the annealed coatings still possessed an obvious laminar structure and oxide inclusions. Laser remelting helps to eliminate the laminar structure formed during TS deposition and defects such as pores and oxide inclusions, and obtain a uniform and dense coating. The $FeCoCrNiAl_{0.5}Si_x$ coating was prepared by Jin et al. [44] using APS and subjected to laser remelting treatment, which significantly improved the tissue morphology of the coating, eliminated the laminar structure of the coating, and produced an obvious grain refinement effect on the coating. The treated coatings showed good wear resistance. However, the method requires optimization of the laser parameters to reduce the resulting heat-affected zone, solidification cracks, and substrate dilution.

With fast deposition rates and low coating-to-substrate dilution, the coatings prepared by the TS technique exhibit excellent performance in the high-temperature domain, making them ideal replacements for conventional thermal barrier coatings. However, due to the inherent characteristics of the thermal spray process, such coatings usually present a typical layered structure with limited layer bonding and more or less defects such as pores, oxide inclusions, microcracks, etc., so that the coatings present a non-uniform and anisotropic structure and properties, which need to be optimized by controlling the coating compositions and process parameters, as well as by appropriate post-treatments [45].

Physical Vapor Deposition

Physical vapor deposition (PVD) is a process in which the source material is transported to the substrate as atoms or molecules by some physical process and forms a solid-state film. The principle of the PVD technique is schematically shown in Fig. 7. According to the preparation method, the PVD technique can be divided into vacuum deposition (VD), magnetron sputtering (MS), and arc ion plating (AIP).

VD utilizes resistance or electron beam to heat the target material, and the surface of the target material rapidly warms up and evaporates to form a coating on the surface of the substrate. The method has simple equipment, fast deposition rate, but it is difficult to effectively control the current, easy to lead to the rapid evaporation of

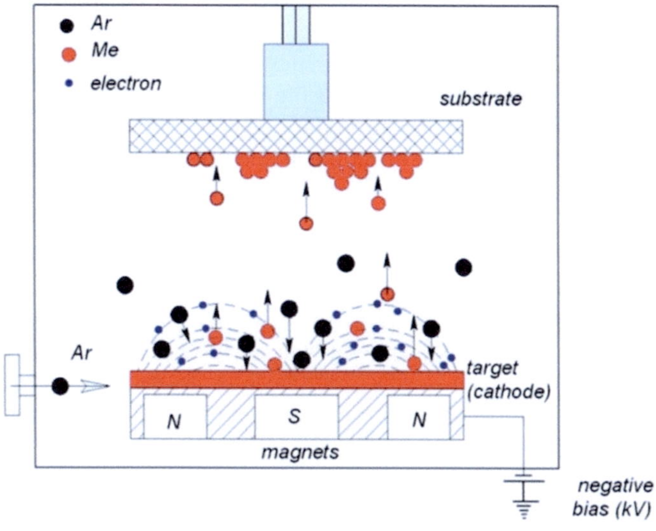

Fig. 7 Schematic diagram of of magnetron sputtering system [46]. Reprinted (adapted) with permission from M. I. Mejía et al. Magnetron-Sputtered Ag Surfaces. New Evidence for the Nature of the Ag Ions Intervening in Bacterial Inactivation. ACS Applied Materials & Interfaces 2, 2010, 230–235. Copyright 2010 American Chemical Society

low-melting point materials, and due to the low energy of the evaporated particles, the adhesion of the coating on the substrate is poor, so its application has greater limitations [47].

The mechanism of MS is that electrons collide with argon atoms under the action of an electric field and ionize a large number of argon ions to bombard the target material, next the sputtered atoms of the target material are gradually reduced in energy after multiple collisions with other atoms and ions, and ultimately deposited on the substrate to form a solid-state film. According to the different power supplies, MS can be divided into DC magnetron sputtering, pulsed magnetron sputtering, and RF magnetron sputtering. Among them, pulse sputtering and RF sputtering techniques sputter a wide range of materials and can avoid target surface poisoning, thereby suppressing the target surface arc problem. MS technique is the ideal first choice for the production of compound thin films. The coating prepared by this method is smooth and dense with few defects but has the disadvantages of low coating deposition rate, high coating stress, and weak bonding between the coating and the substrate [48–50].

AIP uses arc discharge to evaporate metal directly on the cathode target, forming high-energy ions in the substrate, forming a solid film on the surface of the substrate. AIP has one of the highest target material ionization rate among various ion coating techniques, as high as 60–90%. AIP has a high utilization rate of the target, a fast deposition rate, and a good combination of the coating and the substrate. Therefore, it occupies a dominant position in the field of hard coating preparation technology, especially in various kinds of molds and important parts. The main disadvantage of this method is that the coating surface is prone to large particle defects, resulting in roughness of the film layer and loose structure, which reduces the overall performance of the coating. Such defects can be improved by magnetic filtration, adding baffles, increasing the substrate bias voltage, or using a pulsed arc power supply. Among them, the pulsed power supply, instead of a DC power supply, can effectively improve the surface quality of the coating under the premise of ensuring the deposition rate of the coating and strengthening the coating performance, which is the mainstream method of controlling micro-defects in AIP coatings at present [51–53].

The PVD techniques have been successfully applied in the preparation of high-entropy ceramics (HEC) coatings. Different HEC materials such as nitrides, oxides, and carbides can be flexibly prepared by changing the chemical composition of a given target or introducing reactive gases such as N_2, O_2, C_2H_2, etc. Moreover, the structure of the coatings can be precisely tuned to construct advanced structures such as nanocomposite, nanogradient, or nanomultilayer structures. Huang et al. deposited AlCrNbSiTiVN coatings using the MS technique with a high hardness of 41 GPa, maintain a superhard level of more than 40 GPa during annealing at up to 900 °C, and the hardness decreases by only about 4 GPa at 1000 °C, demonstrating excellent thermohardening properties [54]. The (TiZrNbTaFe)N coatings prepared by Bachani et al. [55] by high power pulsed MS technique showed excellent wear resistance with hardness up to 36.2 GPa and wear rate of 2.65×10^{-6} mm^3/Nm. Xu et al. [56] prepared (TiCrZrVAl)N coatings using AIP technique, which have good mechanical properties and wear resistance, with a hardness of up to 31 GPa and a wear rate as low as 7.4×10^{-16} mm^3/Nm.

The current research on PVD-prepared HEC coatings mainly focuses on MS and AIP techniques, which usually have nanoscale grain size, more uniform and dense structure, fewer defects, and less tendency of elemental segregation and second-phase precipitation within the coating. These characteristics make the HECs prepared by these techniques have excellent comprehensive performance [57]. PVD technique has greater flexibility compared with other techniques for the preparation of HEC coatings, both in the selection of deposited materials and in the control of the coating structure, and has obvious advantages in coating performance, therefore this technique is the most cutting-edge technology for the preparation of HEC coatings.

1.3 High-Entropy Nitride Coatings

In recent years, high-entropy ceramics materials such as HEN, HEO, and HEC have received wide attention due to their high hardness, high wear resistance, good oxidation resistance, and excellent chemical stability [58]. It is well known that among the traditional ceramic materials, transition metal nitrides have high hardness, high wear resistance, excellent thermal conductivity, and excellent thermal and chemical stability, and are often used as coatings for industrial molds and have a good basis for industrial applications. Therefore, the current research on high-entropy ceramics coatings mainly centers on nitride coatings, especially the HEN coatings prepared by the PVD method, which nowadays are the mainstream direction of the current research on high-entropy ceramics coatings in single layer [59–67] or multilayer form [68, 69].

HEN coatings are mainly divided into two categories, one is composed of weak nitride or non-nitride forming elements such as Mn, Fe, Co, Ni, Cu, etc. These coatings have low hardness, generally less than 20 GPa. The other is composed of nitride forming elements such as V, Ti, Zr, etc., which have a high hardness greater than 20 GPa. Liang et al. [70] used the MS technique to prepare (TiVCrZrHf)N coatings with a hardness of up to 48 GPa, which exceeds many conventional high-hardness nitride coatings. The elastic modulus and hardness of HEN coatings compared with traditional binary nitrides, ternary nitrides, nanocomposite, and nanomultilayer nitride coatings are shown in Fig. 8, where the hardness and elastic modulus of HENs composed of strong nitride forming elements exceed those of most of the ternary nitride coatings and can reach to the level of some of the nanocomposite and nanomultilayer coatings.

The current research direction of HEN coatings mainly includes optimization of coating composition and structure. Nitrogen flow has an important influence on the structure and performance of the coating. Numerous studies have shown that under low nitrogen flow conditions, HEN coatings tend to form an amorphous structure, and with the increase of nitrogen flow, the coating will gradually transform from an amorphous structure to an FCC structure [71–73]. This is mainly attributed to two

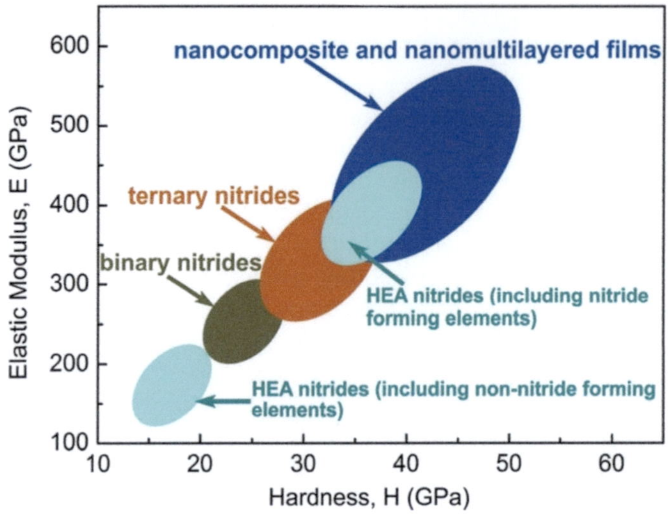

Fig. 8 Trend plots of elastic modulus with hardness for binary nitrides, ternary nitrides, HENs, nanocomposites, and nanomultilayer coatings [6]

factors, one is the formation of more and more binary nitrides with FCC structure with increasing nitrogen flow rate. Secondly, the increase in nitrogen flow increases the mobility of atoms, which makes the grains easy to grow, and this process can effectively improve the mechanical properties of HEN coatings.

In addition to the tuning of N content, multi-component composition is also an important research direction for the HEN coatings. The addition of metal elements can improve the mixing entropy of the material and enhance the solid solution strengthening effect, so that the material has better structural stability and mechanical properties, and still shows comparable or even superior performance to the traditional nitride or HEN coatings under the condition of no bias and heating, which is a better adaptability to many substrate materials with poor temperature resistance and high insulating substrates. Tsai et al. [74] obtained a high hardness of 40.2 GPa of AlMoNbSiTaTiVZrN HEN coating prepared by MS technique without heating and bias. Chang et al. [75] also obtained a high hardness of 34.0 GPa of TiVCrZrNbMoHfTaWAlSiN eleven element HEN coating prepared by MS technique. In addition, the properties of coatings can be optimized by adding non-metallic elements. The addition of C can introduce covalently bonded carbides to form a nanocomposite structure, enhance the lattice distortion of the coating, and improve the hardness of the coating. Lin et al. [76] used MS to prepare AlCrTaTiZrNC coatings, and the hardness of the coatings increased with the increase of bias voltage and C content from 20 to 35 GPa, and their H/E and H^3/E^2 values increased from 0.07 and 0.10 GPa to 0.12 and 0.43 GPa, respectively, showing good resistance to plastic deformation.

Optimization of the coating structure mainly consists of coating densification. The densification of the HEN coating structure is usually realized by controlling the bias voltage of the substrate. Applying bias can increase the ion bombardment energy, strengthen the target material atomic diffusion, and participate in the chemical reaction, which is conducive to improving the coating densification. At the same time, the bias affects the grain growth, orientation, residual stress, defect density, etc., which in turn affects the mechanical properties of the coating. HEN coatings with high hardness and high toughness can be obtained by adjusting the bias [77]. Nanocomposite is mainly used to construct nanocrystalline-amorphous composite structures to strengthen the mechanical properties of HEN coatings. The strengthening mechanism is mainly realized by introducing Si into the coating. The concept of nanocrystalline-amorphous nanocomposite coating was first proposed by Veprek et al. [78] in TiSiN coatings to explain the ultra-high hardness of 70–100 GPa in this coating. The model suggests that the composite structure of amorphous Si_3N_4 encapsulating TiN nanograins (nc-TiN/a-Si_3N_4) can limit the proliferation and movement of dislocations, which leads to the superhard effect. There are two different views on the strengthening mechanism of nanocomposite structures in HEN coatings. One point of view is that there is a similar strengthening mechanism for HEN coatings with conventional nitrides, and the appropriate Si content can make the coating form a nanocrystalline-amorphous composite structure, at which time the hardness of the coating can be strengthened to the maximum extent. Too low content of Si will lead to its solid solution in the form of solute atoms in the FCC lattice, while too high content of Si will lead to the increase of amorphous phases. The results of Hsueh et al. [79] show that due to the high-entropy effect, Si is solidly dissolved in the FCC lattice, the amorphous phase is not precipitated even at very high Si content, and the resulting solid solution strengthening is the main reason for the hardening of the coating.

In addition to the mechanical properties, the high-temperature resistance of HEN coatings has also gained wide attention. Conventional nitrides will cause softening of the coating at high temperatures due to stress relaxation and grain coarsening, reducing the mechanical properties of the coating. The thermal hardness of coatings can be improved in conventional nitrides by limiting grain coarsening through nanocomposite structures or precipitating a second phase, but there are some limitations in this method. For example, the precipitation of the amorphous Si_3N_4 phase along grain boundaries in TiSiN system can limit grain growth to a certain extent, but the coating softens due to the coarsening of grain size at 1200 °C [80]. The age hardening caused by the precipitation of the FCC-AlN phase in TiAlN coatings can compensate for the reduction of hardness of coatings in the process of high-temperature recovery, but the coating is not hardened after annealing at 1000 °C for 2 h due to the precipitation of the AlN phase. After annealing for 2 h, the hardness of the coating decreases significantly due to the transformation of the AlN phase from the FCC structure to the HCP structure accompanied by grain coarsening [81]. Unlike conventional nitride coatings, the grain size of HEN coatings remains almost unchanged at high temperatures [82], and therefore the coatings can still maintain their high hardness after high-temperature annealing. TiVZrNbHfN coatings prepared by Pogrebnjak et al. [83–86] using cathodic arc vapor deposition technique. The coating demonstrated an excellent hardness of up to 60 GPa. The influence of the implantation of heavy ions on the microstructure and mechanical properties of the coating also was studied [87].

The high-temperature resistance of the HEN coating can be attributed to its unique high-entropy effect, lattice distortion effect, and sluggish diffusion effect. The former can counteract the free energy increase caused by different atomic sizes and crystal structures in the coating and inhibit the phase decomposition of the coating, while the latter two can slow down the diffusion of elements and hinder the growth of grains. HEN coatings also have more prominent advantages than conventional ternary and quaternary nitride coatings in resisting high-temperature oxidation, especially the AlCrNbSiTiN system, which has demonstrated quite excellent properties. The $(Al_{23.1}Cr_{30.8}Nb_{7.7}Si_{7.7}Ti_{30.7})_{50}N_{50}$ coating prepared by Hsieh et al. [88] using the MS technique was oxidized at 900 °C for 2 h with an oxidation depth of only 100 nm, while the $(Al_{29.1}Cr_{30.8}Nb_{11.2}Si_{7.7}Ti_{21.2})_{50}N_{50}$ coating not only had a hardness of more than 35 GPa, its oxide film thickness is only 80 nm after 2 h annealing at 900 °C.

High-entropy ceramics coatings show better performance than conventional ceramic coatings. Among them, HEN coatings have become the mainstream research direction of HEA coating materials due to their excellent comprehensive performance and good application basis in the industry. At present, a large number of studies have been carried out on the structure and properties of such coatings, and some excellent results have been achieved, mainly focusing on the study of the coating structure and properties. Multi-alloying and nanocomposites are the current frontiers of HEN coating research. However, the current research suffers from the following shortcomings: lack of theoretical exploration and experimental validation work on the toughening of HEN coatings. In traditional nitride coatings, the construction of nc-TMN/a-Si_3N_4 (TM is the transition group metals) nanocomposite structure and nanomultilayer structure are two classical means of coating toughening. In HEN coatings, the role and applicability of nc-TMN/a-Si_3N_4 nanocomposite structures for hardening model strengthening are still unclear, and the segregation behavior of Si_3N_4 in the coatings needs to be further explored.

2 Deposition of AlCrNbSiTiN Nanocomposite Coatings

AlCrNbSiTiN coatings were deposited using MS [89] and AIP [90] techniques. The schematic diagram of the deposition system is shown in Fig. 9. The system is equipped with one RF magnetron and two DC arc power supplies. Three 6 in. targets of the $Al_{34}Cr_{22}Nb_{11}Si_{11}Ti_{22}$ alloy, pure Ti (99.99%) and pure Cr (99.99%) were used in the deposition process. As substrates, silicon, 304 stainless steel (SS), and YT15 cemented carbide WC-TiC(15%)-Co(6%) were used. The substrates were placed on a rotating sample holder in the deposition chamber. The distance between the targets and the substrates was 150 mm. The chamber was heated to 300 °C to remove the water vapor and was evacuated to a base pressure below 6×10^{-4} Pa. Next, an arc-enhanced glow discharge (AEGD) process was conducted. The substrate surfaces were etched in this process, and the contaminants such as surface oxides could be effectively removed, and the roughness could be increased, thus increasing the adhesion strength. Subsequently, Cr and CrN layers were successively deposited as the interlayers using arc ion plating to improve the adhesion strength of the coating. Then the AlCrNbSiTiN coatings were deposited by RF MS or AIP techniques. The detailed deposition parameters are listed in Table 2.

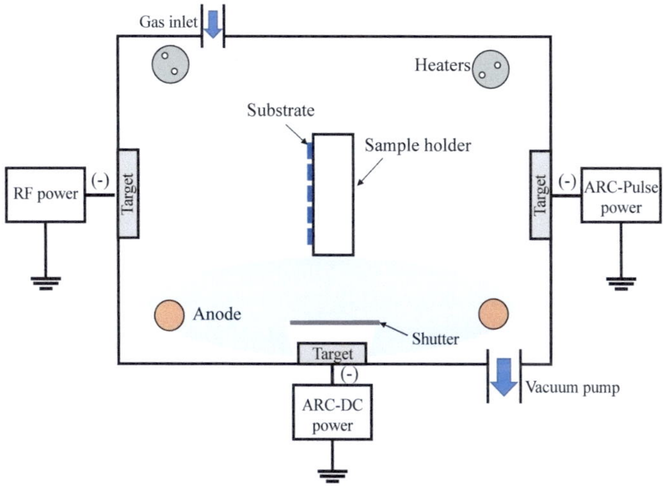

Fig. 9 The schematic diagram of the deposition system

Table 2 Deposition parameters of the AlCrNbSiTiN coating

Layer	Technique	Bias voltage, V	Gas and pressure, Pa	Target/Current or power	Time, min
AEGD	DC arc	−150	Ar/0.5	Ti/100 A	30
Cr	DC arc	−150	Ar/0.5	Cr/100 A	5
CrN	DC arc	−150	Ar:N_2 (1:1)/1	Cr/100 A	5
AlCrNbSiTiN	RF MS	−50, −100,	Ar:N_2 (1:1)/1	AlCrNbSiTi/ 900 W	120
AlCrNbSiTiN	AIP	−150, −200	N_2/3.5	AlCrNbSiTi/ 120 A	45

3 Characterization of the AlCrNbSiTiN Coatings

3.1 AlCrNbSiTiN Coatings Deposited by MS Technique

Morphology and Chemical Composition

The scanning electron microscopy (SEM) and atomic force microscopy (AFM) images of AlCrNbSiTiN coatings deposited under different bias voltages are shown in Fig. 10. At low bias voltage, some particle defects are present on the surface of the coating, while at high bias voltages the particles disappear and craters appear on the surface of the coating. The coating surface roughness increases from 4.16 to 9.21 nm after applying −50 V bias and then decreases with increasing bias. The increase in roughness after applying −50 V bias may be due to the grain growth of the coatings. Further increase of the bias voltage enhances the bombardment effect, the diffusion rate of the adsorbed atoms on the surface is increased, the "peaks" gradually become flat, the coating becomes smoother and flatter, and the surface roughness decreases. The cross-sectional morphology of the coatings shows that when the bias is not applied, the cross-sections of the coatings are in the form of loose scales, and after the bias is applied, the scales disappear and the cross-sections become dense. All the coatings were well bonded to the substrate, and no obvious defects such as holes and cracks were found.

The variation of coating thickness and deposition rate with bias is shown in Fig. 11a. When the bias is not applied, the coating thickness is 2445 nm and the deposition rate is about 20.4 nm/min. At −200 V the coating thickness and the deposition rate are reduced to 1570 nm and 13.1 nm/min, respectively. As the bias voltage increases, ion collisions intensify, and the secondary sputtering effect strengthens, leading to a decrease in coating deposition rate. Figure 11b shows the distribution of elements on the coating surface, and it can be seen that all the elements show uniform distribution. The elemental content of the coating determined by EDS technique is shown in Fig. 11c. It can be seen that with the increase of bias voltage, the N elemental content does not change significantly and fluctuates around 52 at.%, while the Al elemental content decreases slightly from 17 to 14 at.% when the bias voltage increases from 0 to −200 V.

Microstructure

The X-ray diffraction (XRD) patterns of the AlCrNbSiTiN coatings are shown in Fig. 12. It is observed that all the coatings exhibit FCC structure. A strong (200) peak and a weaker (111) peak are observed in the coating deposited at −50 V bias, showing the (200) preferred orientation. As the bias voltage increases, the intensity of the (200) peak decreases, and a (220) peak is detected. At −150 V, the intensity of the (220) peak is further enhanced and the (111) and (200) peaks are almost disappeared, indicating the change of the preferred orientation to (220). It has been

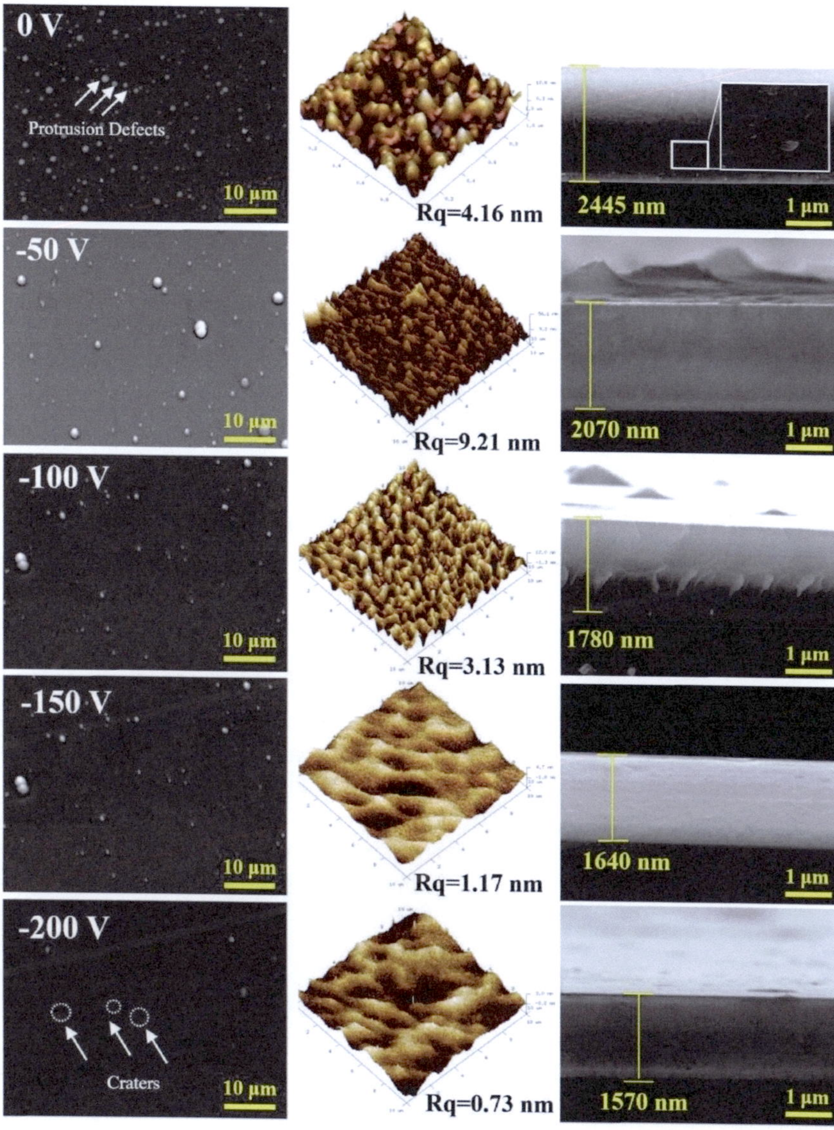

Fig. 10 SEM and AFM images of AlCrNbSiTiN coatings deposited under different bias voltages

reported that the preferred orientation is controlled by the competition between the surface energy (S), strain energy (U), and stopping energy (E). The E is related to the ion bombardment, and the high-energy bombardment can cause high E. When the bias voltage is low, the E is small, and the preferred orientation is mainly controlled by the sum of S and U. It has been reported that the competition between the surface energy and the strain energy is controlled by the ratio of the coating thickness (t)

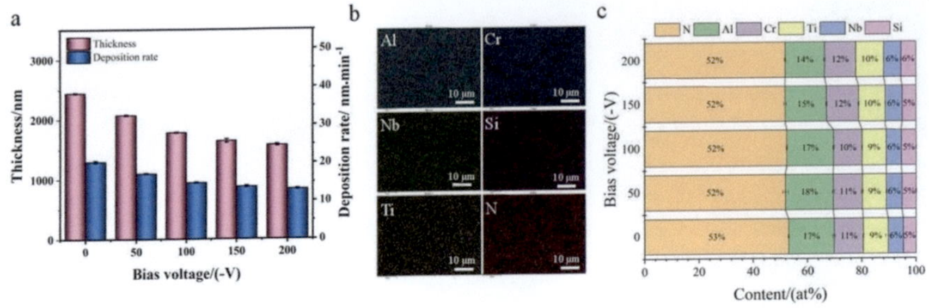

Fig. 11 **a** Thickness and deposition rate **b** elemental distribution and **c** elemental content of AlCrNbSiTiN coatings prepared at different bias voltages

to the bias voltage (V), and there is a critical value of t/V which causes the dominant factor of preferred orientation to change from strain energy to surface energy. The (200) preferred orientation at −50 V exhibits the lowest surface energy and can be attributed to the t/V ratio below the critical value [91]. With increasing the bias voltage, the stopping energy becomes the dominant factor, thus the (220) plane which exhibits the lowest stopping energy is expected to be the preferred orientation [92]. Further increase in the bias voltage results in a decrease in the intensity and broadening of the (220) peak, which can be attributed to the grain refinement. Similar results have been obtained for other coatings deposited by RF MS [93].

X-ray photoelectron spectroscopy (XPS) analysis of the coating deposited at −150 V was conducted to investigate the chemical bonding, and the result is presented in Fig. 13. The Al 2p signal can be assigned to two peaks, Fig. 13a, AlN at 73.4 eV and Al_2O_3 at 74.3 eV [94]. The Cr 2p spectrum is shown in Fig. 13b. The peak at 574.9 eV corresponds to CrN, and the peak at 576.9 eV belongs to Cr_2O_3 [94]. The Nb 3d spectrum reveals the existence of NbN and Nb_2O_5, Fig. 13c, located at 204.9 and

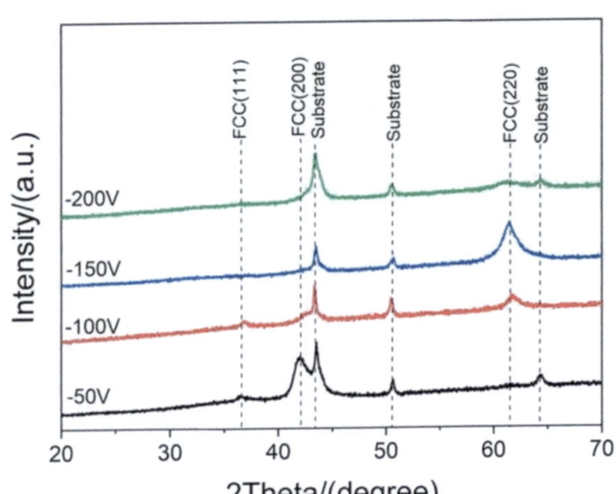

Fig. 12 XRD patterns of the AlCrNbSiTiN coatings deposited at different bias voltages [89]. Reprinted from Surface and Coatings Technology, 433, Liu, Y. et al., Structure and mechanical properties of multi-principal-element (AlCrNbSiTi)N hard coating, 128,113, Copyright (2022), with permission from Elsevier

207.5 eV, respectively [95]. Two peaks located at 101.1 eV and 102.7 eV correspond to Si_3N_4 and SiO_2, Fig. 13d [96]. The Ti 2p spectrum is presented in Fig. 13e, and it can be decomposed into two peaks located at 456.3 and 458.4 eV, which are identified as TiN and TiO_2, respectively [97]. The XPS spectrum of the N 1 s can be decomposed into two peaks, Fig. 13f. The peak centered at 396.5 eV can be attributed to the AlN/CrN/NbN/TiN [98, 99]. The binding energies of these nitrides are very close, thus it is difficult to resolve them. The peak at 398 eV is identified as Si_3N_4 [100].

In order to further investigate the microstructure of the coatings and to clarify the form of Si_3N_4, the coatings deposited at the bias of 0, −50, and −150 V were selected for transmission electron spectroscopy (TEM) analysis. The TEM cross-sectional morphology of the three sets of coatings is shown in Fig. 14. The cross-sectional morphology of the coatings, when no bias was applied, showed loose fibrous grains along the growth direction of the coatings, with the width of the fibrous crystals of about 20 nm, Figure 14a and b. Three strong diffraction rings were detected in the selected area electron diffraction (SAED) patterns, corresponding to FCC(111), FCC(200), and FCC(220). The TEM of the −50 V coating is shown in Fig. 14c and d. The coating cross-section near the substrate shows morphology with inverted cones and the part near the surface shows columns. The average width of the columns is about 99 nm. The SAED pattern of the coatings deposited at −50 V

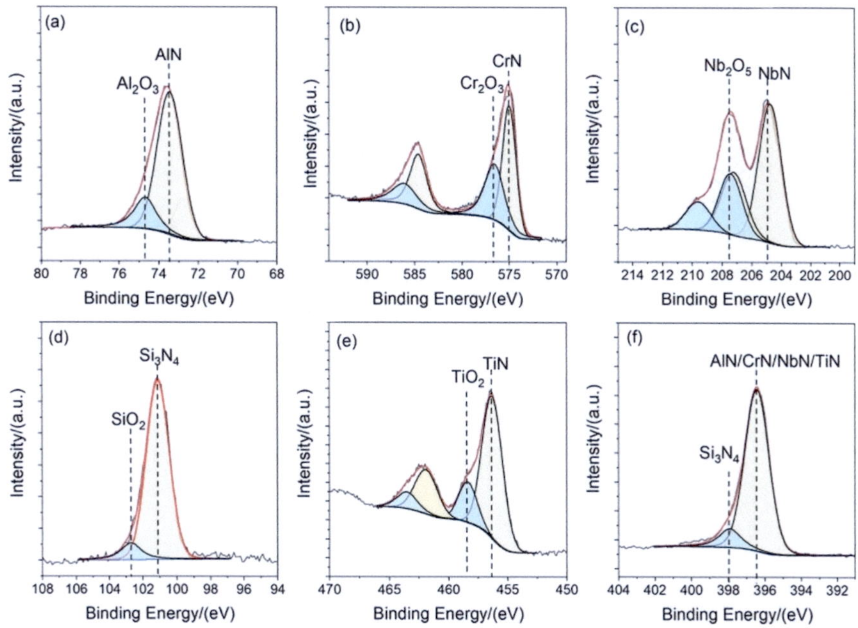

Fig. 13 XPS energy spectra of AlCrNbSiTiN coatings deposited at −150 V with **a** Al 2p, **b** Cr 2p, **c** Nb 2d, **d** Si 2p, **e** Ti 2p, and **f** N 1s [89]. Reprinted from Surface and Coatings Technology, 433, Liu, Y. et al., Structure and mechanical properties of multi-principal-element (AlCrNbSiTi)N hard coating, 128,113, Copyright (2022), with permission from Elsevier

shows three diffraction rings corresponding to FCC(111), FCC(200), and FCC(220). The diffracted ring with the highest intensity is (200), which is consistent with the XRD result. According to Fig. 14e and f, the cross-section of the coatings prepared at −150 V presents a dense fine-fiber-like shape with an average fiber width of about 30 nm. The SAED pattern of the coatings under −150 V bias transforms from diffracted rings to circular arcs corresponding to FCC(111), FCC(200), and FCC(220), where (220) has the highest intensity, which is in agreement with the XRD result.

The coating has a loose fibrous crystal structure at low bias, and with the increase of bias, the coating structure transforms into denser inverted conical grains and finally forms a dense fine fibrous crystal structure. According to the structural zone diagram (SZD) of thin film growth proposed by Thornton [101] and Anders [102], the structure and morphology of the film can be divided into four zones, as shown in Fig. 15.

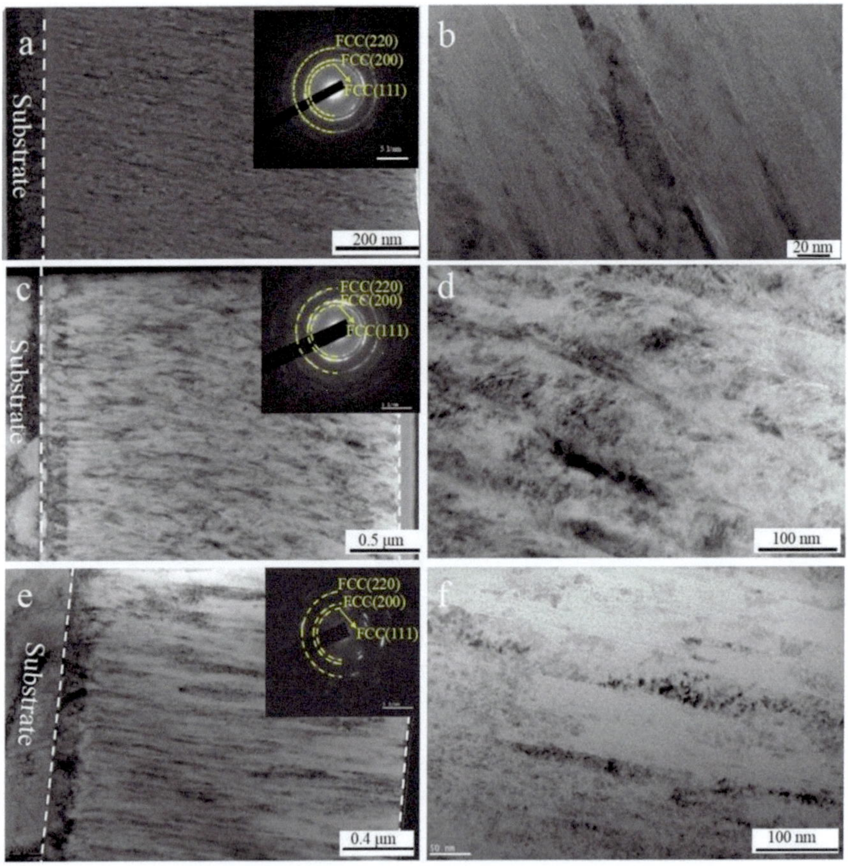

Fig. 14 TEM cross-sectional images and SAED patterns of AlCrNbSiTiN coatings deposited at different bias voltage: **a** and **b** 0 V, **c** and **d** −50 V, **e** and **f** −150 V

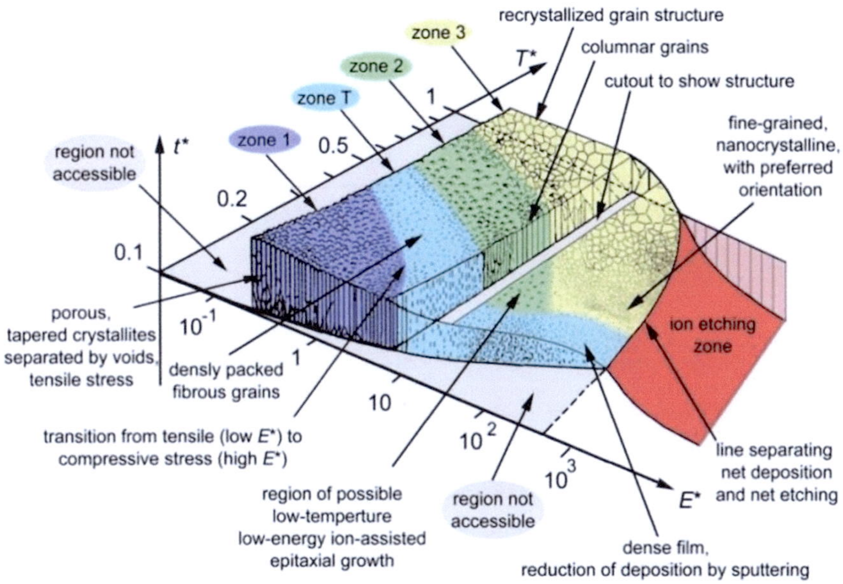

Fig. 15 Structural region modeling of thin film growth [102]. Reprinted from Thin Solid Films, 518(15), Anders, A., A structure zone diagram including plasma-based deposition and ion etching, 4087–4090, Copyright (2010), with permission from Elsevier

In Zone 1, the diffusion rate of adsorbed atoms on the substrate surface is very low, leading to continuous nucleation of grains, which results in the formation of fibrous crystals parallel to the growth direction of the coating. The film structure is relatively loose, with a high density of lattice defects and pores at the grain boundaries. In Zone T, there is a high diffusion rate of the adsorbed atoms but the diffusion at grain boundaries is strongly restricted, so the grains compete for growth and form an inverted conical structure. In Zone 2, the diffusion rate of atoms on the substrate surface and at grain boundaries are high, and a uniform columnar crystal structure is formed. The size of the columnar crystals becomes larger with increasing temperature. In Zone 3, the increased diffusion of adsorbed atoms and recrystallization at high temperatures lead to the formation of large crystals in the coating, and the coating becomes denser. According to the SZD model, at 0 V the coating structure is in zone 1 with a loose fibrous crystal structure, and at −50 V the coating can be classified as zone T with relatively dense inverted conical grains. According to the prediction of the SZD model, with the effect of high-energy bombardment brought about by the increase of deposition bias and the rise of substrate temperature, the grain boundaries that were originally restricted by diffusion gained energy to migrate, and coarser columnar crystals will be formed in the coating. In the experiment, the columnar crystals in the coating under higher bias are not coarsened but further densified, and there are two main reasons for this: first, it is related to the lattice distortion and sluggish diffusion effect of the high-entropy coating, and the activation energy of the atomic

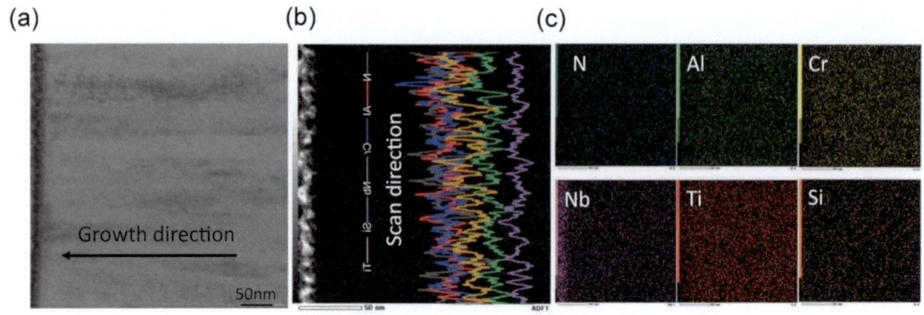

Fig. 16 **a** TEM image of the AlCrNbSiTiN coating deposited at −150 V, **b** EDS line, and **c** mapping analysis of the region shown in (**a**) [89]. Reprinted from Surface and Coatings Technology, 433, Liu, Y. et al., Structure and mechanical properties of multi-principal-element (AlCrNbSiTi)N hard coating, 128,113, Copyright (2022), with permission from Elsevier

migration in the coating is higher, so the grain boundaries are more difficult to move. The second is the grain refinement produced by the introduction of a large number of defects in the coating under the bombardment of high-energy ions, which become the nucleation points of the grains.

TEM energy dispersive X-ray spectrometry (EDS) of the coating deposited at the bias of −150 V was performed and the results are shown in Fig. 16. The test area contains multiple columns. The morphology of the columnar crystals is almost indistinguishable in the TEM dark field image, while all the coating elements in the EDS mapping are uniformly distributed in these columnar crystals without any obvious elemental segregation.

High-resolution transmission electron microscope (HR-TEM) images of the coatings deposited at different bias voltages are shown in Fig. 17. When no bias is applied, the lattice stripes inside the columnar crystals are continuous through the coating, and the amorphous phase is found near the grain boundaries, i.e., the coating structure is nanocrystalline-amorphous. The inverse fast Fourier transform (IFFT) image of the selected region shows clear, distortion-free, continuous lattice fringes. For the coating deposited at −50 V, there are many nanocrystalline grains separated by boundaries inside the columnar crystals, and the lattice striations of the grains can continue through the boundaries, but there are some orientation differences between the neighboring grains. Its IFFT image also shows that the lattice fringes can continuously pass through several nanograins and the interface, indicating that the interface and the grains are co-lattice grown, but the lattice distortion at the interface is serious, resulting in the growth orientation of the grains deflected by a certain angle after they pass through the interface. The HR-TEM image of the coating deposited at −150 V is similar to that deposited at −50 V, but its HR-TEM and IFFT images show that the distribution of boundary phases in the coating under this condition is more dispersed, showing a network-like encapsulation of grains. Combined with the XPS results, we can conclude that the Si exists in the form of Si_3N_4, and Si_3N_4 and TiN, CrN, and NbN have a strong phase separation, so the boundary of the columnar crystal should be Si_3N_4. In the absence of bias, the energy of the ion bombardment is small, and the

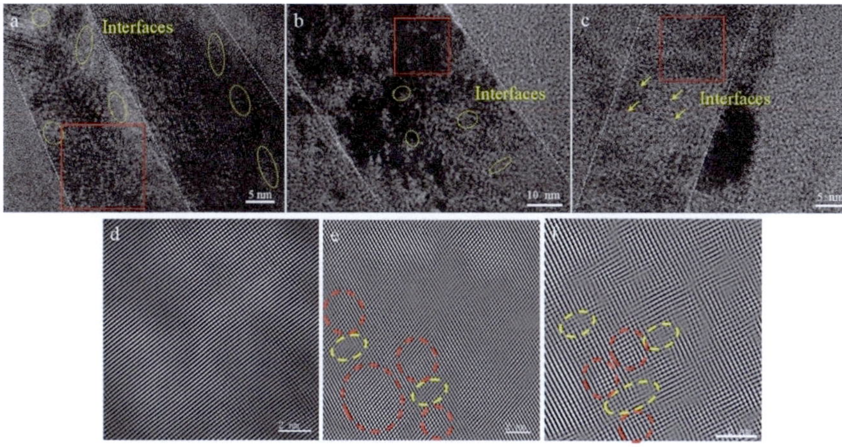

Fig. 17 HR-TEM images of AlCrNbSiTiN coatings, **a** 0 V, **b** −50 V, **c** −150 V, and IFFT images corresponding to the red boxed areas **d** 0 V, **e** −50 V, **f** −150 V. The red circles in the figures mark the grains in the coatings, and the yellow circles mark the interfacial phases

rate of diffusion of Si_3N_4 is limited, so it tends to stay at the boundary of the columnar crystals showing an amorphous state. The coating has a nanocrystalline-amorphous (nc-AlCrNbTiN/a-Si_3N_4) structure. After applying bias voltage, the bombardment energy is enhanced, which improves the diffusion rate of Si, and Si_3N_4 is no longer concentrated at the grain boundaries but distributed in FCC columnar crystals. The distribution of Si_3N_4 is more dispersed and the thickness is thinner, and under the template effect, it forms the crystalline state and eutectic interface with the neighboring grains. The coating forms the nanocomposite structure of nc-AlCrNbTiN/nc-Si_3N_4. Subsequent grains can continue to grow on the Si_3N_4 boundary, but due to the severe distortion of the Si_3N_4 lattice, the post-growth grains will produce small orientation differences, which ultimately result in a continuous deflection of the lattice striations of the columnar crystals. With the increase of bombardment energy, the distribution of Si_3N_4 becomes more dispersed, reticular boundaries wrapping AlCrNbTiN grains can be observed, and the grains within the columnar crystals are further refined. No crystalline Si_3N_4 was detected in XRD and SAED because of the thin Si_3N_4 boundaries, severe lattice distortion, and very dispersed distribution, which made it difficult to be detected by XRD or SAED techniques [103].

Mechanical Properties

The variation of residual stress, nanohardness, and elastic modulus of the coatings deposited at different bias voltages are shown in Fig. 18a. All coatings exhibit compressive residual stresses. The residual stress in the coating is small at about 1 GPa when no bias was applied. The residual stress increased to 3.7 GPa when the bias was −50 V. The residual stress in the coating did not change much when

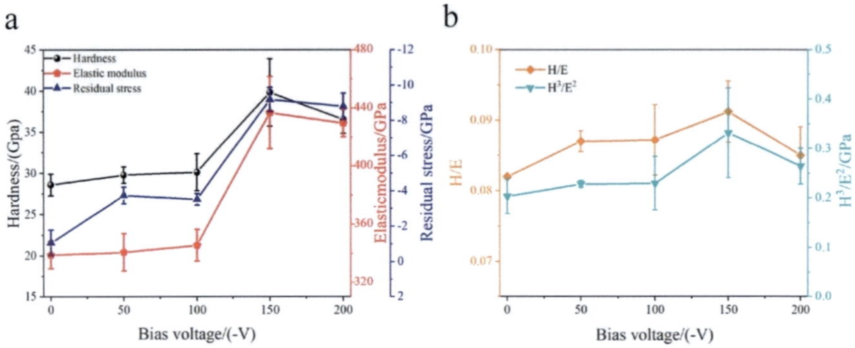

Fig. 18 Variation of **a** residual stress, hardness, and elastic modulus and **b** H/E and H^3/E^2 ratios with bias voltage

the bias was increased to −100 V. A significant increase to 9.2 GPa was observed when applying a bias voltage of −150 V, exhibiting a higher stress level. Further, increase in bias voltage slightly decreased the residual stress to 8.8 GPa. The high-stress level of the coating is due to the fact that particles reaching the coating surface, including sputtered atoms and backscattered carrier ions, typically have energies of 10 to 100 eV [104], which exceeds the threshold energy required to displace the atoms in the coating [105], leading to the displacement of already deposited atoms as well as the entry of newly arrived atoms into the lattice as interstitial atoms. At the same time, the relatively low deposition temperature does not allow the coating to provide enough energy for structural relaxation through atomic movement, resulting in large residual compressive stress. With an increase in bias voltage, the bombardment energy increases, and the defect density in the coating rises, thus increasing the residual stress in the coating [106]. The hardness and modulus of elasticity of the coatings increased from 28.6 GPa and 338 GPa to 39.8 GPa and 436.7 GPa, respectively, as the bias voltage was increased from 0 to −150 V, and decreased slightly to 36.5 and 429.5 GPa at a bias voltage of −200 V. The lower hardness of the coatings at low bias voltages is due to the fact that there are many defects within the coatings, the structure is not sufficiently dense, and the amorphous phase is distributed around the grain boundaries, which cannot interrupt the continuous growth of columnar crystals, and it is difficult to hinder the continuous movement of dislocations within the columnar crystals. The high hardness of the coatings under high bias can be attributed to the densification of the coatings and the presence of nc-AlCrNbTiN/nc-Si_3N_4 nanocomposite structure within the columnar crystals of the coatings. The fine-grain reinforcement resulting from grain refinement by the networked Si_3N_4 boundary phase and the co-lattice interface resulting from the coherent growth of Si_3N_4 and AlCrNbTiN can effectively impede the dislocation motion and thus strengthen the mechanical properties of the coatings [107]. The H/E and H^3/E^2 ratios of the coatings are shown in Fig. 18b. Similar to the hardness and modulus of elasticity, the H/E and H^3/E^2 ratios increased from 0.082 and 0.203 GPa to 0.091 and 0.331 GPa, respectively, with the increase of bias from 0 to −150 V, and then decreased to 0.085 and 0.264 GPa at −200 V, respectively. The coatings had the highest hardness and optimal toughness at −150 V.

The topography of the abrasion wear scars of the coating is shown in Fig. 19. At a bias voltage of 0 V, the width of the wear scar is 276 μm, and it can be seen that the scar shows a bright white color compared with the surrounding coatings, which indicates that the coating is worn through here, exposing the substrate material. The line scan results of the abrasion scar show a strong signal of the substrate element W

Fig. 19 The wear scar morphology of AlCrNbSiTiN coatings prepared under different bias voltages and the corresponding EDS line scan results

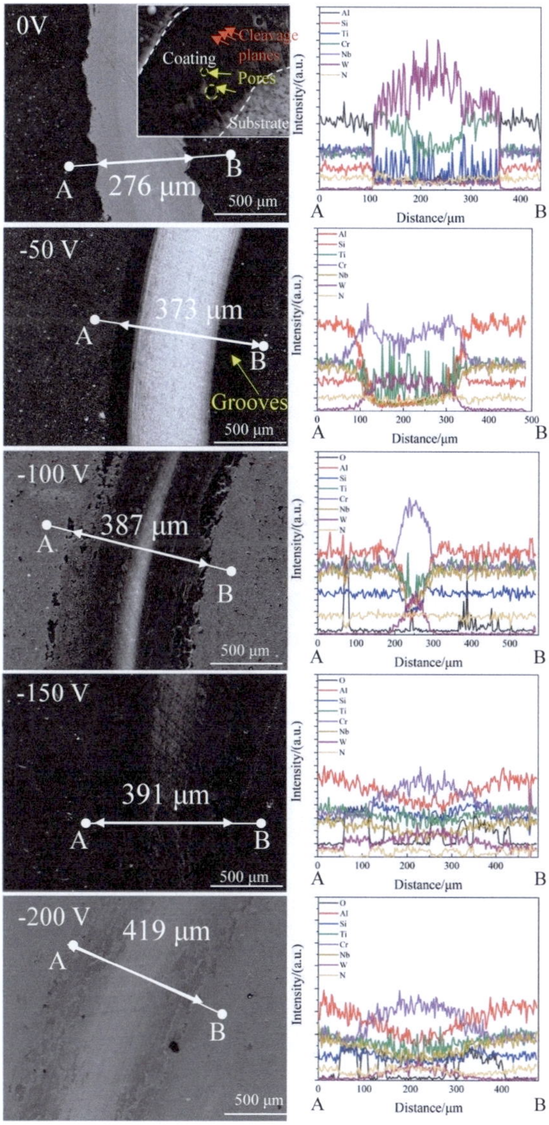

at the abrasion scar, while the coating element has a reduced signal intensity, which further confirms the failure of the coating. From the cross-sectional image of the coating abrasion scar, it can be seen that there are many pores in the coating, and cleavage planes along the growth direction of the coating are observed, indicating that the coating is brittle and fractured. The coating is not dense enough without bias, and defects such as columnar inter- and intra-crystalline defects and holes in the coating are the main reasons for the friction failure of the coating. At −50 V, the width of the abrasion scar was 380 μm and the width of the coating flaking area was 223 μm. At −100 V, the width of the abrasion scar decreases slightly to 371 μm, and the width of the flaking area of the coating is as low as 65 μm. In addition, the EDS results of the abrasion scar show a significant increase in the oxygen content in the area of the abrasion flakes, which indicates that oxidation occurs during the friction process. When the bias voltage was increased to −150 V, the width of the abrasion scar of the coating was 375 μm, and an enhanced region of about 180 μm wide of the signal of the substrate elements was observed, but the enhancement of the signal was smaller compared with that in the low bias voltage condition, indicating that the coating wear degree was relatively low. The width of the abrasion scar for the coating prepared at −200 V increased to 419 μm, and a slight enhancement of the matrix element signal was found in the middle of the abrasion scar in an area of about 250 μm in length.

The coefficient of friction (COF) values and wear rates of the coatings are shown in Fig. 20. The COF of the coatings is 0.44 at 0 V, increases to 0.76 at −50 V, decreases to about 0.6 at −100 V, and then rises with increasing bias to reach 0.67 at −200 V. The coatings have the lowest COF at 0 V can be attributed to the premature flaking failure of the coatings. The coating wear rate decreases and then increases with increasing bias voltage, and the coating deposited at −150 V has the lowest wear rate of 6.7×10^{-6} mm^3/Nm, which exceeds that of many advanced wear-resistant materials such as TiAlN [108] and TiAlSiN [109] coatings. The optimal wear resistance of the −150 V coating is mainly attributed to the high hardness and toughness of the coating.

Thermal Stability and Oxidation Resistance

The coatings prepared at −150 V with the highest hardness and best wear resistance were selected for 800–1100 °C vacuum annealing test. The surface morphology and composition of the coatings after annealing are shown in Fig. 21. It can be seen that the surface of the coating did not change significantly after annealing. The N content in the coating remained unchanged when the annealing temperature was between 800 and 900 °C. When the annealing temperature was increased to 1000 °C, the N content decreased to 35 at.%. A further increase in the annealing temperature to 1100 °C resulted in a rapid decrease in the N content to 4.6 at.%, implying that severe decomposition of the coating occurred at this temperature.

The XRD patterns of the coatings after vacuum annealing are shown in Fig. 22. It can be seen that there is no change in the phase structure of the coating after annealing at 800–900 °C, and only the diffraction peaks are found to be shifted to a high angle.

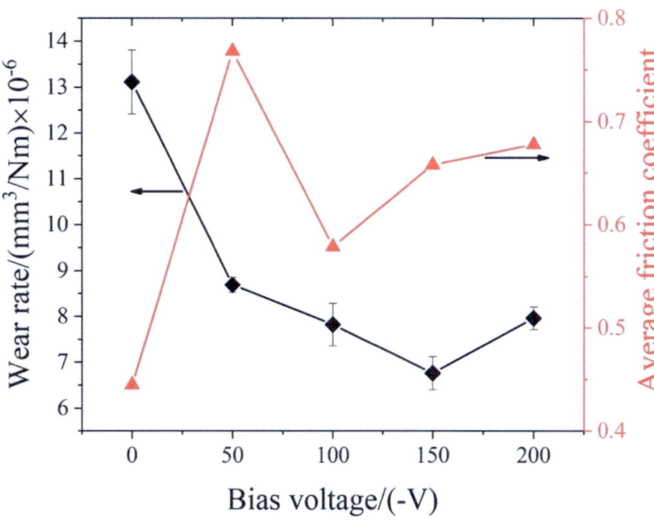

Fig. 20 COF and wear rate of AlCrNbSiTiN coatings at different bias voltages

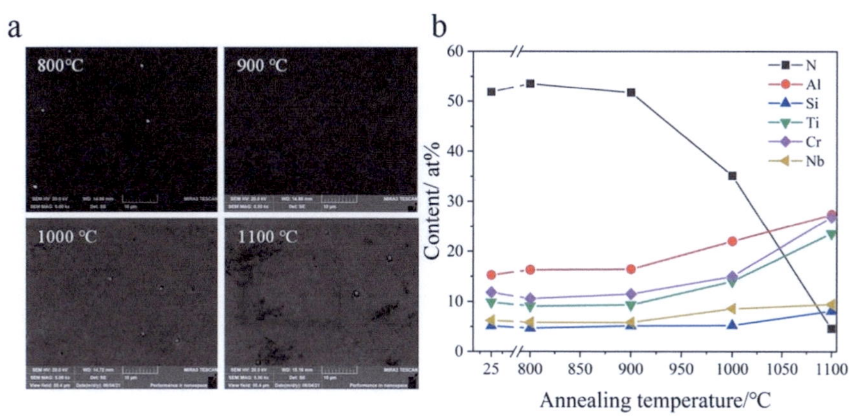

Fig. 21 **a** Surface morphology and **b** compositional changes after vacuum annealing of AlCrNb-SiTiN coating

This is caused by the reduction of the lattice constant due to the elimination of defects such as vacancies during the annealing process. The diffraction peaks of the coating were shifted to a lower angle after annealing at 900 °C. The diffraction peaks of hexagonal AlN and Cr were detected at 1000 °C. The low angle shift of the diffraction peaks is mainly due to two reasons, one is that the residual compressive stress in the coating gradually decreases during the annealing process, and on the other hand, the precipitation of AlN and Cr in the coating makes the average mixing radius of the metal elements in the coating increase, and both of these factors will

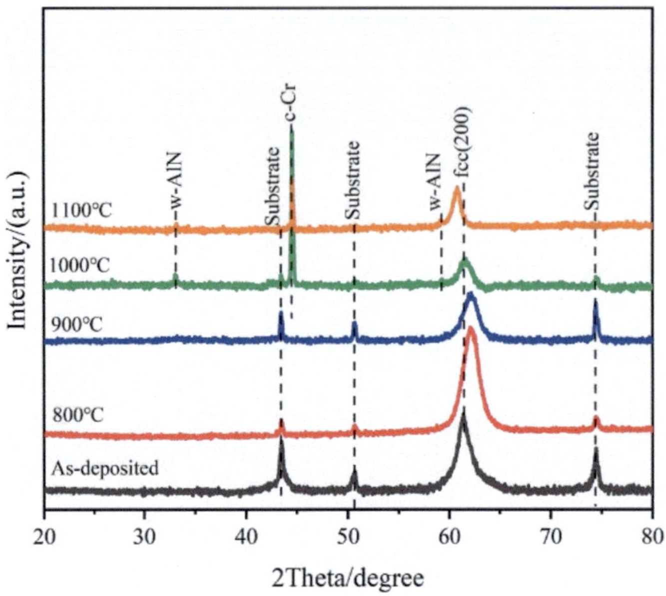

Fig. 22 XRD patterns of AlCrNbSiTiN coatings after annealing in vacuum at different temperatures

lead to an increase in the lattice parameter, and according to the Bragg's formula, the diffraction peaks will be shifted to the low angle direction.

Figure 23 shows the mechanical property changes of AlCrNbSiTiN coatings after annealing. As shown in Fig. 23a, the hardness decreases with the increase of the annealing temperature, and the coating still has a high hardness of 28 GPa after annealing at 900 °C. The hardness of the coating decreases to 26 GPa at 1100 °C. The softening of the coating at high temperatures is, on the one hand, due to the fact that high temperatures promote the movement of the atoms in the coating, which is favorable to the annihilation of point defects within the coating structure, thus generating stress relaxation [110] and reducing the coating hardness, and on the other hand the precipitation of h-AlN and Cr deteriorates the mechanical properties of the coating. The variation of the ratio of H/E and H^3/E^2 of the coatings is shown in Fig. 23b, which shows that the proper annealing temperature can strengthen the coating toughness, and the coating has the highest toughness at 900 °C. When the coating decomposes to form AlN and Cr after the annealing temperature is higher than 900 °C, the toughness of the coating reduces.

The AlCrNbSiTiN coatings prepared at −150 V were air oxidized at 700–1000 °C, and the surface morphology and compositional changes of the coatings after oxidation are shown in Fig. 24. From Fig. 24a, it can be seen that the surface morphology of the coating was basically unchanged after oxidation at 700–900 °C, and many fine oxide particles were detected on the surface of the coating after annealing at 1000 °C. According to Fig. 24b, the N content in the coating decreases with the increase of annealing temperature, while the O content increases to 42.4 at.% at 1000 °C.

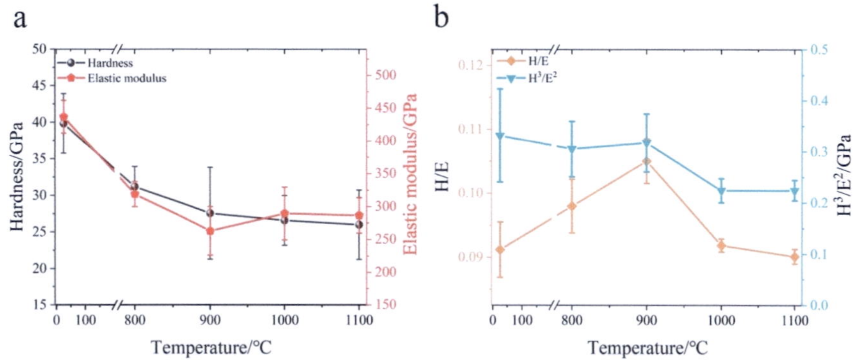

Fig. 23 Variation of **a** hardness, modulus of elasticity and **b** H/E and H^3/E^2 with annealing temperature for AlCrNbSiTiN coatings

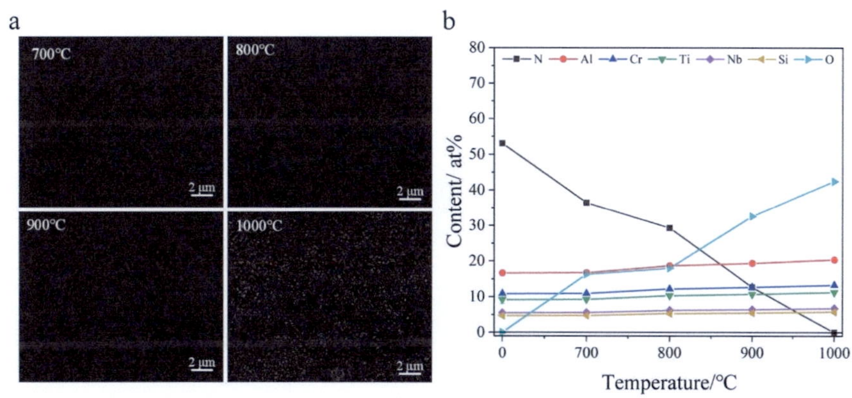

Fig. 24 **a** Surface morphology and **b** compositional changes of AlCrNbSiTiN coatings after oxidation

In order to further study the oxidation behavior of the coating, the cross-sectional morphology of the oxidized coatings was analyzed, and the results are shown in Fig. 25. No obvious delamination was detected on the cross-sectional surface of the coating after oxidation at 700–800 °C, and no obvious enhancement of the O signal was detected by EDS line scanning, which indicated that the oxidized layer in the coating was very. An oxidized layer with a thickness of about 100 nm was detected on the surface after annealing at 900 °C, and this layer was a Cr-rich oxidized layer according to the line-scan results. In the oxidation of pure metals it is observed that Al diffuses preferentially to the surface to form Al_2O_3, whereas in AlCrNbSiTiN it is shown that a Cr-rich oxide layer is preferentially formed on the coating surface. For Cr and Al containing nitride coatings, such as CrAlN, CrAlSiN, etc., there is no consistent conclusion on the diffusion pattern of the elements under oxidation conditions. Hofmann et al. [111] concluded that the oxidation process of CrAlN

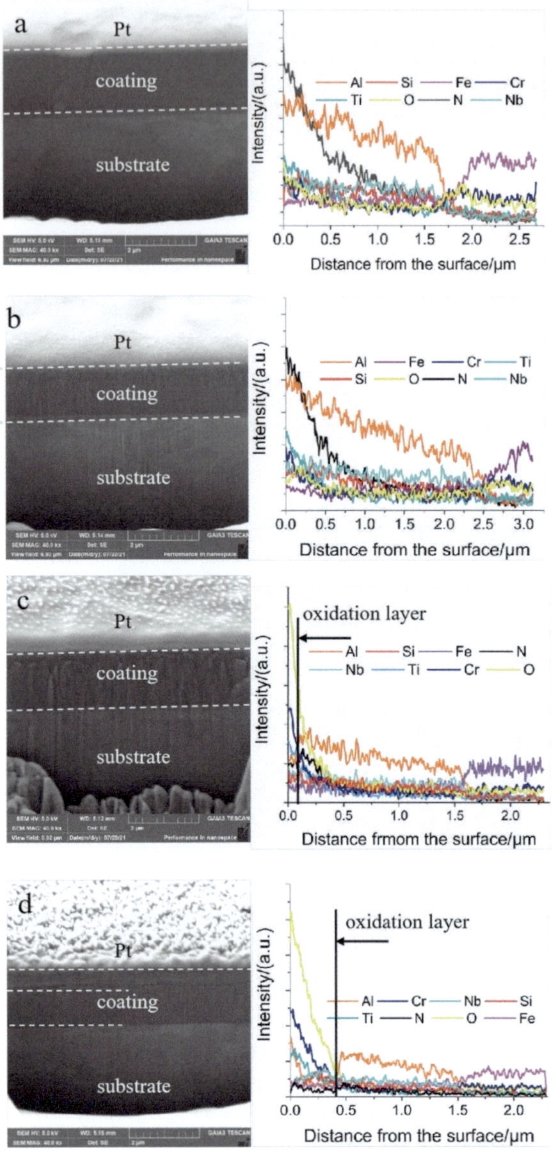

Fig. 25 Cross-sectional morphology and corresponding EDS line analysis of AlCrNbSiTiN coatings after oxidation at different temperatures: **a** 700 °C, **b** 800 °C, **c** 900 °C, and **d** 1000 °C

coatings at 500–900 °C is mainly controlled by the diffusion of Cr. A similar pattern was found by Ichimura and Tytko et. al [112, 113] in AlN/CrN multilayer coatings, and the authors suggested that the lower activation energy of diffusion of Cr in Cr_2O_3 compared to that of Al in Al_2O_3 was the reason why the oxidation process of the coatings was dominated by the outward diffusion of Cr. The results of Kawate et. al [114] showed that when oxidized above 900 °C the surface of the coating will preferentially form an Al-rich oxide layer. The results of Fukumoto and Chang et al.

[115, 116] concluded that the composition of the oxide layer of CrAlN coatings varies with depth, with a thin layer of Al-rich oxide on the surface of the oxide, Cr-rich (Cr, Al) oxide in the middle, and Al-rich (Cr, Al) nitrogen oxide on the bottom. Therefore, the formation of a Cr-rich oxide layer on the surface of the AlCrNbSiTiN coating at 900 °C can be explained as follows: at the early stage of oxidation, Al is preferentially oxidized because the Gibbs free energy for the formation of Al_2O_3 from Al is relatively low, and with the increase in oxidation time or the increase in oxidation temperature, Cr diffuses outward to form Cr_2O_3 and impedes the outward diffusion of Al due to the even lower diffusion activation energy of Cr. Finally, Al stays on the inside of the oxide layer.

At 1000 °C, the thickness of the oxide layer on the coating surface increases to 330 nm. The bonding between the oxide layer and the coating is poor, and continuous cracks are detected at the interface. The formation of cracks is mainly caused by the fractional denitrification of the coating. The oxidation of nitride coatings can be divided into three processes: nitrogen loss due to the breaking of nitride bonds; outward diffusion of metal cations as well as inward diffusion of oxygen ions; and formation of oxides by the combination of metal cations and oxygen ions. For AlCrNbSiTiN coatings, the process of denitrification at high temperatures consists of two main mechanisms, i.e., decomposition of CrN and oxidation of CrN, Cr_2N, AlN, TiN, NbN, and Si_3N_4. Initially, denitrification produces N atoms, which form N_2 bubbles during random diffusion. The bubbles are initially small and located close to the reaction interface, and as N atoms continue to be produced, the N_2 bubbles further absorb isolated N atoms and grow. The accumulation of unreleased N_2 bubbles will lead to the formation of pores at the reaction interface [117].

Figure 26 shows the XRD patterns of the coatings after oxidation at different temperatures. It can be seen that the diffraction peaks of the oxidized coatings at different temperatures are shifted to high angles, which is mainly caused by the elimination of vacancies at high temperatures. In addition, no diffraction peaks of oxides were detected after oxidation from 700 to 900 °C. After oxidation at 1000 °C, the coatings still had the FCC structure, and diffraction peaks of Al_2O_3, Nb_2O_5, and Cr_2O_3 were detected. The grain size of the coating after oxidation was calculated and the results are also represented in Fig. 26, where it can be seen that there is almost no change in the grain size of the coating after oxidation. Grain coarsening is an important factor in the high-temperature oxidation failure of nanocrystalline coatings. The AlCrNbSiTiN coatings show excellent grain size thermal stability during high-temperature oxidation due to three main reasons: first, the lattice distortion effect and the retarded diffusion effect of high-entropy coatings impede the diffusion of the elements and the growth of the grains, second, the networked Si_3N_4 boundary phases inside the coatings restrict the nanocrystalline grain growth, and third, the formation of dense oxide layer on the coating surface can hinder the diffusion of oxygen into the coating interior and inhibit the oxidation.

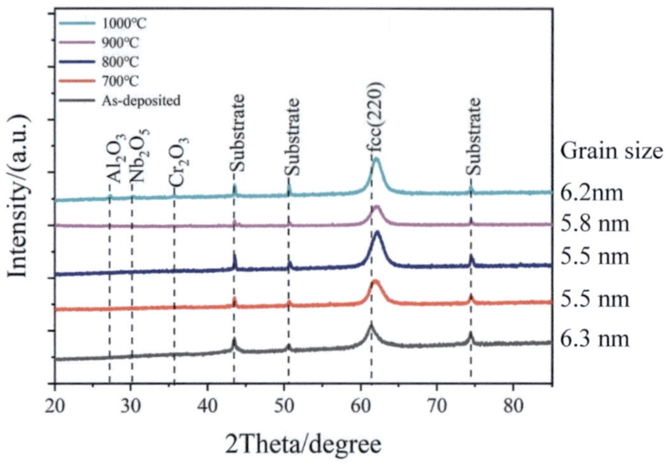

Fig. 26 XRD patterns of AlCrNbSiTiN coatings prepared under −150 V bias after oxidation at different temperatures

3.2 *AlCrNbSiTiN Coatings Deposited by AIP*

Morphology and Chemical Composition

The surface and cross-sectional morphology of AlCrNbSiTiN coatings prepared by AIP technique under different bias voltages are shown in Fig. 27. A large number of particle defects with particle diameters of about 600–1000 nm were observed on the surface of the coating when bias was not applied, and the number of particle defects were significantly decreased after bias was applied. The droplet particle defects are typical defects of AIP-prepared coatings. The decrease of number particle defects under high bias voltage is due to the fact that the application of negative bias will form a negatively charged sheath layer near the substrate, which forms a reaction force on the negatively charged large particles, and the higher the bias voltage is, the stronger the repulsion of the negatively biased substrate on the surface of the negatively charged droplets is [118]. Furthermore, the bias voltage affects not only the number of droplet particles but also their size. Droplet sizes follow a log-normal distribution, and an increase in bias causes the droplet size distribution to become narrower and shifts its maximum towards smaller diameters [119]. After applying the bias voltage, the surface roughness of the coatings decreased with increasing bias voltage from 13.9 nm at -50 V to 3.36 nm at -200 V, and did not change much when the bias voltage was increased above −150 V [120].

The cross-sectional morphology shows that when no bias is applied, the coating has a loose and rough morphology, with holes and large particle defects in the cross-section. In addition many V-shaped growth regions are observed, which originate from particle defects at their tips. During the deposition process, the particles are deposited on the surface, and the subsequent coating layers continue to be deposited

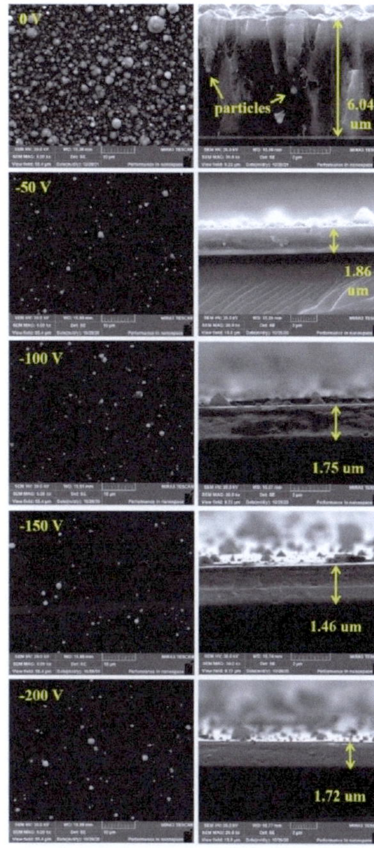

Fig. 27 SEM images of AlCrNbSiTiN coatings deposited at different bias voltages

on the particles. Due to the masking effect, the ion bombardment intensity near the particles is lower, and the coatings in this region are loose and poorly bonded to the surrounding regions, resulting in a more obvious V-shaped boundary. After applying the bias voltage the cross-section was dense and smooth, with no obvious holes and large particle defects observed. The increase in bias voltage improves the coating density, resulting in a decrease in its thickness and deposition rate. As shown in Fig. 28, the thickness of the coating was 6 μm at a bias of 0 V. The thickness of the coating was significantly reduced to 1.86 μm after the bias was increased to −50 V. The thickness of the coating was reduced to 1.46 μm after the bias was further increased to −150 V. Correspondingly, the deposition rate of the coating was drastically reduced from 134 to 32.2 nm/min with the increase of the bias from 0 to −150 V. The deposition rate of AIP is 2–3 times higher than that of MS at the same bias voltage. The variation of elemental content in the coatings with bias voltage is shown in Table 3. The N content slightly increased from 49 to 51 at.% after applying bias voltage, which is attributed to the enhancement of bombardment energy at high bias voltage that can promote the N doping [93]. The Al content decreased after

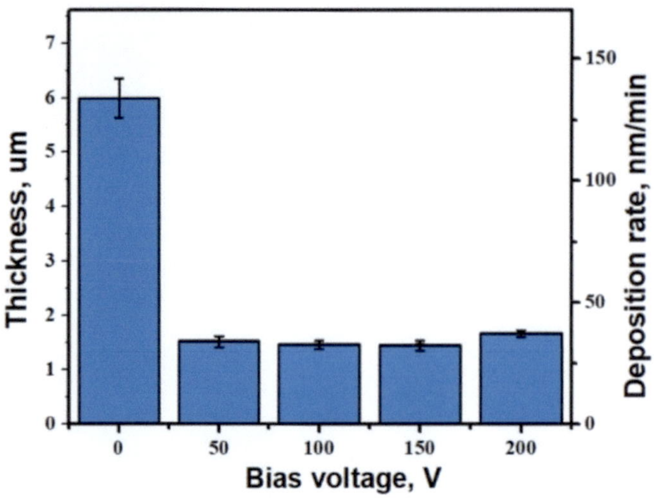

Fig. 28 Variation of thickness and deposition rate of AlCrNbSiTiN coatings deposited at different bias voltages

Table 3 Element composition of AlCrNbSiTiN coatings deposited at different bias voltages

Bias, V	N, at.%	Al, at.%	Cr, at.%	Ti, at.%	Nb, at.%	Si, at.%
0	49	16	12	12	5	5
-50	51	15	13	11	6	4
-100	51	15	13	11	6	4
-150	51	14	14	10	6	5
-200	51	14	13	12	6	5

applying bias voltage, which is due to the enhancement of the re-sputtering effect [121].

Microstructure

Figure 29 shows the XRD patterns of AlCrNbSiTiN coatings deposited at different bias voltages. The diffraction peaks of FCC(111), FCC(200) and FCC(220) are detected in the coating at a bias voltage of 0 V. The coating exhibits a strong (200) preferred orientation. The intensity of the (200) diffraction peaks decreases and the intensity of the (220) diffraction peaks increases with increasing bias voltage, and the diffraction peaks broaden when the bias voltage is increased to −200 V, indicating that the grain refinement occurs. These results indicate that the increase in bias voltage causes a gradual shift of the coating from (200) preferred orientation to

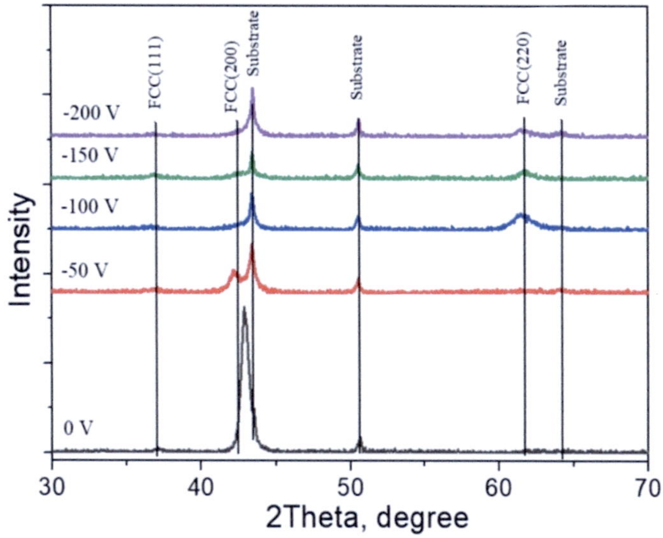

Fig. 29 XRD patterns of AlCrNbSiTiN coatings at different bias voltages

(220), i.e., the growth of the coating gradually shifts from surface energy dominated to stopping energy dominated with the increase in bias voltage.

To further analyze the structure of the coatings, TEM analysis was performed on the coatings prepared at −150 V [120]. The cross-section of the coating exhibits the growth of fiber crystals, and the average width of the fiber crystals is about 40.7 nm. SEAD diffraction arcs of FCC (111), (200), and (220) were detected. A similar microstructure to the AlCrNbSiTiN coating prepared by MS was observed in the HR-TEM of the coating, where Si_3N_4 was distributed inside the columnar crystals in a network-like manner, separating the columnar crystals into fine grains. The Si3N4 showed a crystalline state under the template effect and grew in a co-cellular manner with the surrounding grains to form the nc-AlCrNbTiN/nc-Si_3N_4 nanocomposite structure. The AlCrNbTiN grains and the crystallized Si_3N_4 boundary are observed in the IFFT image, Fig. 30, and the AlCrNbTiN grains can continue to grow through the Si_3N_4 boundary, but the lattice distortion at the Si_3N_4 boundary is severe and there are more dislocations. The TEM-EDS results of the coating show that the elements in the coating are uniformly distributed in the coating [120]. The reason for not detecting the network-like distribution of Si elements is that the distribution of Si_3N_4 is very dispersed and it is difficult to accurately measure its distribution with the TEM instrument used.

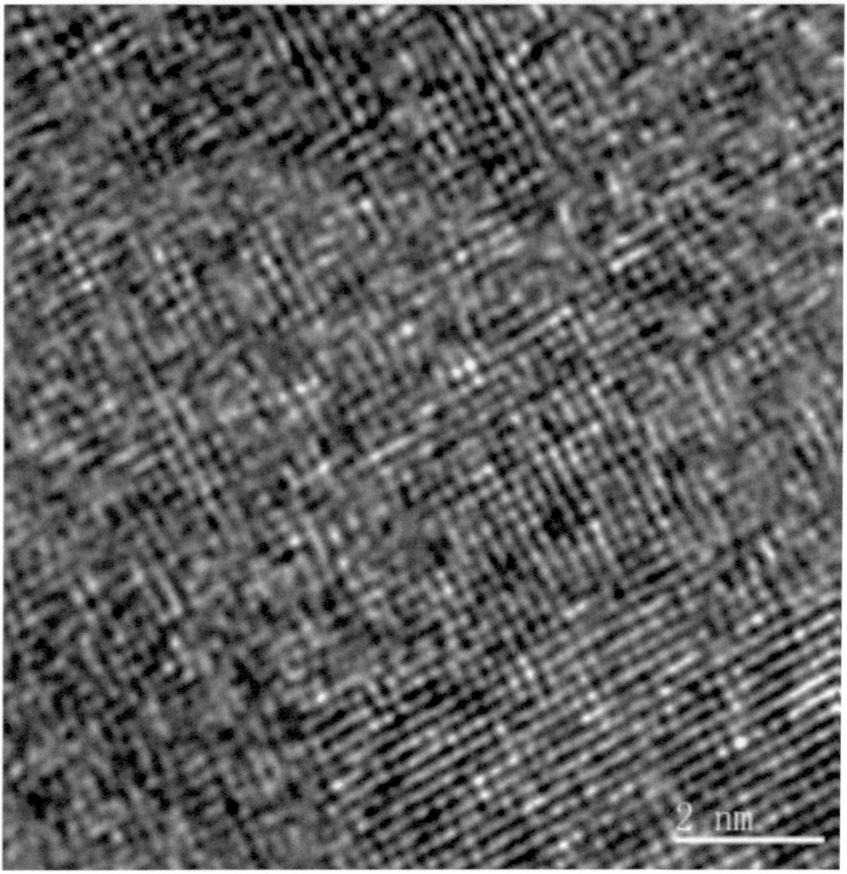

Fig. 30 HR-TEM IFFT image of AlCrNbSiTiN coating prepared at the bias of −150 V

Mechanical Properties

The AlCrNbSiTiN coatings present residual tensile stresses of 450 MPa when no bias is applied [120], which is attributed to the rough surface and loose organization of the coatings, where on the one hand, the tendency of the in-plane contact between the surfaces of the neighboring grains promotes the generation of the tensile stresses, and on the other hand, the ion bombardment effect or the diffusion-based compression mechanism is suppressed with the growth of the rough surface [122]. Upon application of bias voltage, the residual stresses in the coatings were transformed into compressive stresses, which increased from −247 to −585 MPa as the bias voltage increased from −50 to −200 V. Compared to MS, the AIP-prepared AlCrNbSiTiN coatings exhibited considerably less residual stresses, with a bias voltage of −150 V, and residual stresses in the AlCrNbSiTiN coatings of 0.44 GPa, which is only 4.8% of the specimen prepared by magnetron sputtering under the same conditions. The reason for this is that the high ion kinetic energy and high deposition temperature of the AIP technique can increase the surface diffusion of adsorbed atoms, and the intrinsic structural defects in the coating can be partially eliminated during the growth process, which leads to the reduction of residual stresses.

The hardness of the coating at 0 V is low, only 20.3 GPa, Fig. 31(a) which is mainly due to the high number of holes and large particle defects inside the coating, which reduces the coating quality. The hardness of the coating increased rapidly to 32.5 GPa after applying a bias voltage of −50 V. The enhancement of the hardness was attributed to the densification of the coating on one hand, and to the strengthening of

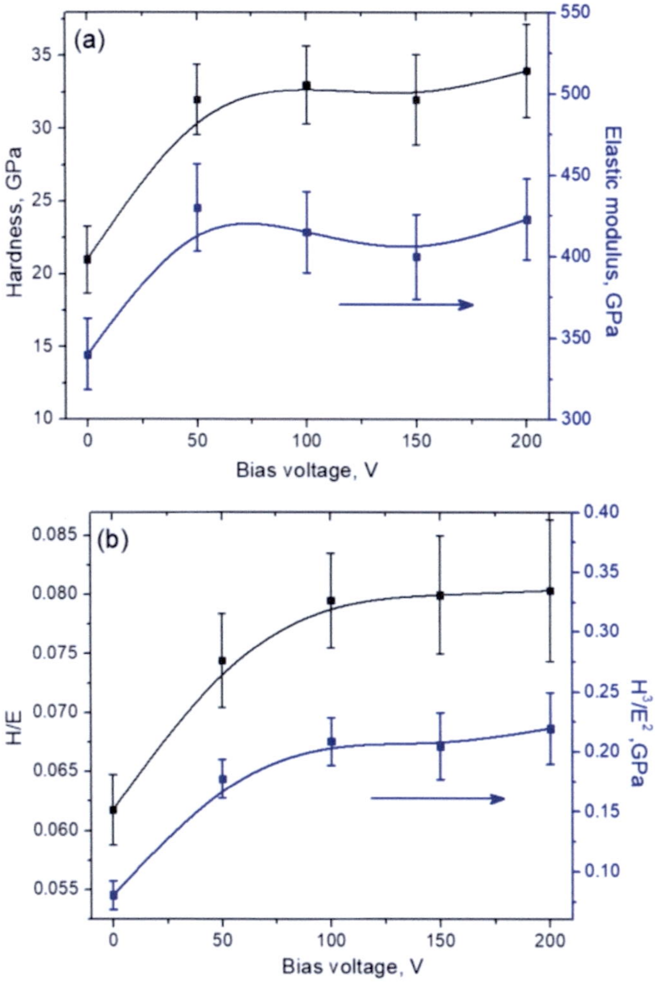

Fig. 31 Variation of hardness and modulus of elasticity **a** and H/E and H^3/E^2 ratios **b** as functions of bias voltage

the fine grains on the other hand. The hardness of the coatings did not change much when the bias voltage was increased to −150 V. The hardness and elastic modulus of the coatings increased further at −200 V, and the highest hardness and elastic modulus were obtained, which were 34.76 GPa and 425 GPa, respectively. The H/E and H^3/E^2 ratios of the coatings showed a tendency of increasing with the increase of bias voltage, and the highest H/E and H^3/E^2 ratios were obtained at −200 V, which were 0.082 and 0.228 GPa Fig. 31(b).

Figure 32 shows the friction test results of AlCrNbSiTiN coating. As can be seen from the figure, without bias voltage, the coating failed during the friction process, and the width of the abrasion scar was about 532 μm. The topography of the abrasion scar section shows that there is a brittle fracture of the coating and there are a lot of pores and large particles defects distributed on the surface, so that a large number of droplets and hole defects in the coating are the main reasons for the abrasion failure of the coating when the bias voltage is not applied. A sharp increase in the intensity of substrate elements in the area of the abrasion scar can be seen in the EDS line scan results of the coating abrasion scar [120]. At −50 V, the width of the coating abrasion scar decreases to 336 μm, and furrows are observed at the abrasion scar, indicating the presence of abrasive wear in the friction. Its line scan shows the presence of a zone of intensity enhancement of the substrate elements with a width of about 150 μm in the center region of the abrasion scar, indicating more severe wear here, but the signal enhancement is much smaller compared to the case when no bias is applied [120]. The wear condition of the coating at −100 V is similar to that at −50 V, with a wear scar width of 337 μm and a matrix signal enhancement zone of approximately 180 μm in width. When the bias voltage was increased to −150 V, the width of the abrasion scar decreased to 252 μm, and no enhancement of matrix elemental signals was detected in the line scan results. At −200 V, the width of the abrasion scar further decreases to 123 μm. These results indicate that the coating has better wear resistance at high bias.

Figure 33 shows the variation of the average COF and wear rate of the coatings deposited with bias voltage. It can be seen that the average COF is low at 0 V, about 0.6, and the COF increases after applying the bias voltage. The specimens at −150 and −200 V achieve the highest COF, about 0.72. This is because the coating wears heavily at low bias voltage, and thus the COF contains the properties of the substrate, and the coating does not shed at high bias voltage, and thus the average COF embodies the characteristics of the coating. The wear rate of the coatings shows a decreasing trend with increasing bias. The coatings have the lowest wear rate of 0.82×10^{-9} mm^3/Nm at −200 V. The wear rate is reduced by four orders of magnitude when compared to the AlCrNbSiTiN coatings prepared by magnetron sputtering. The relief of high stress, reduction of structural defects, and improvement of scratch toughness are responsible for the excellent wear resistance of AlCrNbSiTiN coatings prepared by AIP technique.

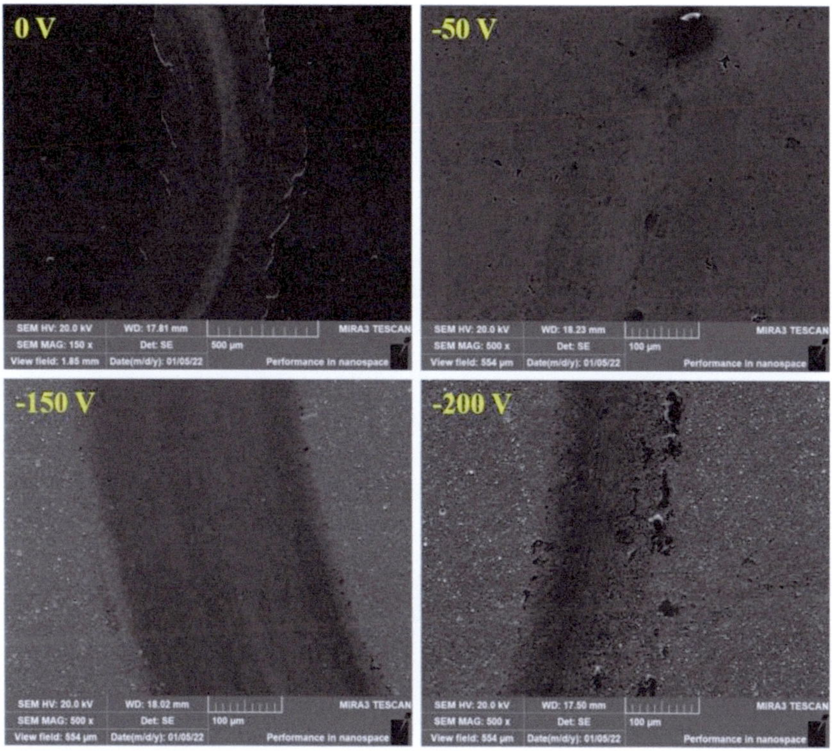

Fig. 32 Wear track morphology of AlCrNbSiTiN coatings deposited at different bias voltages

Thermal Stability and Oxidation Resistance

The coatings prepared at −150 V were subjected to vacuum annealing treatment, and the surface morphology and compositional changes of the coatings after annealing are shown in Fig. 34. From Fig. 34a, it can be seen that there is no obvious change in the surface morphology of the coating after annealing at 1100 °C. According to Fig. 34b, the N content in the coating decreases slowly with the increase of the annealing temperature, and 35.5 at.% N is still detected in the coating at a high temperature of 1100 °C. Compared with the MS technique, the AlCrNbSiTiN coating prepared by AIP has a better compositional stability at high temperatures.

The XRD patterns of the coating after annealing are shown in Fig. 35. It can be seen that the phase structure of the coating can remain stable up to 1000 °C. When the temperature is further increased to 1100 °C, h-AlN is detected in the coating and no diffraction peaks of Cr are detected. The heat resistance temperature is increased by 100 °C compared to the MS technique. The grain size change of the coating after annealing is also represented in Fig. 35. It can be seen that the grain size in the coating does not increase significantly with the increase of annealing temperature, and the

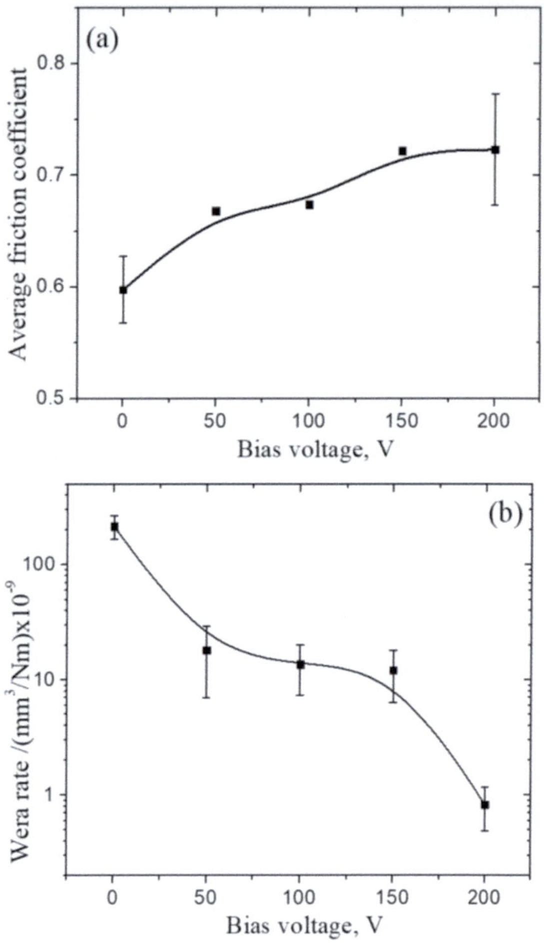

Fig. 33 Variation of average COF **a** and wear rate **b** of AlCrNbSiTiN coatings deposited at different bias voltages

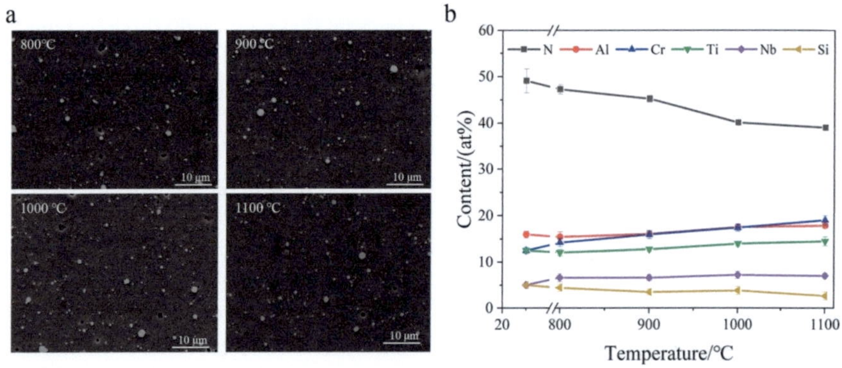

Fig. 34 **a** Surface morphology and **b** compositional changes of AlCrNbSiTiN coatings after annealing

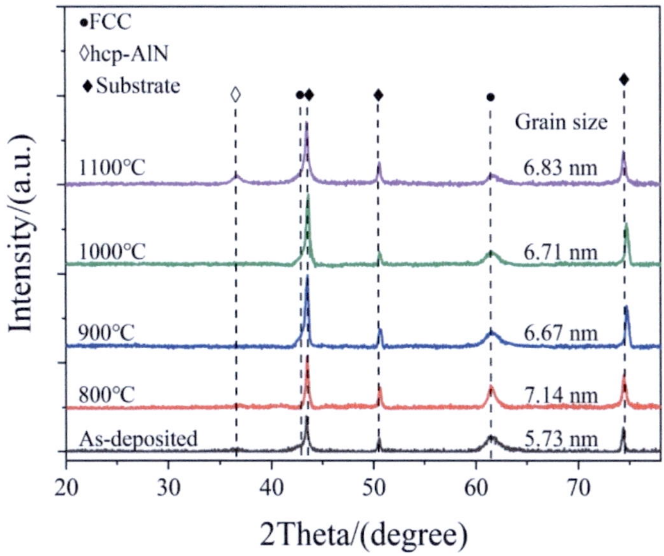

Fig. 35 XRD patterns of of AlCrNbSiTiN coatings after vacuum annealing at different temperatures

coating still maintains its nanostructure at a high temperature of 1100 °C, showing excellent grain size thermal stability.

The changes of hardness, elastic modulus and H/E and H^3/E^2 ratios of the coatings after annealing at different temperatures are shown in Fig. 36. It can be seen that the hardness and elastic modulus of the coatings show a slow decreasing trend with the increase of annealing temperature, and the coatings can maintain a hardness of 31 GPa after annealing at 900 °C. The coatings still have a high hardness of 28 GPa at 1100 °C, which indicates that the coatings have an excellent thermal hardness. The decrease in hardness of the coatings is mainly attributed to stress relief and precipitation of h-AlN. The H/E and H^3/E^2 ratios of the coatings did not change much with the annealing temperature, and the coatings still possessed excellent high-temperature toughness with H/E and H^3/E^2 ratios as high as 0.07 and 0.15 GPa, respectively, at 1100 °C. Compared with MS, the AlCrNbSiTiN coatings prepared by AIP technique have superior thermal stability performance, which is mainly due to the fact that the AlCrNbSiTiN coatings prepared by MS have high internal stresses and many point defects, and thus the unstable coating structure is prone to defect recovery process at high temperatures. In contrast, the AIP-prepared coating has low stress, few point defects within the coating structure, and the defect recovery process of the coating is not obvious at high temperature, so the structure and mechanical properties of the coating are more stable [123].

The coatings prepared at −150 V were subjected to air oxidation at 700–1000 °C, and the surface and cross-sectional morphology of the oxidized coatings are shown in Fig. 37. No significant changes in the surface morphology of the coatings were

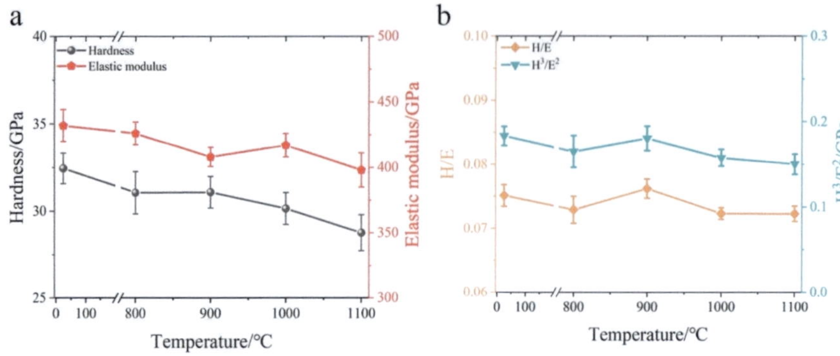

Fig. 36 **a** Hardness and modulus of elasticity and **b** H/E and H^3/E^2 ratios of AlCrNbSiTiN coatings after vacuum annealing

detected after oxidization at 700–800 °C, while fine oxide particles began to appear on the surface at 900 °C, and the oxide particles on the surface of the coatings were further increased at 1000 °C. From the morphology of the coating cross-section, it can be seen that a dense oxide layer with a thickness of about 37 nm was formed on the surface of the coating at 700 °C. The thickness of the oxide layer increases to 280 nm at 1000 °C. The oxide layer formed on the surface of the coating at different temperatures is well bonded with the coating below, and there are no obvious defects such as cracks and pores.

The compositional changes of the coatings after oxidation are shown in Fig. 38, where the N content of the coatings decreases and the O content increases with increasing annealing temperature to ~60 at.% at 1000 °C. Compared with the MS-prepared AlCrNbSiTiN coatings, the AIP-prepared AlCrNbSiTiN coatings exhibited faster oxidation rates at oxidation temperatures in the range of 700–900 °C, and the thickness of the oxide layer and the rate of increase of the oxygen content were higher than that of the MS-prepared specimens. This is due to the fact that the AIP-prepared coatings have many large particles and pore defects, which provide fast channels for the diffusion of oxygen into the interior of the coatings during the oxidation process to accelerate the oxidation. However, at oxidation temperatures higher than 1000 °C, the AIP-prepared specimens showed thinner oxide layer thicknesses because at this temperature, the coating begins to thermally decompose, generating N atoms, and the holes inside the coating provide channels for N release to avoid cracking of the oxide layer due to the agglomeration of N_2 bubbles. The oxide layer, which is well bonded to the coating, continues to provide protection, thus ensuring that the coating has a slower oxidation rate.

The XRD patterns of the coatings after oxidation are shown in Fig. 39. The patterns did not change significantly after oxidation at 800–900 °C, and when the oxidation temperature was increased to 1000 °C, the diffraction peaks of FCC could still be detected, in addition to the diffraction peaks of Al_2O_3. The size changes of the coating grains are also listed in the figure, and it can be seen that the grain size

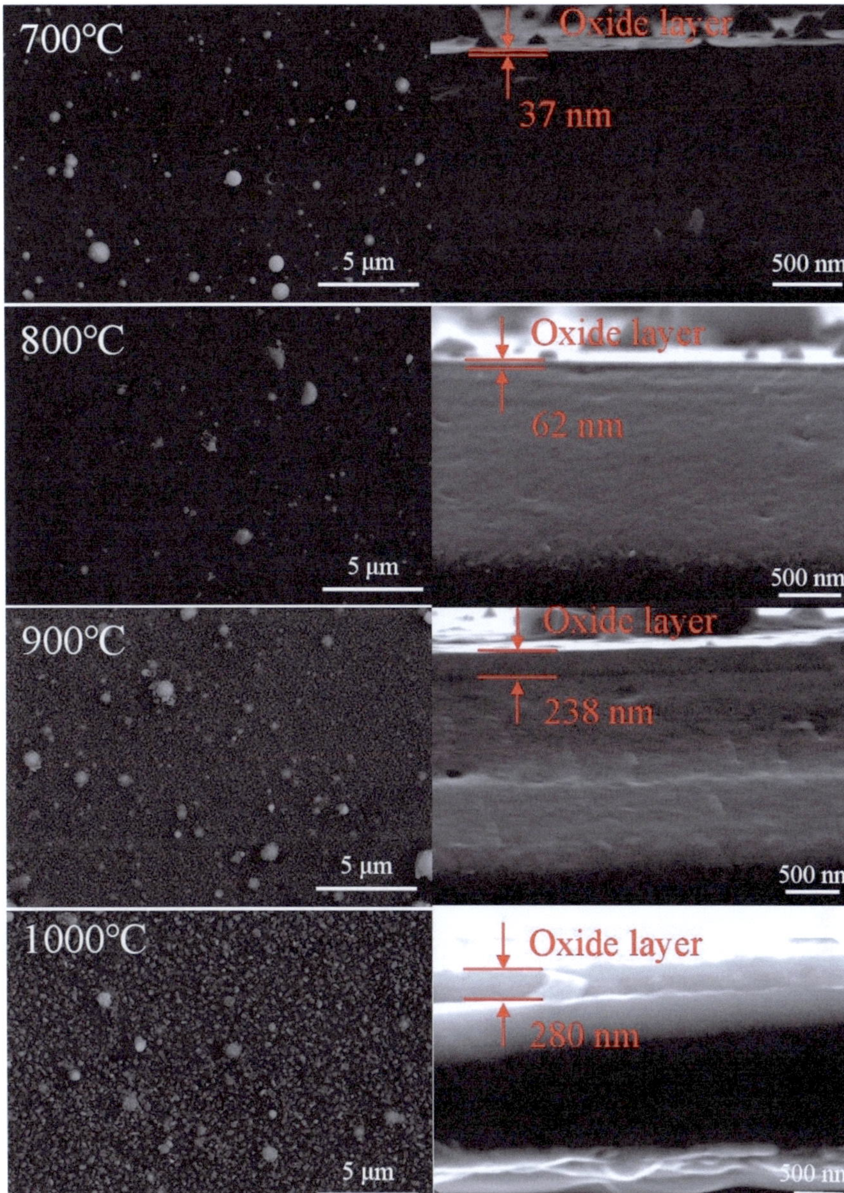

Fig. 37 Surface and cross-sectional images of AlCrNbSiTiN coatings after oxidation at different temperatures

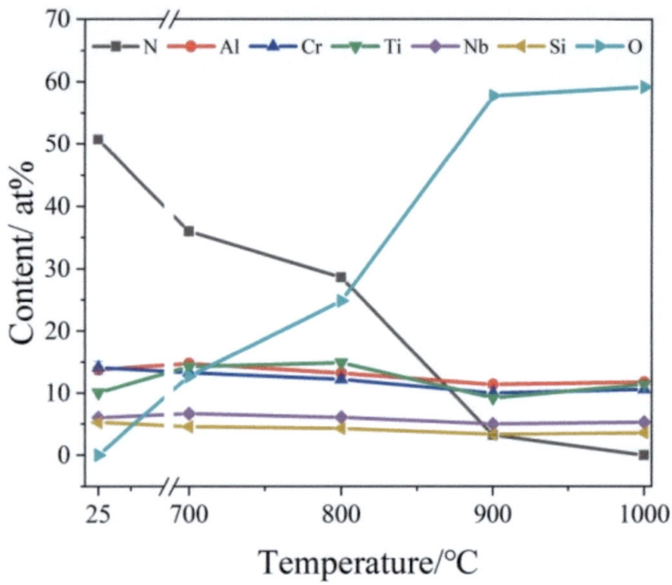

Fig. 38 Variation of elemental content of AlCrNbSiTiN coatings after oxidation at different temperatures

of the coating can be stabilized at about 6.9 nm even at 1000 °C, indicating that the coating has excellent oxidation tolerance and grain size thermal stability.

4 Outlook

In this chapter, results of comparative study of AlCrNbSiTi high entropy alloy nitride coatings prepared by MS and AIP techniques are presented. Microstructure, mechanical properties, wear resistance and high temperature resistance of the coatings were systematically investigated, and the structure and properties of the coatings were optimized by adjusting the bias voltage.

AlCrNbSiTiN coatings deposited by MS technique showed columnar structure. Si_3N_4 in the coatings deposited at low bias existed in amorphous state at the boundary of the columnar crystals. At high bias the Si_3N_4 showed network distribution inside the columnar crystals and grew in a co-lattice with the neighboring AlCrNbTiN grains to form the nanocomposite nc-AlCrNbTiN/nc-Si_3N_4 structure. This structure can strengthen the hardness and toughness of the coating. A high hardness of 39.8 GPa and high H/E and H^3/E^2 ratios of 0.091 and 0.331 GPa, respectively, were obtained for the coatings deposited at a bias voltage of −150 V. The wear mechanism of the coatings was mainly characterized by abrasive and oxidative wear. The coating at a bias of −150 V showed optimal wear resistance due to its high hardness and toughness, and

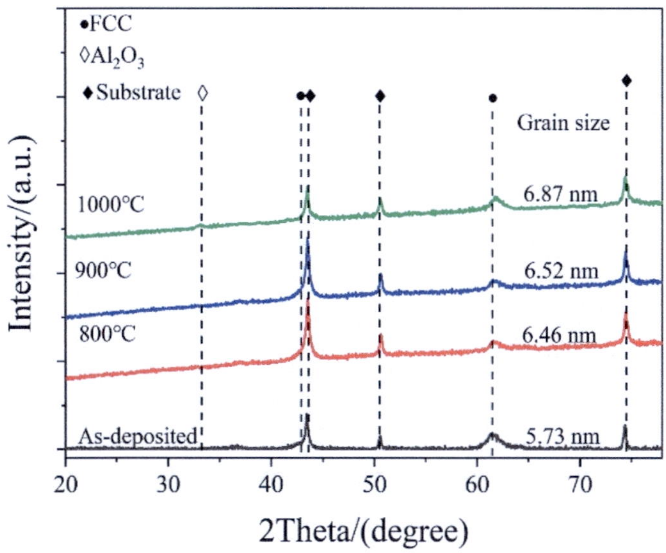

Fig. 39 XRD variation pattern of the coatings after oxidation at different temperatures

its wear rate of 6.7×10^{-6} mm^3/Nm exceeded that of many advanced wear-resistant materials. The coating exhibited excellent thermal stability and oxidation resistance at 900 °C. After annealing at 900 °C, the coating still had a high hardness of 28 GPa, and the thickness of the oxide layer formed by oxidizing at this temperature for 2 h is only 100 nm. Thanks to the lattice distortion effect of the coating, the sluggish diffusion effect, and the grain confinement by the Si_3N_4 network in the nano-composite structure, the coating had an excellent grain size. The coating was thermally stable and could maintain its nanostructure without grain coarsening in vacuum at 1100 °C and oxidizing environment at 1000 °C.

The AIP technique with high ion kinetic energy effectively reduced the point defects inside the coating, relieved the internal stress, and strengthened the hardness of the coating, so that the coating achieved a high hardness of up to 34.8 GPa under the condition of low stress. The wear resistance of the coating was significantly enhanced, and the coating obtained a minimum wear rate of 0.82×10^{-9} mm^3/Nm, which was nearly four orders of magnitude lower than that of the MS technique. The coating prepared by the AIP technique had fewer point defects, which effectively enhanced the thermal stability of the coating. The phase structure of the coatings maintained stable up to 1000 °C, and after annealing at 900 °C, the hardness of the coatings was still 31 GPa. The coatings still had a high hardness of 28 GPa after 1100 °C annealing, and high H/E and H^3/E^2 ratios of 0.07 and 0.15 GPa, respectively, which show excellent thermal stability, thermal hardness, and high-temperature toughness. The thickness of the oxide layer formed after 2 h oxidation of the coating at 1000 °C was is only 280 nm, showing excellent oxidation tolerance. The lattice distortion effect and Si_3N_4 boundary phase restriction of the coating made the coating have excellent grain size thermal stability, and the coating was coarsened in vacuum at 1100 °C and in air at 1000 °C.

The results in this chapter indicate that AlCrNbSiTiN coatings deposited using MS and AIP techniques exhibit good performance, especially from the perspective of oxidation resistance. In this chapter we present results for coating with only one composition. Optimization of both content of elements and types of elements require additional research. However, due to the large number of elements in the coating the composition optimization problem becomes complex. It can be expected that in the coming years, the resolution of these issues will lead to the synthesis of better performance HEA ceramic coatings.

Acknowledgements This work was supported by the Key R&D program of Hubei Province (Grant No. 2022BAA049), the National Key Research and Development Program of China (Grant No. 2022YFB4601000), the Open Fund of Hubei Key Laboratory of Electronic Manufacturing and Packaging Integration (Wuhan University) (Grant No. EMPI2023018), and the Fundamental Research Funds for the Central Universities, grant No. 4207-413100036.

References

1. Yeh, J.W., Chen, S.K., Lin, S.J., Gan, J.Y., Chin, T.S., Shun, T.T., Chang, S.Y.: Nanostructured high-entropy alloys with multiple principal elements: novel alloy design concepts and outcomes. Adv. Eng. Mater. **6**(5), 299–303 (2004)
2. Pogrebnjak, A.D., Bagdasaryan, A.A., Yakushchenko, I.V., Beresnev, V.M.: The structure and properties of high-entropy alloys and nitride coatings based on them. Russ. Chem. Rev. **83**, 1027–1061 (2014)
3. Deng, Y., Tasan, C.C., Pradeep, K.G., Springer, H., Kostka, A., Raabe, D.: Design of a twinning-induced plasticity high entropy alloy. Acta Mater. **94**, 124–133 (2015)
4. Chuang, M.H., Tsai, M.H., Wang, W.R., Lin, S.J., Yeh, J.W.: Microstructure and wear behavior of AlxCo1. 5CrFeNi1. 5Tiy high-entropy alloys. Acta Materialia **59**(16), 6308–6317 (2011)
5. Chen, Y.Y., Duval, T., Hung, U.D., Yeh, J.W., Shih, H.C.: Microstructure and electrochemical properties of high entropy alloys—a comparison with type-304 stainless steel. Corros. Sci. **47**(9), 2257–2279 (2005)
6. Li, W., Liu, P., Liaw, P.K.: Microstructures and properties of high-entropy alloy films and coatings: a review. Mater. Res. Lett. **6**(4), 199–229 (2018)
7. Schinhammer, M., Hänzi, A.C., Löffler, J.F., Uggowitzer, P.J.: Design strategy for biodegradable Fe-based alloys for medical applications. Acta Biomater. **6**(5), 1705–1713 (2010)
8. Inoue, A., Zhang, W., Zhang, T., Kurosaka, K.: High-strength Cu-based bulk glassy alloys in Cu–Zr–Ti and Cu–Hf–Ti ternary systems. Acta Mater. **49**(14), 2645–2652 (2001)
9. Inoue, A., Kong, F., Zhu, S., Liu, C.T., Al-Marzouki, F.: Development and applications of highly functional Al-based materials by use of metastable phases. Mater. Res. **18**, 1414–1425 (2015)
10. Manivasagam, G., Suwas, S.: Biodegradable Mg and Mg based alloys for biomedical implants. Mater. Sci. Technol. **30**(5), 515–520 (2014)
11. Geetha, M., Singh, A.K., Asokamani, R., Gogia, A.K.: Ti based biomaterials, the ultimate choice for orthopaedic implants–a review. Prog. Mater Sci. **54**(3), 397–425 (2009)
12. Lee, M.H., Lee, J.Y., Bae, D.H., Kim, W.T., Sordelet, D.J., Kim, D.H.: A development of Ni-based alloys with enhanced plasticity. Intermetallics **12**(10–11), 1133–1137 (2004)

13. Kumar, A., Singh, A., Suhane, A. Mechanically alloyed high entropy alloys: existing challenges and opportunities. Journal of Materials Research and Technology, 17, 2431–2456 (2022)
14. Cantor, B., Chang, I.T.H., Knight, P., Vincent, A.J.B.: Microstructural development in equiatomic multicomponent alloys. Mater. Sci. Eng., A **375**, 213–218 (2004)
15. Li, Z., Zhao, S., Ritchie, R.O., Meyers, M.A.: Mechanical properties of high-entropy alloys with emphasis on face-centered cubic alloys. Prog. Mater Sci. **102**, 296–345 (2019)
16. Jien-Wei, Y.E.H.: Recent progress in high entropy alloys. Ann. Chim. Sci. Mat **31**(6), 633–648 (2006)
17. Tong, C.J., Chen, M.R., Yeh, J.W., Lin, S.J., Chen, S.K., Shun, T.T., Chang, S.Y.: Mechanical performance of the AlxCoCrCuFeNi high-entropy alloy system with multiprincipal elements. Metall. Mater. Trans. A. **36**, 1263–1271 (2005)
18. Zhao, S., Stocks, G.M., Zhang, Y.: Stacking fault energies of face-centered cubic concentrated solid solution alloys. Acta Mater. **134**, 334–345 (2017)
19. Shun, T.T., Hung, C.H., Lee, C.F.: Formation of ordered/disordered nanoparticles in FCC high entropy alloys. J. Alloy. Compd. **493**(1–2), 105–109 (2010)
20. Ranganathan, S.J.C.S.: Alloyed pleasures: multimetallic cocktails. Curr. Sci. **85**(5), 1404–1406 (2003)
21. Ye, Y.F., Wang, Q., Lu, J., Liu, C.T., Yang, Y.: High-entropy alloy: challenges and prospects. Mater. Today **19**(6), 349–362 (2016)
22. Arif, Z.U., Khalid, M.Y., ur Rehman, E., Ullah, S., Atif, M., Tariq, A.: A review on laser cladding of high-entropy alloys, their recent trends and potential applications. J. Manuf. Process. **68**, 225–273 (2021)
23. Klocke, F., Brecher, C., Wegener, M., Heinen, D., Fischer, B., Do-Khac, D. Scanner-based Laser Cladding. Physics Procedia, **39**, 346–353 (2012)
24. Wang, H., Liu, Q., Guo, Y., Lan, H.: MoFe1.5CrTiWAlNbx refractory high-entropy alloy coating fabricated by laser cladding. Intermetallics 115, 106613 (2019)
25. Guo, Y., Liu, Q.: MoFeCrTiWAlNb refractory high-entropy alloy coating fabricated by rectangular-spot laser cladding. Intermetallics **102**, 78–87 (2018)
26. Yi, M., Wang, J., Bai, X., Xu, Y., Xiao, G., Chen, Z., Xu, C.: Synthesis and properties characterization of an Al_2O_3-based ceramic materials with intragranular self-lubricating nanostructure. Int. J. Refractory Metals Hard Mater. **104**, 105800 (2022)
27. Wan, H., Song, D., Shi, X., Cai, Y., Li, T., Chen, C.: Corrosion behavior of Al0.4CoCu0.6NiSi0.2Ti0.25 high-entropy alloy coating via 3D printing laser cladding in a sulphur environment. J. Mater. Sci. Technol. **60**, 197–205 (2021)
28. Arif, Z.U., Khalid, M.Y., Al Rashid, A., ur Rehman, E., Atif, M.: Laser deposition of high-entropy alloys: A comprehensive review. Opt. Laser Technol. **145**, 107447 (2022)
29. Tong, Z., Ren, X., Jiao, J., Zhou, W., Ren, Y., Ye, Y., Gu, J.: Laser additive manufacturing of FeCrCoMnNi high-entropy alloy: effect of heat treatment on microstructure, residual stress and mechanical property. J. Alloys Compd. **785**, 1144–1159 (2019)
30. Katakam, S., Joshi, S.S., Mridha, S., Mukherjee, S., Dahotre, N.B.: Laser assisted high entropy alloy coating on aluminum: microstructural evolution. J. Appl. Phys. **116**(10), 104906 (2014)
31. Yue, T.M., Xie, H., Lin, X., Yang, H.O., Meng, G.H.: Solidification behaviour in laser cladding of AlCoCrCuFeNi high-entropy alloy on magnesium substrates. J. Alloy. Compd. **587**, 588–593 (2014)
32. Shon, Y., Joshi, S.S., Katakam, S., Rajamure, R.S., Dahotre, N.B.: Laser additive synthesis of high entropy alloy coating on aluminum: corrosion behavior. Mater. Lett. **142**, 122–125 (2015)
33. Chao, Q., Guo, T., Jarvis, T., Wu, X., Hodgson, P., Fabijanic, D.: Direct laser deposition cladding of AlxCoCrFeNi high entropy alloys on a high-temperature stainless steel. Surf. Coat. Technol. **332**, 440–451 (2017)
34. Satyavathi Yedida, V.V., Vasudev H. A review on the development of thermal barrier coatings by using thermal spray techniques. Materials Today: Proceedings, **50**, 1458–1464 (2022)

35. Ang, A.S.M., Berndt, C.C., Sesso, M.L., Anupam, A., Kottada, R.S., Murty, B.S.: Plasma-sprayed high entropy alloys: Microstructure and properties of AlCoCrFeNi and MnCoCrFeNi. Metall. Mater. Trans. A. **46**, 791–800 (2015)
36. Fauchais, P.: Understanding plasma spraying. J. Phys. D Appl. Phys. **37**(9), R86 (2004)
37. Sidhu, T.S., Prakash, S., Agrawal, R.D.: State of the art of HVOF coating investigations—a review. Mar. Technol. Soc. J. **39**(2), 53–64 (2005)
38. Oksa, M., Turunen, E., Suhonen, T., Varis, T., Hannula, S.P.: Optimization and characterization of high velocity oxy-fuel sprayed coatings: techniques, materials, and applications. Coatings **1**(1), 17–52 (2011)
39. Meghwal, A., Anupam, A., Murty, B.S., Berndt, C.C., Kottada, R.S., Ang, A.S.M.: Thermal spray high-entropy alloy coatings: a review. J. Therm. Spray Technol. **29**, 857–893 (2020)
40. Ashokan, V., Gaurav, V., Babu, S.K., Murali, M., Manivannan, S.: Microstructure evolution and erosion behaviour of thermally sprayed AlCoCrNiMo0.1 high entropy alloy coating. Ceram. Int. **49**(19), 32331–32336 (2023)
41. Huang, P.K., Yeh, J.W., Shun, T.T., Chen, S.K.: Multi-principal-element alloys with improved oxidation and wear resistance for thermal spray coating. Adv. Eng. Mater. **6**(1–2), 74–78 (2004)
42. Bhattacharya, R., Senkov, O.N., Rai, A.K., Ma, X., Ruggiero, P.: High entropy alloy coatings for application as bond coating for thermal barrier coating systems. In: ITSC 2016, pp. 279–285. ASM International (2016)
43. Wang, L., Zhang, F., Yan, S., Yu, G., Chen, J., He, J., Yin, F.: Microstructure evolution and mechanical properties of atmosphere plasma sprayed AlCoCrFeNi high-entropy alloy coatings under post-annealing. J. Alloy. Compd. **872**, 159607 (2021)
44. Jin, B., Zhang, N., Guan, S., Zhang, Y., Li, D.: Microstructure and properties of laser re-melting FeCoCrNiAl0.5Six high-entropy alloy coatings. Surf. Coatings Technol. **349**, 867–873 (2018)
45. Cheng, K.C., Chen, J.H., Stadler, S., Chen, S.H.: Properties of atomized AlCoCrFeNi high-entropy alloy powders and their phase-adjustable coatings prepared via plasma spray process. Appl. Surf. Sci. **478**, 478–486 (2019)
46. Mejía, M.I., Restrepo, G., Marín, J.M., Sanjines, R., Pulgarín, C., Mielczarski, E., Mielczarski, J., Kiwi, J.: Magnetron-sputtered Ag surfaces. New evidence for the nature of the Ag ions intervening in bacterial inactivation. ACS Appl. Mater. Interf. **2**, 230–235 (2010)
47. Mehran, Q.M., Fazal, M.A., Bushroa, A.R., Rubaiee, S.: A critical review on physical vapor deposition coatings applied on different engine components. Crit. Rev. Solid State Mater. Sci. **43**(2), 158–175 (2018)
48. Arnell, R.D., Kelly, P.J.: Recent advances in magnetron sputtering. Surf. Coat. Technol. **112**(1–3), 170–176 (1999)
49. Ruske, M., Bräuer, G., Pistner, J., Pfäfflin, U., Szczyrbowski, J.: Properties of SnO_2 films prepared by DC and MF reactive sputtering. Thin Solid Films **351**(1–2), 146–150 (1999)
50. Gudmundsson, J.T.: Physics and technology of magnetron sputtering discharges. Plasma Sources Sci. Technol. **29**(11), 113001 (2020)
51. Sanders, D.M., Anders, A.: Review of cathodic arc deposition technology at the start of the new millennium. Surf. Coat. Technol. **133**, 78–90 (2000)
52. Karpov, D.A.: Cathodic arc sources and macroparticle filtering. Surf. Coat. Technol. **96**(1), 22–33 (1997)
53. Boxman, R.L., Goldsmith, S.: Macroparticle contamination in cathodic arc coatings: generation, transport and control. Surf. Coat. Technol. **52**(1), 39–50 (1992)
54. Huang, P.K., Yeh, J.W.: Effects of nitrogen content on structure and mechanical properties of multi-element (AlCrNbSiTiV)N coating. Surf. Coat. Technol. **203**(13), 1891–1896 (2009)
55. Bachani, S.K., Wang, C.J., Lou, B.S., Chang, L.C., Lee, J.W.: Fabrication of TiZrNbTaFeN high-entropy alloys coatings by HiPIMS: effect of nitrogen flow rate on the microstructural development, mechanical and tribological performance, electrical properties and corrosion characteristics. J. Alloy. Compd. **873**, 159605 (2021)
56. Xu, W., Liao, M., Liu, X., Ji, L., Ju, P., Li, H., Chen, J.: Microstructures and properties of (TiCrZrVAl)N high entropy ceramics films by multi-arc ion plating. Ceram. Int. **47**(17), 24752–24759 (2021)

57. Kim, Y.S., Park, H.J., Mun, S.C., Jumaev, E., Hong, S.H., Song, G., Kim, K.B.: Investigation of structure and mechanical properties of TiZrHfNiCuCo high entropy alloy thin films synthesized by magnetron sputtering. J. Alloy. Compd. **797**, 834–841 (2019)
58. Pogrebnjak, A.D., Shpak, A.P., Azarenkov, N.A., Beresnev, V.M.: Structures and properties of hard and superhard nanocomposite coatings. Phys. Usp. **52**, 29–54 (2009)
59. Firstov, S.A., Gorban', V.F., Danilenko, N.I., Karpets, M.V., Andreev, A.A., Makarenko, E.S.: Thermal stability of superhard nitride coatings from high-entropy multicomponent Ti–V–Zr–Nb–Hf alloy. Powder Metall. Metal Ceram. **52**, 560–566 (2014)
60. Cheng, K.H., Lai, C.H., Lin, S.J., Yeh, J.W.: Structural and mechanical properties of multi-element (AlCrMoTaTiZr)Nx coatings by reactive magnetron sputtering. Thin Solid Films **519**(10), 3185–3190 (2011)
61. Huang, P.K., Yeh, J.W.: Effects of substrate temperature and post-annealing on microstructure and properties of (AlCrNbSiTiV)N coatings. Thin Solid Films **518**(1), 180–184 (2009)
62. Pogrebnjak A.D., Beresnev V.M., Smyrnova K.V., Kravchenko, Y.O., Zukowski P.V., Bondarenko G.G.: The influence of nitrogen pressure on the fabrication of the two-phase superhard nanocomposite (TiZrNbAlYCr)N coatings. Mater. Lett. **211**, 316–318 (2018)
63. Pogrebnjak A.D., Yakushchenko I.V., Sobol' O.V., Beresnev V.M., Kupchishin A.I., Bondar O.V., Lisovenko M.A., Amekura H., Kono K., Oyoshi K., Takeda Y.: Influence of residual pressure and ion implantation on the structure, elemental composition, and properties of (TiZrAlYNb)N nitrides. Tech. Phys. **60**, 1176–1183 (2015)
64. Pogrebnjak, A., Yakushchenko, I., Bondar, O., Beresnev, V., Oyoshi, K., Ivasishin, O., Amekura, H., Takeda, Y., Opielak, M., Kozak, C.: Irradiation resistance, microstructure and mechanical properties of nanostructured (TiZrHfVNbTa)N coatings. J. Alloy. Compd. **679**, 155–163 (2016)
65. Pogrebnjak, A.D., Yakushchenko, I.V., Bagdasaryan, A.A., Bondar, O.V., Krause-Rehberg, R., Abadias, G., Chartier, P., Oyoshi, K., Takeda, Y., Beresnev, V.M., Sobol, O.V.: Microstructure, physical and chemical properties of nanostructured (Ti-Hf-Zr-V-Nb)N coatings under different deposition conditions. Mater. Chem. Phys. **147**, 1079–1091 (2014)
66. Zhang X., Pelenovich V., Liu Y., Ke X., Zhang J., Yang B., Ma G., Li M., Wang X.: Effect of bias voltages on microstructure and properties of (TiVCrNbSiTaBY)N high entropy alloy nitride coatings deposited by RF magnetron sputtering. Vacuum **195**, 110710 (2022)
67. Zhang, X., Zeng, X., Liu, Y., Liu, J., Pogrebnjak, A., Pelenovich, V., Wan, Q., Liu, X., Wang, H., Lei, Y., Yang, B.: Effects of RF magnetron sputtering power on the structure and nanohardness of high-entropy alloys (TiVCrNbSiTaBY)N hard coatings. Ceram. Int. **49**, 33418–33424 (2023)
68. Bagdasaryan, A.A., Pshyk, A.V., Coy, L.E., Konarski, P., Misnik, M., Ivashchenko, V.I., Kempiński, M., Mediukh, N.R., Pogrebnjak, A.D., Beresnev, V.M., Jurga, S.: A new type of (TiZrNbTaHf)N/MoN nanocomposite coating: Microstructure and properties depending on energy of incident ions. Compos. B Eng. **146**, 132–144 (2018)
69. Bagdasaryan, A.A., Pshyk, A.V., Coy, L.E., Kempiński, M., Pogrebnjak, A.D., Beresnev, V.M., Jurga, S.: Structural and mechanical characterization of (TiZrNbHfTa)N/WN multilayered nitride coatings. Mater. Lett. **229**, 364–367 (2018)
70. Liang, S.C., Chang, Z.C., Tsai, D.C., Lin, Y.C., Sung, H.S., Deng, M.J., Shieu, F.S.: Effects of substrate temperature on the structure and mechanical properties of (TiVCrZrHf)N coatings. Appl. Surf. Sci. **257**(17), 7709–7713 (2011)
71. Chang, Z.C., Liang, S.C., Han, S., Chen, Y.K., Shieu, F.S.: Characteristics of TiVCrAlZr multi-element nitride films prepared by reactive sputtering. Nucl. Instrum. Methods Phys. Res., Sect. B **268**(16), 2504–2509 (2010)
72. Liang, S.C., Tsai, D.C., Chang, Z.C., Sung, H.S., Lin, Y.C., Yeh, Y.J., Shieu, F.S.: Structural and mechanical properties of multi-element (TiVCrZrHf)N coatings by reactive magnetron sputtering. Appl. Surf. Sci. **258**(1), 399–403 (2011)
73. Lai, C.H., Lin, S.J., Yeh, J.W., Chang, S.Y.: Preparation and characterization of AlCrTaTiZr multi-element nitride coatings. Surf. Coat. Technol. **201**(6), 3275–3280 (2006)

74. Tsai, M.H., Lai, C.H., Yeh, J.W., Gan, J.Y.: Effects of nitrogen flow ratio on the structure and properties of reactively sputtered (AlMoNbSiTaTiVZr)Nx coatings. J. Phys. D Appl. Phys. **41**(23), 235402 (2008)
75. Chang, Z.C.: Structure and properties of duodenary (TiVCrZrNbMoHfTaWAlSi)N coatings by reactive magnetron sputtering. Mater. Chem. Phys. **220**, 98–110 (2018)
76. Lin, S.Y., Chang, S.Y., Huang, Y.C., Shieu, F.S., Yeh, J.W.: Mechanical performance and nanoindenting deformation of (AlCrTaTiZr)NCy multi-component coatings co-sputtered with bias. Surf. Coat. Technol. **206**(24), 5096–5102 (2012)
77. Lin, Y.C., Hsu, S.Y., Song, R.W., Lo, W.L., Lai, Y.T., Tsai, S.Y., Duh, J.G.: Improving the hardness of high entropy nitride (Cr0.35Al0.25Nb0.12Si0.08V0.20)N coatings via tuning substrate temperature and bias for anti-wear applications. Surf. Coatings Technol. **403**, 126417 (2020)
78. Vepřek, S., Reiprich, S.: A concept for the design of novel superhard coatings. Thin Solid Films **268**(1–2), 64–71 (1995)
79. Hsueh, H.T., Shen, W.J., Tsai, M.H., Yeh, J.W.: Effect of nitrogen content and substrate bias on mechanical and corrosion properties of high-entropy films (AlCrSiTiZr)100−xNx. Surf. Coat. Technol. **206**(19–20), 4106–4112 (2012)
80. Männling, H.D., Patil, D.S., Moto, K., Jilek, M., Veprek, S.: Thermal stability of superhard nanocomposite coatings consisting of immiscible nitrides. Surf. Coat. Technol. **146**, 263–267 (2001)
81. Mayrhofer, P.H., Hörling, A., Karlsson, L., Sjölén, J., Larsson, T., Mitterer, C., Hultman, L.: Self-organized nanostructures in the Ti–Al–N system. Appl. Phys. Lett. **83**(10), 2049–2051 (2003)
82. Huang, P.K., Yeh, J.W.: Effects of substrate bias on structure and mechanical properties of (AlCrNbSiTiV)N coatings. J. Phys. D Appl. Phys. **42**(11), 115401 (2009)
83. Pogrebnjak, A.D.: Structure and properties of nanostructured (Ti-Hf-Zr-V-Nb)N coatings. J. Nanomater. **2013**, 780125 (2013)
84. Pogrebnjak, A.D., Yakushchenko, I.V., Abadias, G., Chartier, P., Bondar, O.V., Beresnev, V.M., Takeda, Y., Sobol', O.V., Oyoshi, K., Andreyev, A.A., Mukushev, B.A.: The effect of the deposition parameters of nitrides of high-entropy alloys (TiZrHfVNb)N on their structure, composition, mechanical and tribological properties. J. Superhard Mater. **35**, 356–368 (2013)
85. Krause-Rehberg, R., Pogrebnyak, A.D., Borisyuk, V.N., Kaverin, M.V., Ponomarev, A.G., Bilokur, M.A., Oyoshi, K., Takeda, Y., Beresnev, V.M., Sobol', O.V.: Analysis of local regions near interfaces in nanostructured multicomponent (Ti-Zr-Hf-V-Nb)N coatings produced by the cathodic-arc-vapor-deposition from an arc of an evaporating cathode. Phys. Metals Metallogr. **114**, 672–680 (2013)
86. Grigoriev, S.N., Sobol, O.V., Beresnev, V.M., Serdyuk, I.V., Pogrebnyak, A.D., Kolesnikov, D.A., Nemchenko, U.S.: Tribological characteristics of (TiZrHfVNbTa)N coatings applied using the vacuum arc deposition method. J. Frict. Wear **35**, 359–364 (2014)
87. Pogrebnjak, A.D., Yakushchenko, I.V., Bondar, O.V., Sobol', O.V., Beresnev, V.M., Oyoshi, K., Amekura, H., Takeda, Y.: Influence of implantation of Au- ions on the microstructure and mechanical properties of the nanostructured multielement (TiZrHfVNbTa)N coating. Phys. Solid State **57**, 1559–1564 (2015)
88. Hsieh, M.H., Tsai, M.H., Shen, W.J., Yeh, J.W.: Structure and properties of two Al–Cr–Nb–Si–Ti high-entropy nitride coatings. Surf. Coat. Technol. **221**, 118–123 (2013)
89. Liu, Y., Zhang, X., Tu, M., Hu, Y., Wang, H., Zhang, J., Yang, B.: Structure and mechanical properties of multi-principal-element (AlCrNbSiTi)N hard coating. Surf. Coat. Technol. **433**, 128113 (2022)
90. Liu, J., Zhang, X., Zeng, X., Xiong, Z., Liu, Y., Lei, Y., Yang, B.: Effect of bias voltages on microstructure and mechanical properties of (AlCrNbSiTi)N high entropy alloy nitride coatings deposited by arc ion plating. Surf. Eng. **39**(4), 495–505 (2023)
91. Zhao, J.P., Wang, X., Chen, Z.Y., Yang, S.Q., Shi, T.S., Liu, X.H.: Overall energy model for preferred growth of TiN films during filtered arc deposition. J. Phys. D Appl. Phys. **30**(1), 5 (1997)

92. Oh, U.C., Je, J.H.: Effects of strain energy on the preferred orientation of TiN thin films. J. Appl. Phys. **74**(3), 1692–1696 (1993)
93. Aouadi, K., Tlili, B., Nouveau, C., Besnard, A., Chafra, M., Souli, R.: Influence of substrate bias voltage on corrosion and wear behavior of physical vapor deposition CrN coatings. J. Mater. Eng. Perform. **28**, 2881–2891 (2019)
94. Geng, D., Li, H., Zhang, Q., Zhang, X., Wang, C., Wu, Z., Wang, Q.: Effect of incorporating oxygen on microstructure and mechanical properties of AlCrSiON coatings deposited by arc ion plating. Surf. Coat. Technol. **310**, 223–230 (2017)
95. Lubenchenko, A.V., Iachuk, V.A., Krause, S., Pavolotsky, A.B., Ivanov, D.A., Lubenchenko, O.I., Pavlov, O.N.: Interface layers of niobium nitride thin films. J. Phys. Conf. Ser. **1410**(1), 012124 (2019). IOP Publishing
96. Matsuoka, M., Isotani, S., Sucasaire, W., Zambom, L.D.S., Ogata, K.: Chemical bonding and composition of silicon nitride films prepared by inductively coupled plasma chemical vapor deposition. Surf. Coat. Technol. **204**(18–19), 2923–2927 (2010)
97. Greczynski, G., Hultman, L.: Self-consistent modelling of X-ray photoelectron spectra from air-exposed polycrystalline TiN thin films. Appl. Surf. Sci. **387**, 294–300 (2016)
98. Tam, P.L., Zhou, Z.F., Shum, P.W., Li, K.Y.: Structural, mechanical, and tribological studies of Cr–Ti–Al–N coating with different chemical compositions. Thin Solid Films **516**(16), 5725–5731 (2008)
99. Leith, S.B., Vogel, M., Jiang, X., Seiler, E., Ries, R.: Deposition parameter effects on niobium nitride (NbN) thin films deposited onto copper substrates with DC magnetron sputtering. In: 19th International Conference on RF Superconductivity (SRF 2019), pp. 947–951
100. Lan, Y., Li, X., Luo, J., Zhou, L., Wei, X., Yin, C.: Direct nitridation synthesis of quasi-spherical β-Si_3N_4 powders with CaF2 additive. Materials **12**(18), 2870 (2019)
101. Thornton, J.A.: Influence of apparatus geometry and deposition conditions on the structure and topography of thick sputtered coatings. J. Vac. Sci. Technol. **11**(4), 666–670 (1974)
102. Anders, A.: A structure zone diagram including plasma-based deposition and ion etching. Thin Solid Films **518**(15), 4087–4090 (2010)
103. Hultman, L., Bareño, J., Flink, A., Söderberg, H., Larsson, K., Petrova, V., Petrov, I.: Interface structure in superhard TiN-SiN nanolaminates and nanocomposites: film growth experiments and ab initio calculations. Phys. Rev. B **75**(15), 155437 (2007)
104. Le Meur, Y., Meyer, F., Pellet, C., Schwebel, C., Möller, P., Buxbaum, A., Eizenberg, M.: Stress in Si1−xGex films prepared by ion sputtering: origin and relaxation. Thin Solid Films **222**(1–2), 180–183 (1992)
105. Lefèvre, J., Costantini, J.M., Esnouf, S., Petite, G.: Silicon threshold displacement energy determined by photoluminescence in electron-irradiated cubic silicon carbide. J. Appl. Phys. **105**(2), 023520 (2009)
106. Pfeiler, M., Kutschej, K., Penoy, M., Michotte, C., Mitterer, C., Kathrein, M.: The influence of bias voltage on structure and mechanical/tribological properties of arc evaporated Ti–Al–V–N coatings. Surf. Coat. Technol. **202**(4–7), 1050–1054 (2007)
107. Cheng, K.H., Tsai, C.W., Lin, S.J., Yeh, J.W.: Effects of silicon content on the structure and mechanical properties of (AlCrTaTiZr)–Six–N coatings by reactive RF magnetron sputtering. J. Phys. D Appl. Phys. **44**(20), 205405 (2011)
108. Kılınc, B., Cegil, O., Sen, S., Sen, U.: Wear behavior of TiAlN and CrAlN coatings deposited by TRD process on AISI D2 Steel. Acta Phys. Pol., A **125**(2), 362–364 (2014)
109. Cao, F., Munroe, P., Zhou, Z., Xie, Z.: Mechanically robust TiAlSiN coatings prepared by pulsed-DC magnetron sputtering system: scratch response and tribological performance. Thin Solid Films **645**, 222–230 (2018)
110. Loopstra, O.B., Van Snek, E.R., de Keijser, T.H., Mittemeijer, E.J.: Model for stress and volume changes of a thin film on a substrate upon annealing: application to amorphous Mo/Si multilayers. Phys. Rev. B **44**(24), 13519 (1991)
111. Hofmann, S., Jehn, H.A.: Oxidation behavior of CrNx and (Cr, Al) Nx hard coatings. Mater. Corros. **41**(12), 756–760 (1990)

112. Tytko, D., Choi, P.P., Raabe, D.: Oxidation behavior of AlN/CrN multilayered hard coatings. Nano Convergence **4**, 1–5 (2017)
113. Ichimura, H., Kawana, A.: High-temperature oxidation of ion-plated TiN and TiAlN films. J. Mater. Res. **8**(5), 1093–1100 (1993)
114. Kawate, M., Hashimoto, A.K., Suzuki, T.: Oxidation resistance of Cr1−xAlxN and Ti1−xAlxN films. Surf. Coat. Technol. **165**(2), 163–167 (2003)
115. Fukumoto, N., Ezura, H., Suzuki, T.: Synthesis and oxidation resistance of TiAlSiN and multilayer TiAlSiN/CrAlN coating. Surf. Coat. Technol. **204**(6–7), 902–906 (2009)
116. Chang, Y.Y., Hsiao, C.Y.: High temperature oxidation resistance of multicomponent Cr–Ti–Al–Si–N coatings. Surf. Coat. Technol. **204**(6–7), 992–996 (2009)
117. Liu, J., Cui, Z., Ma, D., Lu, J., Cui, Y., Li, C., Yun, D.: Investigation of oxidation behaviors of coated Zircaloy as accident-tolerant fuel with CrAlN and CrAlSiN coatings in high-temperature steam. Corrosion Sci. **175**, 108896 (2020)
118. Keidar, M., Aharonov, R., Beilis, I.I.: Influence of an electrical field on the macroparticle size distribution in a vacuum arc. J. Vac. Sci. Technol., A: Vac., Surf. Films **17**(5), 3067–3073 (1999)
119. Warcholinski, B., Gilewicz, A., Ratajski, J., Kuklinski, Z., Rochowicz, J.: An analysis of macroparticle-related defects on CrCN and CrN coatings in dependence of the substrate bias voltage. Vacuum **86**(9), 1235–1239 (2012)
120. Liu, J., Zhang, X., Zeng, X., Xiong, Z., Liu, Y., Lei, Y., Yang, B.: Effect of bias voltages on microstructure and mechanical properties of(AlCrNbSiTi)N high entropy alloy nitride coatings deposited by arc ion plating. Surf. Eng. **39**, 495–505 (2023)
121. Warcholinski, B., Gilewicz, A., Myslinski, P., Dobruchowska, E., Murzynski, D., Kochmanski, P., Raaen, S.: Effect of nitrogen pressure and substrate bias voltage on the properties of Al–Cr–B–N coatings deposited using cathodic arc evaporation. Tribol. Int. **154**, 106744 (2021)
122. Detor, A.J., Hodge, A.M., Chason, E., Wang, Y., Xu, H., Conyers, M., Hamza, A.: Stress and microstructure evolution in thick sputtered films. Acta Materialia **57**(7), 2055–2065 (2009)
123. Chen, L., Chang, K.K., Du, Y., Li, J.R., Wu, M.J.: A comparative research on magnetron sputtering and arc evaporation deposition of Ti–Al–N coatings. Thin Solid Films **519**(11), 3762–3767 (2011)

Antiferromagnetic Spintronic Oscillators: Fundamentals and Applications

Denys Slobodianiuk, Oleh Shtanko, and Oleksandr Prokopenko

Abstract Antiferromagnetic spintronic oscillators are nanoscale multilayer devices consisting of one or more antiferromagnetic layers and capable of generating an AC output signal due to the current-driven (sub-) terahertz frequency dynamics of magnetic sublattices in an antiferromagnet when the device is biased by an electric DC current. Antiferromagnetic oscillators are now considered as promising robust and compact signal sources for many everyday and high-tech applications, including medical imaging, communications, security, military technologies, materials science, etc., although such devices have not yet been studied experimentally. In this chapter, we formulate the basic principles of antiferromagnetic spintronics with a focus on oscillator problems, discuss the key difference between the novel antiferromagnet-based oscillators and conventional ferromagnet-based oscillators, and also review recent advances, trends, and problems in the physics and applications of antiferromagnetic spintronic oscillators.

Keywords Spintronics · Antiferromagnet · Oscillator · Magnetic dynamics · Synchronisation of oscillations · Neuromorphic computing

1 Introduction

Magnetic materials are widely used in modern science, engineering, and technology. In particular, they are widely used in a variety of electronic devices and equipment that can operate at low frequencies and microwaves [1–5].

For quite a long time, magnetic materials (mainly ferromagnets—see Sect. 2.1) were used mainly in low-frequency electronic devices or in information storage devices (e.g., hard disks) [3–7]. With the advent of ferrites in the 1960s, microwave magnetic electronics began to develop rapidly, allowing the study of a number of

D. Slobodianiuk · O. Shtanko (✉) · O. Prokopenko
Educational and Scientific Institute of High Technologies, Taras Shevchenko National University of Kyiv, 64/13 Volodymyrs'ka Str., Kyiv 01601, Ukraine
e-mail: olegshtanko2000@gmail.com

A. D. Pogrebnjak et al. (eds.), *Nanocomposite and Nanocrystalline Materials and Coatings*, Advanced Structured Materials 214,
https://doi.org/10.1007/978-981-97-2667-7_3

unusual physical effects in ferrites and the creation of a number of unique microwave devices based on them [3–13].

Interest in magnetic materials and magnet-based electronics was revived 30–40 years ago with the emergence of such a promising area of modern nanoelectronics as spintronics [13–18]. In contrast to traditional electronic systems, whose principle of operation is based on the transfer of electric charges, spintronic systems work not with electric charges but with spins (projections of magnetic moments of moving particles along a certain dedicated axis). It turns out that creating a non-equilibrium configuration of spins, transferring such an ensemble of spins in space, and detecting the spin configuration of such an ensemble requires significantly less energy than similar processes in conventional electronic devices [14, 15, 17–20]. This, in turn, opens the way for the design and creation of energy-efficient electronics that can use less energy to perform key operations, which of course simplifies the task of heat dissipation from such a device and potentially improves its performance.

To date, various researchers and engineers have proposed and practically implemented many different spintronic devices based on ferromagnetic materials. The corresponding results are reflected in hundreds of original papers (including numerous patents), many review articles [11, 13–16, 18, 20] and books [4, 5, 12, 17, 18]. At the same time, it has been found that the characteristic operating frequencies of spintronic devices based on ferromagnetic materials depend significantly on the magnitude of the external bias DC magnetic field and typically do not exceed 20–50 GHz. Since the presence of a bias DC magnetic field for proper device operation is a rather unpleasant drawback, the last 10–15 years have demonstrated the development and creation of spintronic devices based on magnetic materials with strong anisotropy [18–24], in particular $L1_0$ materials [25], which are capable of operating in the absence of a bias DC magnetic field. It should be noted, however, that the operating frequencies of such devices are usually lower than those of magnetic field-controlled devices, usually not exceeding 20–30 GHz (typically 1–10 GHz).

A significant improvement in the performance of spintronic devices is currently associated with the use of antiferromagnetic (AFM) rather than ferromagnetic materials (see Sect. 2.1). The corresponding branch of spintronics—antiferromagnetic spintronics [26–33]—was emerged about 20 years ago and is now developing rapidly. Dynamic processes in spintronic devices based on AFM spintronic nanostructures are realised at frequencies close to the antiferromagnetic resonance frequency, which can reach $\sim 50 - 10^4$ GHz in different AFM materials and thus significantly exceed the typical frequencies of ferromagnetic resonance in ferromagnets and ferrites. Also, the key factors that determine the prospects of antiferromagnetic spintronics are the following: (i) antiferromagnetic devices are capable of operating at $\sim 50 - 10^4$ GHz in the absence of a bias DC magnetic field; (ii) antiferromagnets are almost insensitive to DC magnetic fields, which makes them easy to use even in the presence of strong DC magnetic fields; (iii) the ground state of antiferromagnets is usually characterised by zero net magnetisation, which avoids the parasitic effect of the magnetic subsystem of antiferromagnets on the surrounding elements, and thus facilitates the integration of antiferromagnetic elements into various (e. g., semiconductor) nanoelectronic devices.

Although the history of antiferromagnetic spintronics began not so long ago [26, 27], more than a dozen different devices based on antiferromagnetic spintronic nanostructures have been proposed so far, among which the most famous are memory devices [34–38], antiferromagnetic oscillators [39–51], whose properties will be discussed in this chapter, and antiferromagnetic signal detectors [52], whose properties will be discussed in details in Chap. 4. Unfortunately, the practical realisation of such devices currently faces a number of serious problems, primarily related to the difficulty of fabricating high-quality AFM spintronic nanostructures (typically based on monocrystalline single-domain AFM films with a certain type of magnetic anisotropy) [19, 20, 28–33, 40–43]. At the same time, the first experiments for the simplest AFM systems demonstrated a good agreement between theory and experiment [53, 54]. This opens up the possibility of finding new physical effects in AFM spintronic systems, developing new AFM spintronic devices, optimising their technical properties, etc. using the methods of theoretical physics and numerical modelling.

2 Basic Physics of Antiferromagnetic Spintronic Devices

2.1 Basic Physics of Magnetic Materials

A characteristic feature of all magnets is the presence of a large number of elementary magnetic moments $\boldsymbol{\mu}_i$ ($i = 1, 2, 3, \ldots$) inside them, oriented in a certain way in space. These magnetic moments have a quantum nature (magnetic moments of "free" electrons, ions, atoms, molecules, etc.) and can interact in different ways with each other and with an external bias DC magnetic field with a strength of $\mathbf{H}$ [2–7, 55, 56]. This interaction leads to a certain ordering of the magnetic moments inside the magnetic material. Depending on the type of ordering, magnets usually can be divided into three classes: diamagnets, paramagnets, and magnetically ordered materials, which include ferromagnets, ferrites, and antiferromagnets (Fig. 1).

The magnetic moments $\boldsymbol{\mu}_i$ in diamagnets are mainly oriented against the field direction $\mathbf{H}$, and in paramagnets the magnetic moments $\boldsymbol{\mu}_i$ are mainly oriented parallel to the field $\mathbf{H}$ direction (Fig. 1a). However, in both cases, the influence of the field is rather weak, and therefore the magnetic permeability μ of diamagnets and paramagnets, i.e. the ratio of the magnetic induction $|\mathbf{B}| = B$ inside the magnet to the magnetic field induction $\mu_0|\mathbf{H}| = \mu_0 H$ in vacuum, $\mu = B/\mu_0 H$ is close to 1 (here μ_0 is the vacuum permeability) [2–7, 55, 56]. For diamagnets $\mu \approx 1 - \chi_d$, and for paramagnets $\mu \approx 1 + \chi_p$, where $\chi_d \sim 10^{-6} - 10^{-4}$, $\chi_p \sim 10^{-4} - 10^{-2}$ at room temperatures.

In ferromagnets, the magnetic properties are much more pronounced because the individual magnetic moments $\boldsymbol{\mu}_i$ in a ferromagnet are bound together by a strong exchange interaction [55, 56]. The presence of this interaction leads to the formation of magnetic domains in ferromagnets—macroscopic regions with a particular

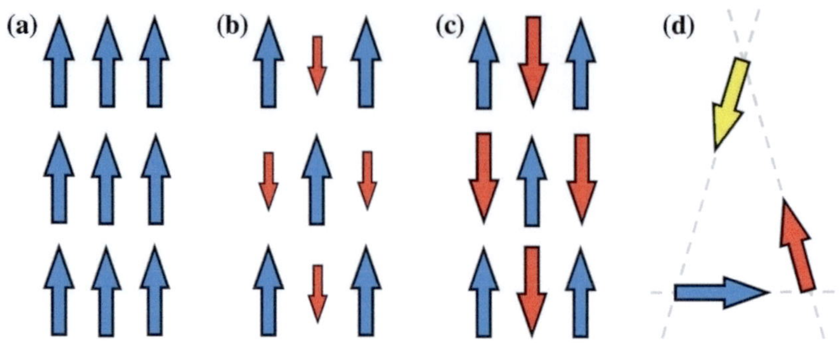

Fig. 1 Type of magnetic ordering in different magnetic materials: **a** paramagnet of ferromagnet, **b** ferrimagnet (ferrite), **c** collinear antiferromagnet, and **d** non-collinear antiferromagnet. Magnetic moments (indicated by arrows) form one (**a**) or more (**b–d**) ordered magnetic sublattices. The same colour indicates the magnetic moments belonging to a particular sublattice. The size of the arrow indicates the relative size of the magnetic moment

dominant direction for all magnetic moments (Fig. 1a). If a ferromagnetic sample has small dimensions $\sim 10-100$ nm, which is typical for spintronic nanostructures, only one domain can exist in it, i.e. a single-domain state is realised [17–20, 25]. At the same time, for a large-sized ferromagnet, it is energetically advantageous to form a set of domains separated by domain walls with different orientations of magnetic moments within each domain. In a certain frequency range, the magnetic permeability of ferromagnets can be significantly larger than 1, $\mu \sim 1 - 10^6$ [2–7, 55, 56].

Ferrimagnets (ferrites) have a more complex magnetic structure than ferromagnets [8, 12, 55, 56]. They contain two or more magnetic sublattices, the magnetic moments in each of which are approximately equally oriented (the case similar to the case of ferromagnets), but the magnetic moments of different magnetic sublattices partially compensate each other (Fig. 1b). The result is a magnetic state with incompletely compensated magnetic moments, for which a single-domain state (in small-sized samples) or a domain structure (in films and bulk samples) can be observed [10–12, 55, 56]. According to their electrical properties, ferrimagnets are semiconductors. Their main advantage over ferromagnets is low losses. Yttrium-iron garnet has the lowest losses among the known ferrimagnets, which makes it possible to study various complex, including nonlinear and parametric, phenomena in it and to design microwave devices with record-breaking characteristics [8–13, 55].

Antiferromagnets are a large group of magnetic materials that have two or more magnetic sublattices whose magnetic moments usually compensate each other (Fig. 1c, d) [2–7, 31, 55, 56]. In collinear antiferromagnets, the magnetic moments of the sublattices are antiparallel (Fig. 1c); in non-collinear antiferromagnets, the magnetic moments of the sublattices can be arbitrarily oriented, but in such a way that the influence of the magnetic moments of all sublattices is mutually compensated (Fig. 1d). The ground state of such an antiferromagnet (the lowest energy state) is a compensated magnetic state with zero net magnetisation.

The magnetic moments in a compensated antiferromagnet are held together by a very strong internal magnetic field B_{ex} of exchange nature. This field typically reaches $B_{ex} \sim 10^2 - 10^3$ T, which is 3–4 orders of magnitude higher than the bias DC magnetic fields B_{dc} commonly used in practice, especially in ferromagnetic spintronics. The consequence of this is the natural insensitivity of antiferromagnets to magnetic fields $B_{dc} \ll B_{ex}$, , since the field is not able to change the compensated ground magnetic state of the antiferromagnet. On the one hand, this property of antiferromagnets is useful because it allows them to be used in the presence of strong magnetic fields, electromagnetic interference, ionising radiation, etc. On the other hand, unlike ferromagnetic and ferrimagnetic spintronic systems, where the operating frequency of the system (ferro- or ferrimagnetic resonance frequency) and some other properties can be controlled by applying an external bias DC magnetic field B_{dc} to the system (device) (this is a typical practice in most known experiments), this control method is not applicable to antiferromagnetic devices and systems. Therefore, to excite magnetic oscillations in antiferromagnets and to control their properties, it is necessary to use alternative experimental techniques that do not rely on the effect of an external DC magnetic field on the antiferromagnet. The relevant physical phenomena useful for the creation of antiferromagnetic spintronic oscillators will be discussed further in Sect. 2.2.

2.2 *The Spin-Transfer Torque, the Spin Hall, and Spin Pumping Effects*

The spin-transfer torque (STT) effect is well known in ferromagnetic spintronics [6, 7, 14–25]. This effect was theoretically predicted independently by Slonczewski [57] and Berger [58] in 1996 and then thoroughly studied theoretically [59, 60] and observed experimentally in layered spintronic nanostructures (nanopillars and nanocontacts) based on ferromagnets [61–66]. This effect is typically manifested in layered spintronic nanostructures with at least two magnetic layers separated by a non-magnetic layer (Fig. 2a). One magnetic layer of the nanostructure is “fixed” or “pinned”—the magnetisation of this layer can be considered stationary. The other layer of the nanostructure is “free”; its magnetisation can move relatively easily under the influence of external influences (current, field, etc.). The STT effect implies the appearance of an additional torque $\mathbf{T}_{STT}$ acting on the magnetisation $\mathbf{M}$ of the free layer when an electric current $I(t)$ flows through the layered spintronic nanostructure. The mechanism of the STT effect is as follows. When a current $I(t)$ flows through the nanostructure, the charge carriers became spin-polarised, as a result of which they are able to carry not only an electric charge but also an uncompensated magnetic moment (a certain selected projection of the spin magnetic moment) [57–60]. This case corresponds to the so-called spin-polarised current. It is also possible for the current to carry an uncompensated magnetic moment without carrying an electric charge. This so-called “pure” spin current is formed either by the counterflow of two spin-polarised currents of equal magnitude with different spin polarisation directions, or by the appearance of a pure spin current perpendicular to the direction of the electric current. In the latter case, typical for spin Hall oscillators (SHOs), it is possible to realise the case of transfer of uncompensated spin moment to the free magnetic layer of the nanostructure without transfer of electric charge through the whole nanostructure (electric current $I(t)$ flows only in one of the layers of the nanostructure) [67].

In antiferromagnetic nanostructures, the physics of the STT effect is somewhat more complicated because the magnetic moments in antiferromagnets are organised as a set of interacting magnetic sublattices $j = \overline{1, 2, \ldots N}$, each of which is characterised by its own magnetisation $\mathbf{M}_j$ [26, 68–72]. At the same time, the linear order of $\mathbf{M}_j$ of the torque $\mathbf{T}_{\mathrm{STT},j}$ acting on the j-th magnetic sublattice has the same form as in ferromagnets:

$$\mathbf{T}_{\mathrm{STT},j} = \frac{\tau}{M_{S,j}} \left[\mathbf{M}_j \times \left[\mathbf{M}_j \times \mathbf{p}\right]\right], \tag{1}$$

where $M_{S,j} = |\mathbf{M}_j|$ is the saturation magnetisation for the j-th magnetic sublattice, τ is the amplitude STT coefficient (given in frequency units), and $\mathbf{p}$ is the spin-polarisation of the current $I(t)$. Since the torque $\mathbf{T}_{\mathrm{STT},j}$ is directly proportional to the current $I(t)$, the coefficient τ is typically expressed as $\tau = \sigma I(t)$ or $\tau = \sigma j(t)$ where

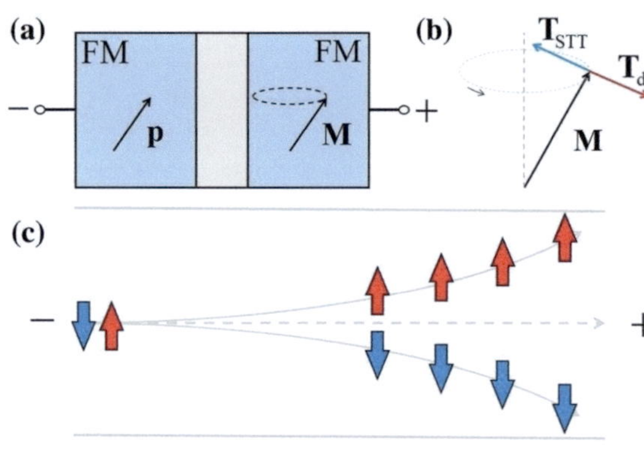

Fig. 2 **a** Typical ferromagnetic spintronic nanostructure consisting of two ferromagnetic (FM) layers, pinned (left) and free (right), separated by a non-magnetic spacer. When electrons move through the structure from the pinned to the free layer, they transfer the spin angular momentum from the pinned to the free layer due to the STT and cause the excitation of a magnetisation precession in the free magnetic layer of the structure. **b** The STT effect manifests itself as a negative damping, characterised by the antidamping torque $\mathbf{T}_{\mathrm{STT}}$, which is able to compensate natural magnetic damping described by the damping torque $\mathbf{T}_{\mathrm{d}}$. In this case a self-sustained magnetisation precession can start. c) Electrons moving in a material with a strong spin–orbit interaction are deflected in opposite directions depending on their spin state (spin “up” or spin “down”). Due to that effect, called the spin Hall effect, an unpolarised electric current flowing in a material with a strong spin–orbit interaction generates a transverse spin current

$j(t)$ is the density of the current $I(t)$, and σ is the STT coefficient, given in frequency/current or frequency/(current density) units, respectively. More information on the STT effect in antiferromagnets can be found in [26, 27, 68–71].

Equation (1) formally characterises the changes of magnetisation, $\partial \mathbf{M}_j/\partial t$, for some j-th magnetic sublattice of an antiferromagnet under the action of a current. These changes can manifest themselves as the so-called antidamping torque (Fig. 2b). Due to this effect, even in compensated antiferromagnets with zero net magnetisation, local oscillations of magnetic sublattices can be excited under the action of current $I(t)$, which is a key effect for the realisation of antiferromagnetic spintronic oscillators.

The second physical effect commonly used to excite oscillations in antiferromagnetic spintronic nanostructures is the spin Hall effect (SHE) [73]. This effect, which has been well studied for ferromagnetic spintronic nanostructures [67, 74–77], manifests itself as the generation of a transverse spin current when an electric current $I(t)$ flows in a conductive nonmagnetic material with a strong spin–orbit interaction. Typically, a heavy metal (HM) like Pt, Au, Pd, Al, Mo, etc. is used as such a material. In such a HM, when an unpolarised electric current $I(t)$ flows, in which the spins of the moving charge carriers (electrons) are initially randomly oriented in space (the spin "up" and spin "down" states), the electrons with spins "up" and spins "down" are deflected in opposite directions due to the strong spin–orbit interaction (Fig. 2c). This effect makes it possible to create a non-equilibrium configuration of magnetic moments (spins) at the interface of AFM/HM thin film structures, which can exert a sufficiently strong local influence on the magnetic sublattices of the adjacent AFM layer. The presence of such an influence leads to the STT effect discussed earlier. The STT efficiency due to the SHE can be estimated by the STT σ (see Eq. (1)). The value of σ (in frequency per current density units) for an AFM/HM thin film structures is typically given in the form:

$$\sigma = \theta_{\text{SHE}} g_r \frac{e\gamma\lambda\rho}{2\pi M_S d_{\text{AFM}}} \tanh\left(\frac{d_{\text{HM}}}{2\lambda}\right) \tag{2}$$

where θ_{SHE} is the spin Hall angle, g_r is the spin-mixing conductance at the AFM/HM interface, λ is the spin diffusion length in the HM, ρ is the HM electric resistivity, M_S is the saturation magnetisation of one of the AFM magnetic sublattices, d_{AFM} and d_{HM} are the thicknesses of the AFM and HM layers, respectively.

It should be noted that there is in fact a whole family of spin Hall spin effects, of which the above-mentioned "direct" spin Hall effect and the inverse spin Hall effect discussed in Chap. 4 are used in antiferromagnetic spintronics. These effects are analysed in more detail in the review [73].

The spin pumping effect is a complementary effect to the SHE [73] and has the same nature for ferromagnetic [78] and AFM devices [70, 79, 80]. This effect leads to the appearance of an alternating spin current at the AFM/HM interface under the influence of the magnetisation oscillations occurring in the AFM layer. The resulting spin current traverses the AFM/HM interface and flowing into the HM layer, which can lead to the appearance of a transverse electric current in the HM layer (conversion of the spin current into an electric current due to the inverse spin Hall effect [23]) and other effects. It should be noted that spin pumping effects have been well studied for ferromagnetic nanostructures [78, 81].

2.3 Antiferromagnetic Spintronic Nanostructures

In contrast to ferromagnetic spintronic nanostructures, for which a large number of device schematics are known—with one (spin Hall devices) [67, 75, 78, 81], two (traditional spin-torque nano-oscillators) [21–24, 75] or even more magnetic layers (unconventional oscillators) [82–84], antiferromagnetic spintronic nanostructures are typically realised as AFM/HM bilayers. A typical AFM nanostructure that works as a spin Hall oscillator is a thin film AFM/HM structure (Fig. 3a). When a DC or AC electric current $I(t)$ is passed through the HM layer, a transverse spin current is generated in the HM layer due to the SHE. This spin current, if its magnitude exceeds a certain threshold, excites the undamped oscillations of the magnetic sublattices in the adjacent AFM layer with characteristic frequencies that depend on the current $I(t)$ magnitude and are typically in the ~$50 - 10^3$ GHz frequency range. Subsequently, the energy of these excited oscillations can be extracted from the oscillator, which allows us to consider such a system as a current-driven source of (sub-)terahertz-frequency signals.

Another type of AFM nanostructure that has been studied recently is the antiferromagnetic tunnel junction (ATJ) [44, 52], which is more complex than a bilayer SHO structure. Such a nanostructure is actually a specific variant of a conventional SHO with additional non-magnetic layers.

A typical ATJ is a four-layer HM/AFM/D/HM structure (Fig. 3b). The first two layers, HM and AFM, separated by the HM/AFM interface, play the same role as in a conventional antiferromagnetic spin Hall oscillator. When an electric current is passed through the HM layer adjacent to the AFM layer, magnetic oscillations are excited in the AFM layer due to the spin Hall and STT effects. While in a conventional SHO, the energy of these oscillations is directly extracted to the external circuit connected to the SHO, the mechanism of energy extraction from the ATJ may be

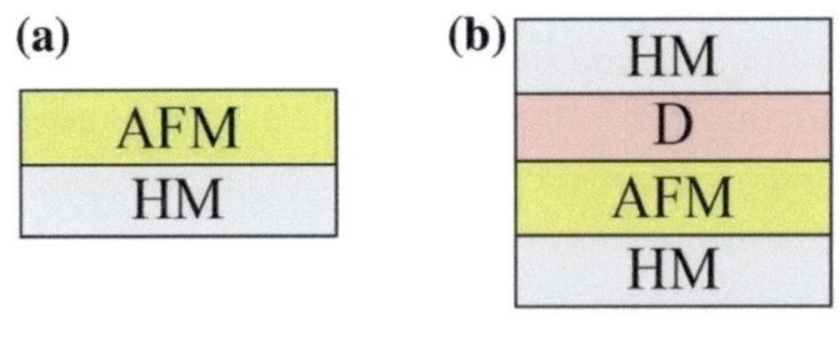

Fig. 3 **a** Schematics of spin Hall oscillator based on heavy metal (HM) and antiferromagnetic (AFM) layers. **b** Schematics of antiferromagnetic tunnel junction based on two heavy metal (HM), antiferromagnetic (AFM), and dielectric (D) layers

different. Such structures are characterised by the tunnelling anisotropic magnetoresistance (TAMR) phenomenon [85–87]. Due to this, the oscillations of the magnetic sublattices of the AFM layer, excited by a DC electric current, lead to the appearance of the structure TAMR oscillations with a frequency close to the frequency of the oscillations of the magnetic sublattices ($\sim 50 - 10^3$ GHz). When a DC current flows through such a nanostructure, the oscillations of the magnetoresistance and the current are mixed. This leads to the appearance of (sub-)terahertz oscillations of the electric voltage between the outer conductive HM layers of the structure. This oscillating voltage is the generated signal whose energy can be transferred to the external circuit.

2.4 Magnetic Dynamics in Antiferromagnets

The dynamics of magnetic sublattices in a thin AFM film in the presence of the STT is governed by the coupled Landau-Lifshits-Gilbert-Slonczewski equations written for each sublattice. For the simplest case of an antiferromagnet with two magnetic sublattices, this system of equations has the form [27, 57, 58, 70]:

$$\begin{aligned} \tfrac{d\mathbf{M}_1}{dt} &= \gamma\left[\mathbf{B}_{\mathrm{eff},1} \times \mathbf{M}_1\right] + \tfrac{\alpha}{M_S}\left[\mathbf{M}_1 \times \tfrac{d\mathbf{M}_1}{dt}\right] + \tfrac{\tau}{M_S}\left[\mathbf{M}_1 \times \left[\mathbf{M}_1 \times \mathbf{p}\right]\right] \\ \tfrac{d\mathbf{M}_2}{dt} &= \gamma\left[\mathbf{B}_{\mathrm{eff},2} \times \mathbf{M}_1\right] + \tfrac{\alpha}{M_S}\left[\mathbf{M}_2 \times \tfrac{d\mathbf{M}_2}{dt}\right] + \tfrac{\tau}{M_S}\left[\mathbf{M}_2 \times \left[\mathbf{M}_2 \times \mathbf{p}\right]\right] \end{aligned} \tag{3}$$

where the first terms in the right-hand parts of these equations describe the conservative motion of the sublattice magnetisations $\mathbf{M}_{1,2}$ about the effective fields $\mathbf{B}_{\mathrm{eff},1,2}$ acting on each sublattice, the second term is the damping term describing the natural magnetic damping in the AFM material, and the last term is the Slonczewski-Berger antidamping term excited by the STT discussed earlier in Sect. 2.2. In Eq. (2) γ is the modulus of the gyromagnetic ratio, α is the effective damping parameter, M_S is the saturation magnetisation (assumed to be the same for both sublattices), and τ is the STT amplitude, which is defined as $\tau = \sigma j(t)$, where the STT coefficient σ is given by Eq. (2).

The bonding between the magnetic sublattices is characterised by a very strong internal field $\mathbf{B}_{\mathrm{ex}}$ of the exchange origin, which is the major contributor to the effective fields $\mathbf{B}_{\mathrm{eff},1,2}$ [27, 31, 70]:

$$\mathbf{B}_{\mathrm{eff},1,2} = -\frac{1}{2M_S}\mathbf{B}_{\mathrm{ex}}\mathbf{M}_{2,1} + \mathbf{B}_{\mathrm{a},1,2}\left(\mathbf{M}_{1,2}\right). \tag{4}$$

Here $\mathbf{B}_{\mathrm{a},1,2}(\mathbf{M}_{1,2})$ is the relatively weak ($|\mathbf{B}_{\mathrm{a},1,2}| \ll |\mathbf{B}_{\mathrm{ex}}|$) anisotropy field for the j-th sublattice depending on its magnetisation vector $\mathbf{M}_j$, $j = \overline{1,2}$.

The system of Eqs. (3) can be simplified in the framework of the σ-model [88] for two vectors, the AFM Neel vector $\mathbf{l} = (\mathbf{M}_1 - \mathbf{M}_2)/2M_S$ and $\mathbf{m} = (\mathbf{M}_1 + \mathbf{M}_2)/2M_S$. Note that for an ideally compensated antiferromagnet, where $\mathbf{M}_1 = -\mathbf{M}_2$, $\mathbf{l}$ is the unit vector along $\mathbf{M}_1$ ($\mathbf{l} = M_1 = M_S$), and $\mathbf{m} = 0$. Even if $\mathbf{M}_1 \neq -\mathbf{M}_2$, but $|\mathbf{m}| \ll |\mathbf{l}|$ (in the context of this assumption, one can write $\mathbf{m}$ as a slave variable $\mathbf{m} = \left[\boldsymbol{l} \times \frac{dl}{dt}\right]/\gamma B_{\text{ex}}$), one can obtain from (3) the simplified equation describing the dynamics in antiferromagnets in the terms of the Neel vector $\mathbf{l}$ [40]:

$$\mathbf{l} \times \left[\frac{1}{\omega_{\text{ex}}}\frac{d^2\mathbf{l}}{dt^2} + \alpha\frac{d\mathbf{l}}{dt} + \widehat{\Omega}\cdot\mathbf{l} + \tau\left[\mathbf{p}\times\mathbf{l}\right]\right] = 0 \tag{5}$$

where $\widehat{\Omega}$ is the term that depends on the magnetic anisotropy of an AFM film.

3 Signal Sources Based on AFM Nanostructures

The idea of exploiting current-induced magnetic dynamics in antiferromagnetic (AFM) materials was first proposed more than a decade ago [26] and inspired a number of theoretical proposals for AFM-based signal sources potentially capable of generating terahertz-frequency (TF) signals.

In one of the first such studies [39], the current-induced magnetisation dynamics in insulating antiferromagnets was considered for a famous spin Hall geometry [73], namely for the spin Hall oscillator (SHO) shown in Fig. 4. Assuming that an AFM material has a monocrystalline structure and using the macrospin description for the investigated dynamics in the AFM, the authors of [39] analysed the dynamics of the Néel $\mathbf{l} = (\mathbf{m}_{\text{A}} - \mathbf{m}_{\text{B}})/2$ and the net magnetisation $\mathbf{m} = (\mathbf{m}_{\text{A}} + \mathbf{m}_{\text{B}})/2$ vectors, where $\mathbf{m}_{\text{A}}$ and $\mathbf{m}_{\text{B}}$ are the unit vectors of the AFM sublattice magnetisations. Using the coupled Landau-Lifshitz-Gilbert-Slonczewski equations for $\mathbf{l}$ and $\mathbf{m}$ vectors, it was shown that sufficiently large in-plane currents, being applied perpendicular to $\mathbf{l}$ (see current density $\mathbf{J}_e$ in Fig. 4), can induce spontaneous TF oscillations. When the amplitude of this current exceeds a threshold, it is possible to achieve steady magnetic dynamics as a result of spin pumping and current-induced torques (see Sect. 2.1). In addition, the amplitudes of these oscillations were found to be controllable via the applied current, providing the opportunity to build AFM sources of TF signals without cryogenic temperatures, bulk and complicated experimental setups.

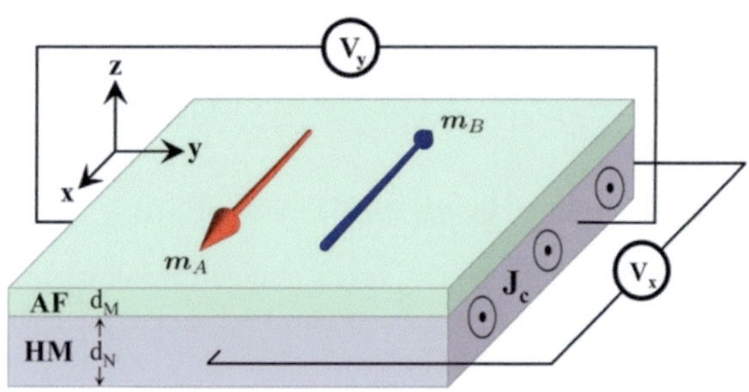

Fig. 4 A SHO consisting of a heavy metal (HM) and an insulating antiferromagnet (AF) layer. $\mathbf{J}_e$ is the applied DC current density, $\mathbf{m}_{\text{A}}$ and $\mathbf{m}_{\text{B}}$ are the unit vectors of the AFM sublattice magnetisations

In the next work [40] dealing with AFM generators, the ability to obtain TF signals was demonstrated for a similar SHO consisting of a heavy metal platinum (Pt) layer and an AFM dielectric NiO layer with biaxial anisotropy (Fig. 5). In the proposed system, the spin current with the polarisation **p** induced due to the spin Hall effect (see Sect. 2.2), flows from a DC current-driven Pt layer into the adjacent AFM layer. If the current polarisation **p** is parallel to the hard anisotropy axis of the AFM layer, the aforementioned spin current excites a non-uniform precession of the AFM magnetisation sublattices as a result of the relatively weak easy-plane AFM anisotropy. The frequencies of the obtained magnetisation precession are in the range of 0.1–2.0 THz and are found to be dependent on the amplitude of the driving current (typical values of the DC current densities vary from 10^8 A/cm^2 to 10^9 A/cm^2). In the proposed generator, output TF signals exceeding the amplitude of 1 V/cm, are collected by means of the inverse spin Hall effect in Pt. It is important to note that the proposed signal source is similar to the cryogenic oscillator based on Josephson junction, but in their case, the energy of easy-plane magnetic anisotropy plays a crucial role instead of Josephson energy in the above-mentioned junctions [89].

The next gradual step towards the implementation of AFM TF generators was made in [41], where SHO with a canted AFM layer is considered (Fig. 6). In this SHO, the AFM magnetic sublattices are considered to be canted via the Dzyaloshinskii-Moriya interaction, resulting in the appearance of a relatively small net magnetisation $\mathbf{m}_{\mathrm{DMI}}$. The spin current $\mathbf{j}_s$ being induced by the spin Hall effect in a heavy metal Pt layer interacts with $\mathbf{m}_{\mathrm{DMI}}$ and causes its TF rotation. This oscillation dynamics around the hard AFM anisotropy axis shows a substantial dependence of the precession frequency on the amplitude of the injected spin current and the value of the AFM

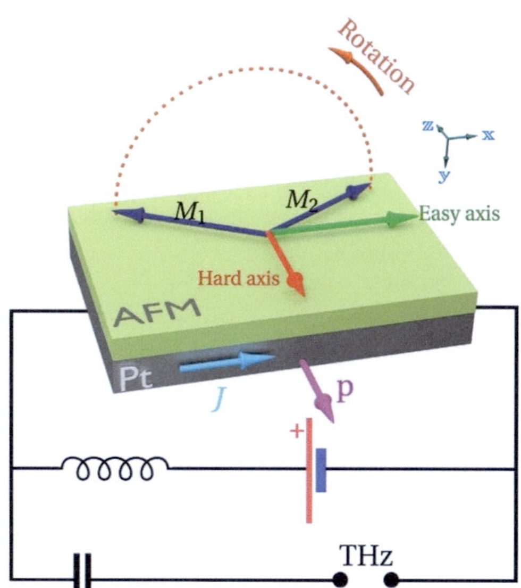

Fig. 5 Schematic of the TF generator based on a Pt/AFM SHO, where **J** is the applied DC current density, **p** is the polarisation of the excited spin current, $\mathbf{M}_1$ and $\mathbf{M}_2$ are the vectors of the AFM sublattices' magnetisations

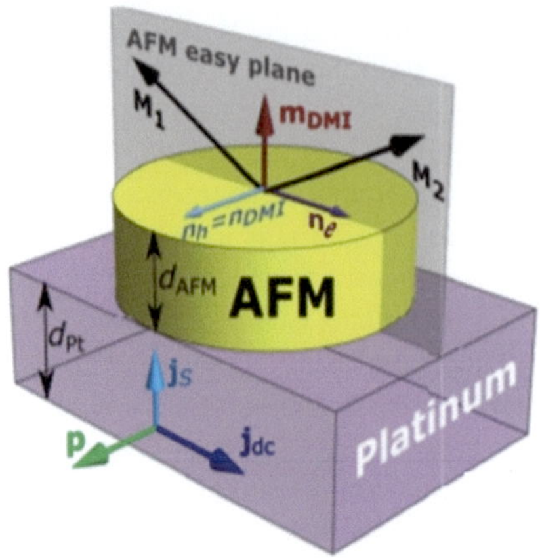

Fig. 6 Schematic of the SHO, where d_{AFM} and d_{Pt} are the thicknesses of the AFM and Pt layers respectively, $\mathbf{j}_{dc}$ is the applied DC current density, $\mathbf{p}$ is the polarisation of the excited spin current with density $\mathbf{j}_s$, $\mathbf{M}_1$ and $\mathbf{M}_2$ are the vectors of the sublattices' magnetisations, $\mathbf{m}_{DMI}$ is the vector of the net magnetisation, $\mathbf{n}_e$ is the unit vector lying in the easy plane, $\mathbf{n}_h$ is the unit vector along the hard anisotropy axis

exchange field. This effect is therefore useful for the development of novel, reliable, and tunable AFM sources of TF signals.

It should be noted that the obtained TF rotation of $\mathbf{m}_{DMI}$ causes a dipolar radiation of the same frequency, which energy could be collected by some adjacent electrodynamic system (a similar case has been analysed for ferromagnet-based spintronic oscillators in [90]). For instance, in [41], it is shown that a high-Q dielectric resonator provides the best efficiency of such a source, among a number of different electrodynamic systems (free space, nano-sized loop, parallel plate line, rectangular and dielectric waveguides, rectangular and parallel plate resonators), even in comparison with the previous generator proposal [40] built on the basis of NiO/Pt SHO (Fig. 7). Using experimentally reachable values of the applied driving current density $\mathbf{j}_{dc}$, it is shown in [41] that such a source can operate effectively in the frequency range from 0.05 THz to 2 THz, with a frequency-dependent output power of up to 1 μW at the 0.5 THz frequency (see Fig. 7).

A completely new design of a TF generator has been proposed in [44]. It is based on the utilisation of the so-called antiferromagnetic tunnel junction (ATJ) [85] shown in Fig. 8. This junction consists of four layers, the first two of which (heavy metal Pt and AFM layers) are similar to those in the spin Hall oscillator geometry. An ATJ also contains a layer of MgO, which is used as a tunnel barrier for moving electrons traversing all the four layers of the structure. The final fourth layer in an ATJ is the top Pt layer. The operating principles of an ATJ-based signal source are partly similar to those of a conventional SHO-based source considered above. In both nanostructures, the spin current acting on an AFM layer is excited via the spin Hall effect in a heavy metal Pt layer (bottom Pt layer in the ATJ shown in Fig. 8). However, the energy

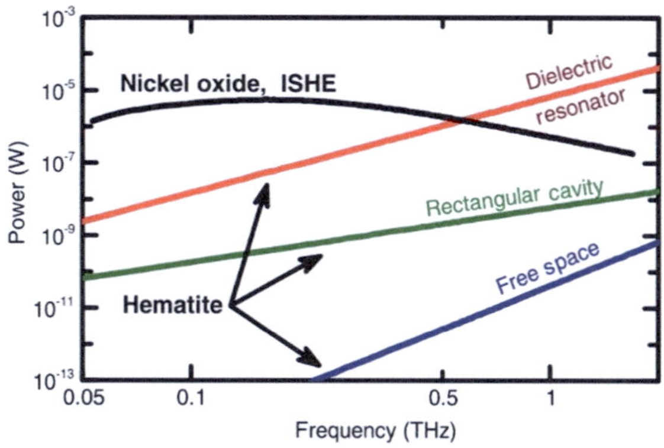

Fig. 7 Frequency dependence of the generated signal power for SHO based on the Pt/canted AFM structure. The generated dipolar radiation is registered by using the different electrodynamic systems: dielectric resonator (red line), rectangular cavity (green line) and free space (blue line), respectively. The black curve shows the frequency dependence of the generated signal power for the case of NiO/Pt SHO using the power extraction mechanism based on the inverse spin Hall effect

extraction mechanism for the ATJ-based source is completely different from that of the conventional SHO-based source.

In an ATJ, the rotation of magnetisation sublattices driven by the drive current I_{drive} leads to the TF variations of the junction tunnelling anisotropic magnetoresistance. When a bias DC electric current I_{dc} is then passed through the junction, as shown in Fig. 8, the mixing of the oscillating magnetoresistance and this DC current results in the generation of the AC voltage across an ATJ.

Using an equivalent electrical scheme for such a Pt/AFM/MgO/Pt junction, the authors of [44] evaluate the output power of such a source and obtain a result of about 1 μW at 0.5 THz frequency for optimised geometrical parameters of the junction, which is similar to the results obtained in [41]. However, the considered ATJ-based signal source is characterised by a strong frequency dependence of the source output power demonstrating a substantial reduction of this power with the increase of frequency. Nevertheless, the main advantage of the proposed generator is its relative simplicity for experimental implementation, which could potentially be used to

Fig. 8 Schematic of a TF signal source based on ATJ, where I_{drive} is a drive DC current, I_{SH} is a spin current, I_{dc} is a bias DC current

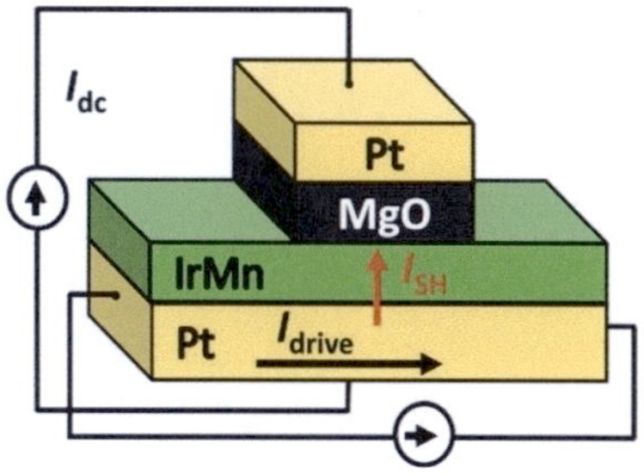

create ATJs compatible with existing fabrication technology and reliable ATJ-based TF signal sources.

In order to increase the output power of the proposed TF sources, systems with multiple generating structures (SHOs or ATJs) have been proposed [91–93]. In [91], two boundary cases are considered: independent and perfectly coupled SHOs embedded in a high-Q dielectric resonator, forming an array of generating SHOs embedded in some electrodynamic system. The theoretical model is based on an approach used in [41] where the calculation of an electromagnetic field inside the dielectric resonator is considered using the magnetic wall boundary conditions. According to the estimates made (Fig. 9), it is possible to generate signals with power up to 1 mW using even a small number ($\leq$10) of phase-locked SHOs placed at optimum locations inside a high-Q dielectric resonator.

Following a promising prediction made in [91], the authors investigated the behaviour of this system by studying the effect of the spatial configuration of the AFM SHO array on the frequency-dependent power of such a source, which is crucial for the development of such devices [92, 93]. The case of perfectly synchronised SHOs was chosen and studied for three different placement schemes (see A, B, and C schemes in Fig. 10), which show that the highest values of output AC power could be guaranteed when all AFM SHOs are placed in locations where the magnetic field is maximum. Therefore, it is predicted that it is possible to obtain 100 μW power at 2 THz frequency for the A-type placement scheme with only three phase-locked AFM SHOs.

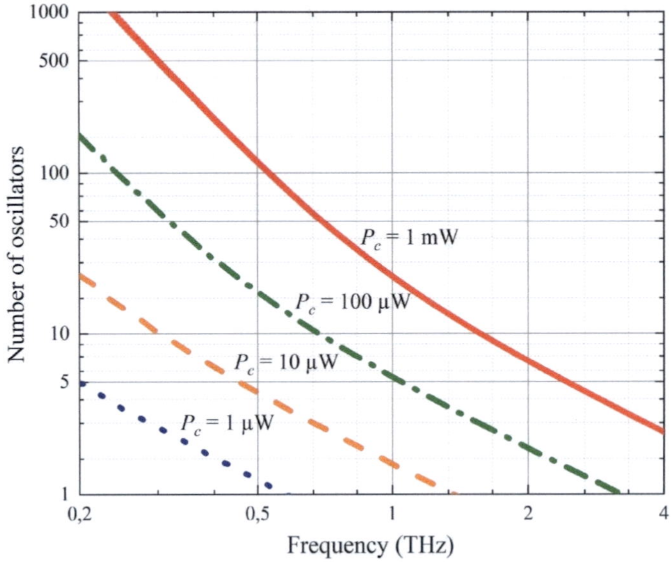

Fig. 9 Frequency dependence of the number of perfectly coupled SHOs required to achieve powers of 1 μW, 10 μW, 100 μW and 1 mW

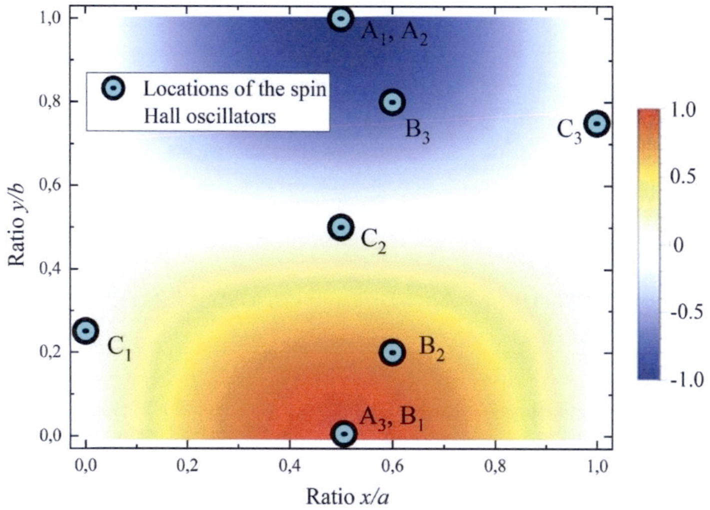

Fig. 10 Density plot for the y-component of the magnetic field in the case of E_{111} mode for A, B, and C placement schemes of three AFM SHOs

Another case has also been investigated, where a single ATJ is embedded in a high-Q dielectric resonator to create an alternative generator design [44]. Using an electrical model of the ATJ developed in [41] and the optimal geometric parameters of the junction, the output power is estimated for different ATJ locations within the dielectric resonator. According to the study, it is possible to achieve 1–10 μW of power in the sub-terahertz frequency range (0.1–1 THz) by placing ATJs at optimal locations inside the high-Q dielectric resonator.

In conclusion, AFM-based sources of TF signals are promising in various fields of science and technology because their use does not require cryogenic temperatures, strong bias currents and/or magnetic fields as well as bulk and complex experimental setups. To date, there are some prospective theoretical proposals, based on conventional AFM SHOs and ATJs, which foresee the possibility of generating TF signals of 1 μW power for a single operating nanostructure and up to 1 mW power for arrays of synchronised oscillators embedded inside an electrodynamic system (e.g. a high-Q dielectric resonator).

4 Synchronisation of a Josephson-Like AFM SHO to an External AC Signal

Synchronisation of oscillators, including their mutual synchronisation and their synchronisation to an external AC signal, is a general phenomenon well known in physics and widely used in modern engineering and technology [94]. Similar problems are also well known and partly solved for a small number (<10) of ferromagnet-based spintronic oscillators [76, 95–99]; there are also numerical and analytical studies for large arrays of interacting oscillators [100–103]. However, analogous problems have hardly been considered for antiferromagnetic oscillators. The only well-studied problem of oscillator synchronisation has been performed for a Josephson-like AFM SHO considered in Sect. 3.

4.1 General Description and Basic Features of SHO Synchronisation

In this section, we consider the case where a Josephson-like AFM oscillator, proposed and considered for the first time in [40], is under the action of the STT caused not only by the DC current τ (written in units of angular frequency; see Eqs. (1), (3)) but also by an additional torque dependent on the harmonic AC current applied to the SHO structure with magnitude τ_f, and frequency ω_f. The Eqs. (3) and (5) in this case can be rewritten in the form [40, 42]:

$$\frac{1}{\omega_{\text{ex}}}\ddot{\phi} + \alpha\dot{\phi} + \frac{\omega_e}{2}\sin(2\phi) - \tau - \tau_f\cos\big(\omega_f t\big) = 0, \tag{6}$$

where ϕ is the rotation angle of the Neel vector $\mathbf{l}$, $\ddot{\phi} = d^2\phi/dt^2$, $\dot{\phi} = d\phi/dt$, $\omega_{\text{ex}} = \gamma B_{ex}$ is the exchange frequency, $B_{ex} = \mu_0 H_{ex}$, α is the effective Gilbert damping parameter, $\omega_e = \gamma B_e$ is the angular frequency that is proportional to the easy-plain anisotropy field $B_e = \mu_0 H_e$, and μ_0 is the vacuum permeability.

To obtain the numerical solution of Eq. (6), the following typical parameters of the NiO/Pt SHO can be used [40, 42]: the Gilbert damping $\alpha = 0.01$, the exchange field $H_{ex} = 9 \cdot 10^6$ Oe, the easy-plane anisotropy field $H_e = 625$ Oe. Using these parameter values, it is easy to investigate how the generation frequency of the AFM oscillator depends on the DC current τ (given in units of angular frequency) and how the previously obtained approximate equation $\omega_g = 2\tau/\alpha$ [40] is fulfilled (Fig. 11). As it can be seen from Fig. 11, the considered AFM oscillator is nonlinear with respect to the dependence of its generation frequency on the applied DC current. Only for relatively large DC currents τ the approximation $\omega_g = 2\tau/\alpha$ is fulfilled. In contrast near the generation threshold, at $\tau/\omega_e = 0.5$, the generation frequency deviates significantly from the linear dependence $\omega_g = 2\tau/\alpha$. This feature plays a fundamental role in synchronising the AFM oscillator to an external AC signal, as discussed below.

By adding a non-zero AC current to the system (6), it is possible to achieve synchronisation of the AFM oscillator at two different AC signal frequencies ω_f as shown in Fig. 12. It can be seen that the AFM SHO is synchronised to an external AC signal and that the synchronisation bandwidths have substantially different values for different AC signal frequencies. As expected, the bandwidth for the 150 GHz signal

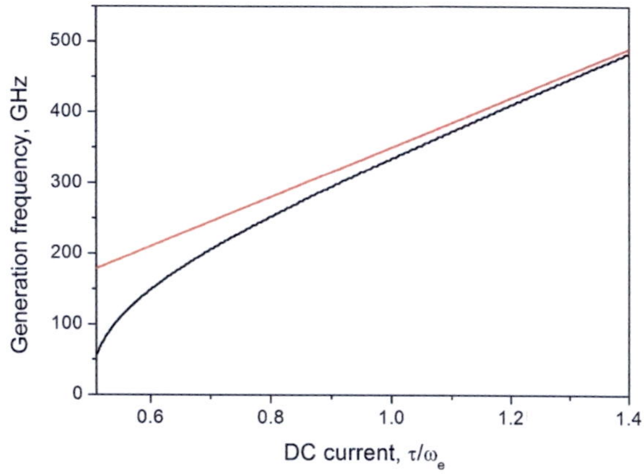

Fig. 11 AFM oscillator generation frequency as a function of normalised DC current. The black curve is the result of the numerical solution of Eq. (6) with the AC current amplitude set to 0 ($\tau_f = 0$) and the red curve is a theoretical approximation $2\tau/\alpha$ [40]

is larger than that for the 400 GHz signal. This discrepancy is even more striking if we assume that the external AC signal amplitude τ_f is the same for both considered cases. It is known from the synchronisation theory [94] that the synchronisation bandwidth of an oscillator is directly proportional to the amplitude of the synchronisation force (τ_f in our case, see Eq. (6)). Therefore, the result shown in Fig. 12 cannot be explained by the linear theory [94]. Considering that the Josephson-like AFM SHO is a strongly nonlinear oscillator [40], the specific nonlinear theory [99] should be used to describe its behaviour.

Nonlinear theory [99] takes into account not only the external signal amplitude F_0 to determine the synchronisation bandwidth BW but also the intrinsic oscillator nonlinearity coefficient N, which leads to the equation for the bandwidth [104]:

$$BW = F_0\sqrt{1 + N^2}. \tag{7}$$

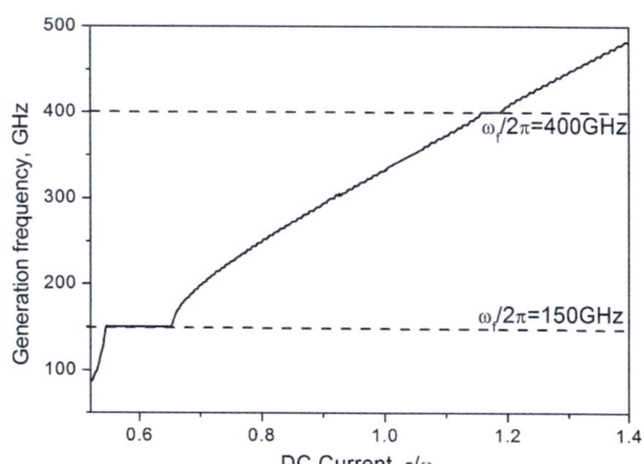

Fig. 12 Dependence of the Josephson-like AFM SHO generation frequency on the normalised DC current τ/ω_e for two different applied AC current frequencies ($\omega_f/2\pi = 150$ GHz and 400 GHz). The black curve is the result of the numerical solution of Eq. (6) with the AC current amplitude $\tau_f/\omega_e = 0.15$ and the dashed lines are a guide for an eye

Obviously, in the case of a linear oscillator $N = 0$, one can obtain a classical linear relationship between the synchronisation bandwidth BW and the external signal amplitude F_0.

The nonlinearity coefficient N can be introduced as follows. Assume that the oscillator frequency ω depends on some governing parameter p. For the simplest case of a nonlinear oscillator, this parameter is typically the power of the oscillations $p \sim a^2$ where a is the oscillation amplitude. Then the nonlinearity coefficient can be introduced as $N = d\omega/dp$. For the case of a spin-torque oscillator considered in [99], the dependence of the oscillator frequency on the oscillator power is $\omega = \omega_0 + Np$.

The numerically calculated nonlinearity coefficient for the AFM oscillator is shown in Fig. 13. As can be seen, the nonlinearity coefficient decreases monotonically with an increase in the applied DC current and consequently with an increase in the generation frequency. At the threshold $\tau/\omega_e = 0.5$ it goes to infinity and for large DC currents $\tau/\omega_e \gg 1$ it approaches a constant value of $N_e = 2\omega_e/\alpha$. This is an important result, indicating that the choice of an appropriate oscillator working point can lead to a significant increase in its synchronisation bandwidth $BW \sim \sqrt{1 + N^2}$, since the nonlinearity coefficient N varies greatly with the change in DC current. To maximise the synchronisation bandwidth using an external AC signal, the oscillator working point should be chosen close to the generation threshold where the nonlinearity coefficient is maximum (see Fig. 11).

So, as we can see, the main conclusion is that the Josephson-like AFM SHO is a strongly nonlinear oscillator. Its generation frequency depends nonlinearly on the applied DC current, and the synchronisation bandwidth depends not only on the external signal amplitude but also on the nonlinearity of the oscillator itself.

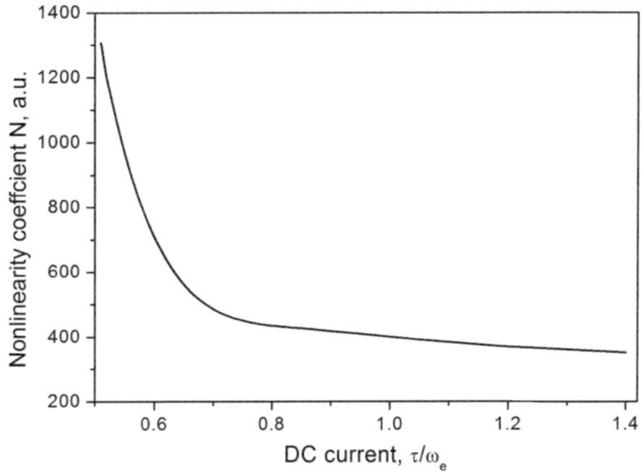

Fig. 13 Dependence of the nonlinearity coefficient N of the Josephson-like AFM oscillator on the normalised DC current τ/ω_e, obtained numerically as a derivative for the data shown in Fig. 11

4.2 Fractional Synchronisation

It is well known from oscillation theory [94] that an oscillator can be synchronised not only by an AC signal whose frequency ω_f coincides with the oscillator frequency ω_g but also by a signal whose frequency is a fraction of the oscillator frequency:

$$\frac{\omega_g}{\omega_f} = \frac{m}{n} = r \tag{8}$$

where $m, n \in \mathbb{N}, r \in \mathbb{Q}$.

To describe the magnetisation dynamics in the Josephson-like AFM SHO, Eq. (6) can be used as it was done before. In this case, it is possible to obtain some analytical results regarding the synchronisation bandwidth BW in the case of an overdamped oscillator, if the term $\omega_{\text{ex}}^{-1}\ddot{\phi}$ can be omitted. With this assumption, one can obtain the following first-order differential equation:

$$\alpha\dot{\phi} + \frac{\omega_e}{2}\phi - \tau - \tau_f\cos(\omega_f t) = 0. \tag{9}$$

Following the calculation procedure from [94], the following expression can be written for the synchronisation BW [105]:

$$BW \approx \left| J_r\left(\frac{\tau_f}{2\alpha\omega_f}\right) \right| \tag{10}$$

where $r = m/n, m, n$ are the oscillation mode numbers, and $J_r(x)$ is a Bessel function of the first kind. m and n have the following meanings: m is a harmonic number with frequency $m\omega_g$ (ω_g is the oscillator generation frequency), n is a harmonic number with frequency $n\omega_f$ (ω_f is the external signal frequency), i.e. if, for instance, we consider an oscillation with the frequency ω_g twice greater than the driving frequency ω_f, we should substitute the value $\frac{m}{n} = r = 2$ in Eq. (5). It should be noted that this equation does not include the nonlinearity coefficient N introduced in Sect. 4.1. However, the nonlinearity is "hidden" in the expression itself by the Bessel function J_r.

As before, one can solve Eq. (6) numerically for the same system parameters as used before (see Sect. 4.1). It has been already shown that the synchronisation bandwidth BW depends on the nonlinearity N of the oscillator, and therefore it is advisable to work in a region close to the threshold where the nonlinearity has a maximum value. On the other hand, this approach usually prevents the fractional synchronisation regime from being observed, since the fractional frequencies $\omega_r = r\omega_g$, $r < 1$ are typically below the fundamental (first harmonic) frequency of the oscillator $\omega_g = \omega_f$. By choosing an external signal frequency far from the threshold at 300 GHz, one can observe numerous synchronisation regions, as shown in Fig. 14 [105].

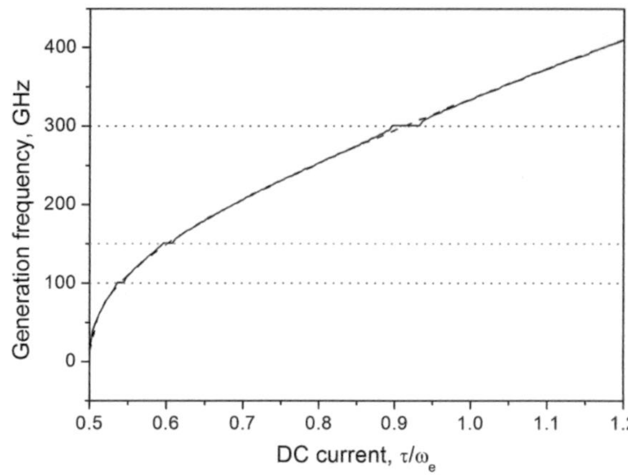

Fig. 14 Frequency of generation of the AFM oscillator as a function of normalised DC current τ/ω_e without (dashed line) and with (solid line) AC signal with amplitude $\tau_f = 0.1\omega_e$ and frequency of 300 GHz. Horizontal dotted lines indicate synchronisation regions at 300 GHz, 150 GHz and 100 GHz

The further increase in the amplitude of the AC signal τ_f leads to the appearance of even more synchronisation regions at more frequencies, as can be seen in Fig. 15 [105].

By fixing an external DC current amplitude τ and varying the AC signal frequency ω_f, it is possible to plot the AC power of the generated oscillations in frequency coordinates (signal frequency ω_f—generation frequency ω_g). In such a dependence, the colour of a point codes the AC power of the generated oscillations. Results for relatively small AC currents $\tau_f/\omega_e = 0.1$ are shown in Fig. 16. Regions of linear proportionality $\omega_g \approx \omega_f$ correspond to the ordinary synchronisation regime of the Josephson-like AFM SHO when the oscillator is driven by an external signal. Furthermore, we can distinguish different synchronisation regions with different proportionality between ω_g and ω_f. In the case of $\tau_f/\omega_e = 0.1$ there are only two such regions: $r = 1$ and $r = 1/2$.

Increasing the amplitude τ_f of the external AC signal results in a much more complex synchronisation pattern, as shown in Fig. 17. As we can see, there are

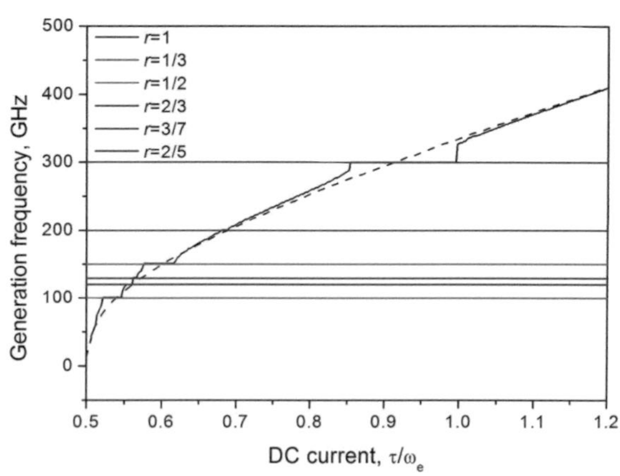

Fig. 15 Frequency of generation of the AFM oscillator as a function of normalised DC current τ/ω_e without (dashed line) and with (solid line) AC signal with amplitude $\tau_f = 0.3\omega_e$ and frequency of 300 GHz. Horizontal colour lines indicate synchronisation regions at different frequencies

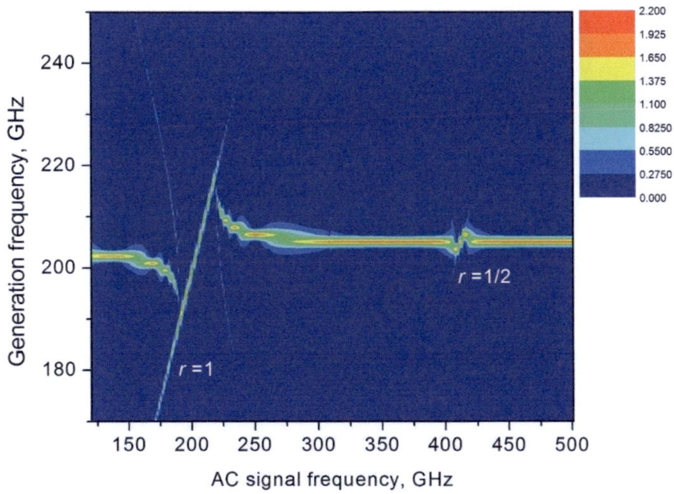

Fig. 16 AFM SHO generation frequency ω_g as a function of AC signal frequency ω_f, calculated numerically for normalised DC $\tau/\omega_e = 0.7$ and normalised AC $\tau_f/\omega_e = 0.1$ currents. Regions of fractional synchronisation (m/n) are labelled on a plot. The coloured areas correspond to different values of the AC power generated by an AFM SHO

numerous synchronisation regions at larger current amplitudes. It should be noted that different synchronisation regions have different synchronisation bandwidths, as can be seen in Fig. 17. The maximum bandwidth BW is obtained for region $r = 1$, the minimum BW—for $r = 2/3$.

To investigate this effect in detail, we use Eq. (10). For the small AC signal amplitude $\tau_f/\omega_e \ll 1$, we can obtain the following approximate expression for the bandwidth:

$$BW \approx \left(\frac{\tau_f}{2\alpha\omega_f}\right)^{\frac{m}{n}}. \tag{11}$$

According to (11), the bandwidth BW depends strongly on the ratio $r = m/n$ and is weaker for $r < 1$ and stronger for $r > 1$ regions. However, the observation of $r > 1$ regions is only possible for AC signals with sufficiently large amplitudes τ_e or sufficiently small frequencies ω_e. These conclusions are in good agreement with the numerical data shown in Fig. 18, where the synchronisation bandwidth is plotted as a function of the signal amplitude on a double-logarithmic scale [105].

As we can see from Fig. 18, the curve for $r = 1$ is situated lower than the curve for $r = 1/2$ and so on, because $BW(r = 1) \approx \tau_e$, $BW(r = 1/2) \approx \sqrt{\tau_e}$, $BW\left(r = \frac{1}{3}\right) \approx \sqrt[3]{\tau_e}$, etc. Note that there is a deviation from the aforementioned asymptotic (11) in a region $\tau_e/\omega_e > 0.3$ where the condition $\tau_f/\omega_e \ll 1$ is not satisfied.

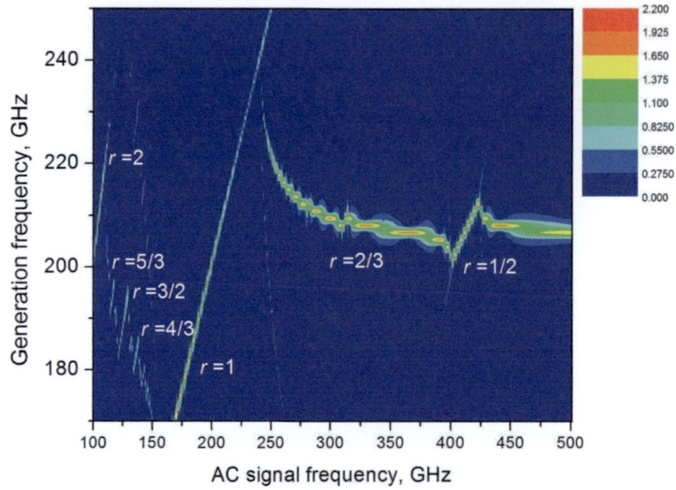

Fig. 17 AFM SHO generation frequency ω_g as a function of AC signal frequency ω_f, calculated numerically for normalised DC $\tau/\omega_e = 0.7$ and normalised AC $\tau_f/\omega_e = 0.3$ currents. Regions of fractional synchronisation (m/n) are labelled on a plot. The coloured areas correspond to different values of the AC power generated by an AFM SHO

Fig. 18 Dependence of AFM oscillator synchronisation bandwidth BW in different synchronisation regions on AC signal amplitude, $\tau_f/\omega_e \ll 1$, $\omega_f/2\pi = 300$ GHz

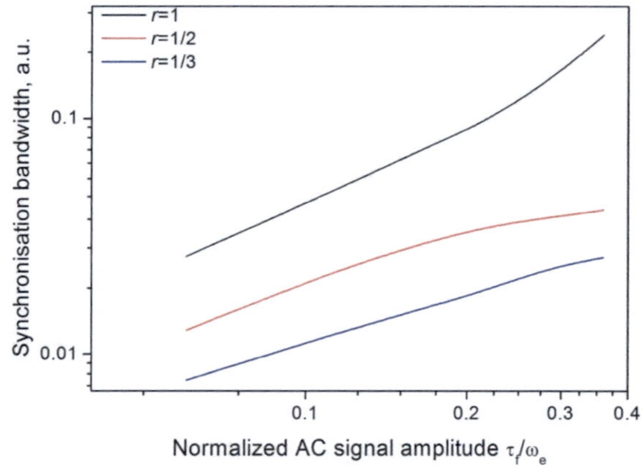

5 Chaotic Dynamics in a Josephson-Like AFM SHO

In the previous sections, we analysed the deterministic dynamics of a Josephson-like AFM SHO. It is clear that the input of an AC or DC signal with a given amplitude and frequency leads to the generation of an output signal which is usually deterministic in nature. However, the Eq. (6) describing the Josephson-like AFM SHO supports another type of dynamics—the stochastic dynamics. In this case, even under the influence of a fixed harmonic AC signal, the resulting output signal has a complicated shape and cannot be described analytically, even approximately. Such behaviour is well known for nonlinear oscillators [94] described by an equation similar to Eq. (6). Probably, the best-known case of a nonlinear oscillator exhibiting chaotic dynamics under the action of a harmonic external force is the Duffing oscillator. The Duffing oscillator is essentially a damped oscillator with cubic nonlinearity. The combination of damping, nonlinearity and external driving force is sufficient to support stochastic dynamics in such a system for some range of driving force parameters.

In the case of a Josephson-like AFM SHO described by Eq. (6), the nonlinearity is "hidden" in the term $\frac{\omega_e}{2}\sin(2\phi)$. Indeed, for relatively small Neel vector angles ϕ one can assume: $\frac{\omega_e}{2}\sin(2\phi) \approx \omega_e\phi$. However, this approximation does not hold as the Neel vector angle ϕ grows and to obtain a good precision for $\frac{\omega_e}{2}\sin(2\phi)$ one must include more and more expansion terms to describe the system: $\frac{\omega_e}{2}\sin(2\phi) \approx \omega_e\phi - \frac{2\omega_e}{3}\phi^3 + \ldots$. Furthermore, equation like the Eq. (6) has been used to describe the stochastic dynamics of Josephson junctions [106–109] under the action of an external AC signal (for this reason, the SHO considered is called a Josephson-like SHO).

To describe the dynamics of the AFM SHO, it is convenient to consider the phase portrait of the system in coordinates $(\phi, \dot{\phi})$. In the case of Eq. (6), the phase portrait is a two-dimensional surface in the coordinates $(\phi, \dot{\phi})$. Due to the cyclic condition, where the angles $\phi = 0$ and $\phi = 2\pi$ correspond to the same state of the system, the phase space of the system can be considered as the surface of a cylinder. The results of the numerical modelling for Eq. (6) are shown in Fig. 19, where we plot typical dependencies $\dot{\phi}(t)$ and phase portraits in coordinates $(\phi, \dot{\phi})$ for the AFM SHO under the action of a harmonic AC current with fixed amplitude $\tau_f = 0.5\omega_e$ but different angular frequencies ω_f.

As can be seen from Fig. 19a, the dynamics of the system is described by relatively small oscillations near the equilibrium $\dot{\phi} = 0$ and a set of periodic large pulses when the AC current frequency is $\omega_f = 2\pi \cdot 50$ GHz. This behaviour corresponds to the phase portrait $(\phi, \dot{\phi})$ in the form of a complex double closed loop (Fig. 19b). Larger loops correspond to a periodic spike-like pulse generation, while smaller loops correspond to the case of small oscillations that take place between the strong pulses (see Fig. 19a, b).

When the AC signal frequency ω_f is increased up to $\omega_f = 2\pi \cdot 150$ GHz while keeping the current amplitude remains constant, the system dynamics undergo drastic changes (Fig. 19c, d). As it can be seen from Fig. 19c, under the action of a 150 GHz AC signal, the system dynamics become stochastic: no (quasi-)periodic pulse generation can be observed. Instead of that we have a pseudo-random output AC signal with approximately constant frequency, but time-dependent pseudo-random amplitude. Accordingly, such a signal can be treated as a quasi-periodic signal with random amplitude modulation. The phase portrait of this stochastic signal generated in the Josephson-like AFM SHO is much more complicated than in the case of regular SHO generation regimes (compare Fig. 19d with Fig. 19b, f). It no longer consists of rather clear closed loop(s) in a relatively small confined region of the parametric space $(\phi, \dot{\phi})$ Instead, the phase portrait occupies the entire plane $(\phi, \dot{\phi})$. This regime is one of the stochastic regimes of AFM SHO operation.

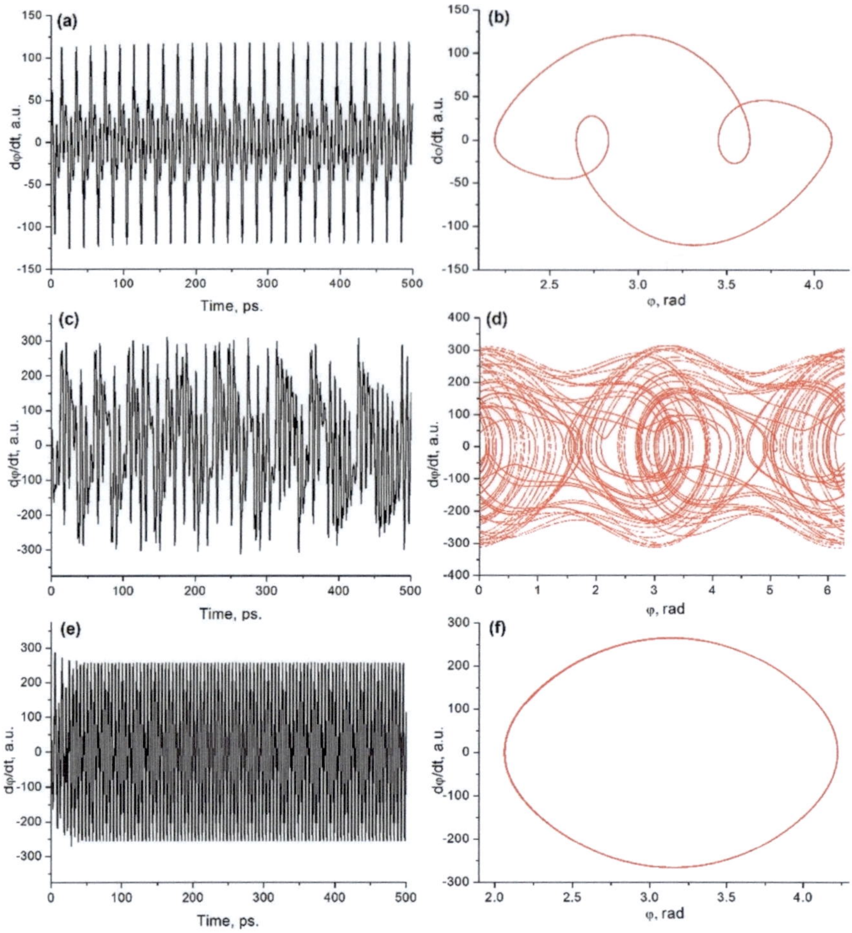

Fig. 19 Time dependence of the Neel vector rotation speed $\dot{\phi}$ (left, black line) and the corresponding phase portrait of the system in $(\phi, \dot{\phi})$ coordinates (right, red line) for the fixed AC signal amplitude $\tau_f/\omega_e = 0.5$ and different AC signal frequencies ω_f: **a–b** $\omega_f = 2\pi \cdot 50$ GHz; **c–d** $\omega_f = 2\pi \cdot 150$ GHz **e–f** $\omega_f = 2\pi \cdot 200$ GHz. All curves are calculated numerically from Eq. (6). Simulation parameters: $\alpha = 0.01$, $H_{\text{ex}} = 9 \cdot 10^6$ Oe, $H_e = 625$ Oe DC current amplitude $\tau = 0$

Finally, by further increasing the AC signal frequency, it can be achieved that at sufficiently high frequencies, e.g. $\omega_f = 2\pi \cdot 200$ GHz, the system again exhibits regular (deterministic) dynamics demonstrated in Fig. 19e, f. Furthermore, in this case, the system dynamics is even simpler than in the case of the smaller frequency $\omega_f = 2\pi \cdot 50$ GHz (see Fig. 19a, b), because $\dot{\phi}$ has a form of regular periodic oscillations around the ground state $\dot{\phi} = 0$. Thus, the phase portrait of the system driven by an AC current with a rather high frequency is the simplest of all the three cases considered and contains only one "almost perfect" closed loop (Fig. 19f).

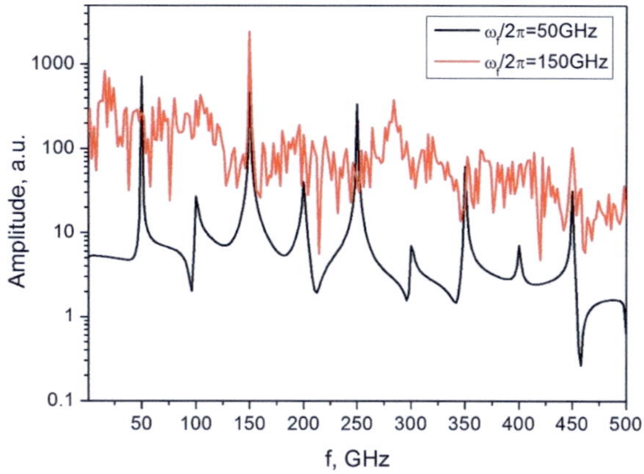

Fig. 20 Output AFM SHO signal spectrum under the action of an AC signal with amplitude $\tau_f/\omega_e = 0.5$ and different AC signal frequencies: 50 GHz (black) and 150 GHz (red)

As one can see, the output signal in the stochastic generation regime retains the input signal frequency but can be considered to be randomly amplitude modulated. To show this, we have calculated the output signal spectrum for two different input signal frequencies ω_f shown in Fig. 20.

Thus, by choosing an appropriate input signal frequency and amplitude, one can drive the Josephson-like AFM SHO into the stochastic generation regime. In Sect. 7, we will discuss some promising applications for this generation regime. It should be noted that the driving AC signal frequency must be in the ~ 10–100 GHz range, while as we have shown before, the output of the AFM SHO under the action of DC current is a (sub-)THz frequency signal, so it can be used to drive the second AFM SHO into the stochastic generation regime.

6 Application: Neuromorphic Computing

Recent studies in various fields of modern science and technology offer the perspective of building a brain-inspired way of processing, storing, and learning information that resembles the actions and interactions of the neurons and synapses that make up a living being's brain [99, 110]. Such an approach is known as neuromorphic computing [111–113] and, compared to traditional computational processes based on von Neumann architecture, it appears to be extremely efficient for various cognitive tasks, including classification of multidimensional signals, prediction, control problems, image and pattern recognition, etc.

To date, this idea has been implemented using neural networks that mimic the behaviour of neurons and synapses using conventional CMOS electronics and software. While this technique can solve many practical problems, it has two major drawbacks. Firstly, it is a famous bottleneck of the von Neumann architecture, and secondly, these neural networks operating in high-performance regimes require an enormous power supply, which drastically reduces the efficiency of such systems

[110]. Therefore, many researchers propose a completely different technology, which does not foresee the aforementioned difficulties, they propose to start from the hardware stage, building artificial neurons, synapses and whole neural networks on the basis of nanoelectronic devices such as quantum dots, nanowires, two-dimensional layered materials, von der Waals heterostructures, etc. [111].

At the same time, various spintronic oscillators appear to be useful for creating these artificial "neurons" and "synapses", which are the cornerstones of neuromorphic computing. Initial proposals in this area consider conventional ferromagnetic spintronic nanostructures because of their non-linearity and their ability to mimic neural connections by influencing each other's behaviour via current and magnetic field without huge power or voltage supply [112, 114]. Since the frequency of ferromagnetic oscillators is relatively low for practical needs (usually 10–30 GHz), the replacement of ferromagnets by antiferromagnets became a natural solution in this field [42, 43, 113], as it was done for THz frequency generators.

A principle scheme of a device that could operate as an ultra-fast analogue of a biological neuron is shown in Fig. 21, using the SHO with biaxial magnetic anisotropy of the AFM layer [40] as the core element. The system envisages the ability to generate ultra-short "Dirac delta-like" pulses (picosecond duration spikes) in response to a driving electric current whose amplitude exceeds a threshold.

To investigate the behaviour of the system shown in Fig. 21, we carry out analytical studies and numerical simulations using the Eq. (6) for the Neel vector [40] and obtain the results presented in Fig. 22, where for different current amplitudes we discover the appearance of three different regimes of no spikes (see Fig. 22b), single spike (see Fig. 22c) and discrete group of spikes generation (bursting mode) (see Fig. 22d). The duration of the spikes is several picoseconds for ordered simulation parameters and is determined by the in-plane magnetic anisotropy and the effective damping of the AFM layer. Another important feature of these results is that the generated output signal depends on the input signal frequency f_{ac}, which can exceed tens of GHz.

The next gradual step towards the realisation of neuromorphic computing was taken in [43], where the authors propose the entire system of an artificial "neuron" based on the spin Hall geometry, as shown in Fig. 23.

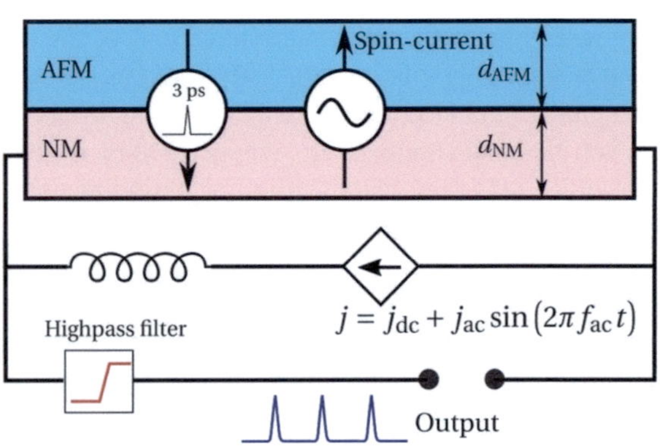

Fig. 21 Principal operating scheme of an ultra-short pulse (spike) spintronic generator consisting of a bilayer structure of AFM and normal metal (NM) with thicknesses d_{AFM} and d_{NM}, respectively. The NM layer is driven by current j with DC j_{dc} and AC $j_{ac}\sin(2\pi f_{ac}t)$ components. A highpass filter is used to pass signals with frequencies $\geq f_{ac}$

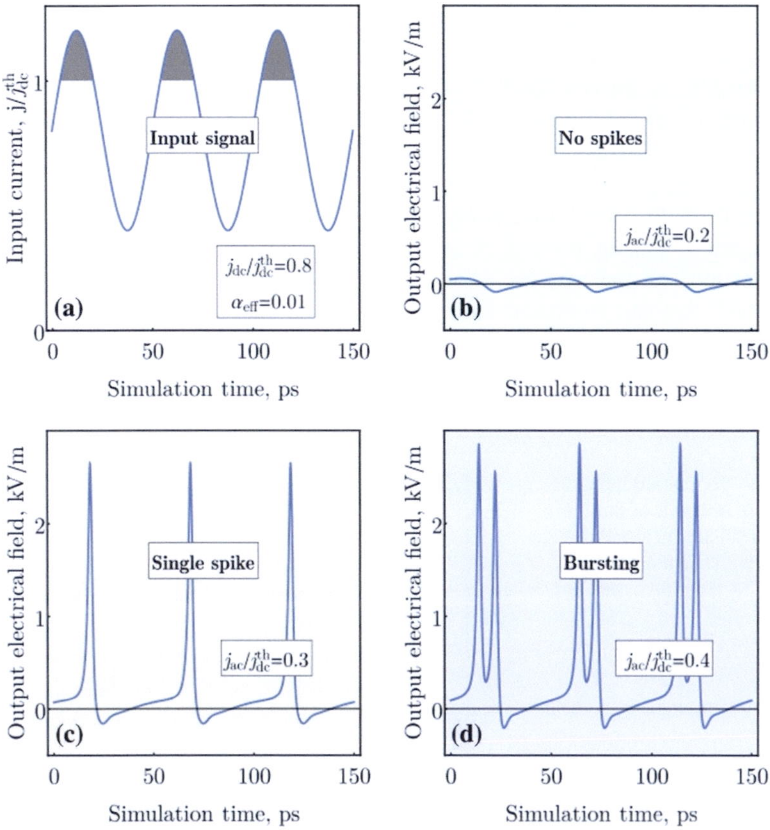

Fig. 22 Numerical simulation of spike generation in different regimes of AFM SHO: **a** Shape of the total input driving current; **b** no generation regime; **c** single spike generation regime; **d** discrete group of spikes generation (bursting mode)

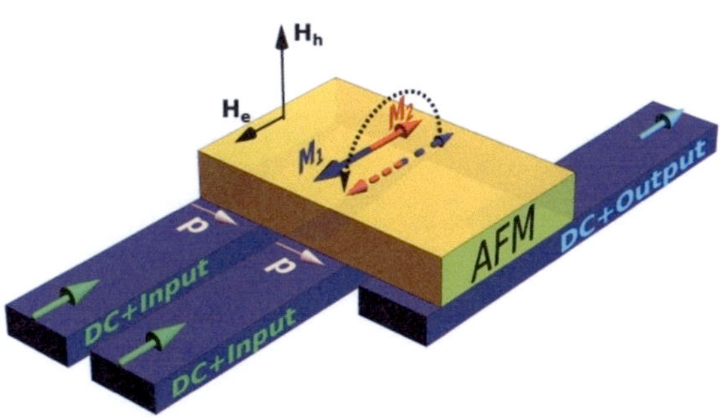

Fig. 23 Schematic of an SHO-based artificial "neuron". The green arrows represent the driving current with polarisation **p** and the dashed arrows show the magnetisation vectors $\mathbf{M}_1$ and $\mathbf{M}_2$ after spike generation

With two input terminals, the artificial "neuron" depicted in Fig. 23 makes it possible to perform AND, OR, MAJORITY, and Q-gate logic functions (Fig. 24) [43]. To extend the progress in this field, the authors also discover that a system consisting of a small number of such "neurons" can also function as a FULL ADDER or even as a dynamic memory loop with variable clock frequency. As already mentioned for the generation of spikes, the clock frequencies of these systems could be tens of GHz, making them effective for many practical applications.

In conclusion, the concept of neuromorphic computing could be successfully implemented on the basis of AFM SHOs, which allow us to construct artificial "neurons" that can work with the ultra-short spike-like pulses and ~10–100 GHz operating frequencies to perform basic logic operations and more complex functions. Therefore, such AFM "neurons" have a good chance of becoming a standard element of future neuromorphic computers.

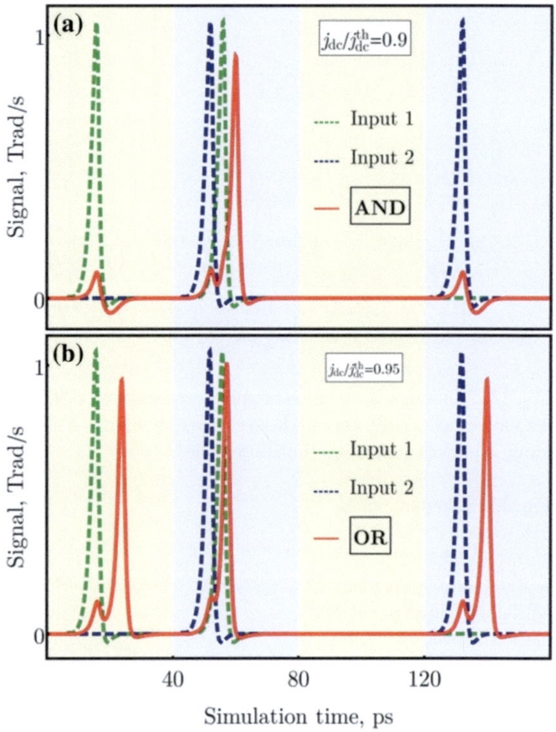

Fig. 24 Normalised input and output signals of the AFM artificial "neuron", which works in the AND **a** and OR **b** regimes

7 Application: Probabilistic Computing

Conventional CMOS logic operates on deterministic systems, with two distinct states: either "0" or "1". However, it is possible to construct a stochastic so-called "p-bit" [115, 116] which operates in the following way: it has a normalised output m_i which takes on the values "0' and "1" with probabilities p_0 and p_1 respectively. Moreover, these probabilities can be controlled with some normalised inputs I_i as follows: if $I_i = 0$ then probabilities are equal ($p_0 = p_1 = 0.5$), large $+I_i$ input pins output to the p_1 state ($p_1 = 1$, $p_0 = 0$), and large $-I_i$ input pins output to the "0" state ($p_1 = 0$, $p_0 = 1$) Most notably, such "p-bit" base systems, possess an invertible nature, a distinctive characteristic not found in conventional digital circuits. When operating in direct mode, the input is stabilised, resulting in an accurate output of the device. In inverted mode, the output is fixed while the network fluctuates through all potential inputs compatible with that output.

The common realisation of "p-bits" with magnetic systems utilises magnetic tunnel junctions (MTJ) [117]. Such systems have two distinct states depending on the orientation of the magnetisations in the layers—ferromagnetic and antiferromagnetic. Typically, the energy barrier between these two states is quite large, so thermal fluctuations cannot switch the system from one state to the other. However, by applying an external voltage, this barrier can be lowered, and the system begins to fluctuate between these states on a millisecond time scale [117].

Moving from MTJ to AFM SHO, we have discussed in Sect. 4, we have shown that this system can also exhibit stochastic behaviour when driven by AC current with some amplitude and frequency. However, there is a key difference between the chaos in MTJ and AFM SHO systems. In MTJ, chaos is caused by thermal fluctuations as the system "hops" between two states. In AFM SHO, the nature of the chaos is different and is not caused by thermal fluctuations. It is so-called dynamic chaos, which occurs in nonlinear systems [94]. Here we cannot speak of two distinct states of the system, but of a complex behaviour of the system dynamics. Such behaviour creates some obstacles for the direct use of the AFM SHO output signal in the stochastic regime (see Fig. 19c) for the construction of the "p-bit". Therefore, we need a way to convert the AFM SHO oscillator output signal into the "p-bit" like signal by switching between two states (for convenience, let us consider the "–1" and "1" states). Mathematically, there are an infinite number of ways to achieve such a conversion, here we propose only one of them, using a Markov-like stochastic signal converter [51].

Let us consider the stochastic AFM SHO output signal as a driving signal for some other system, which operates via a given set of rules. It compares incoming signal amplitude and compares it with two given amplitudes A_- and A_+. If incoming positive amplitude A is smaller than A_+ ($A < A_+$) or incoming negative amplitude A is smaller than A_- $\left(|A| < A_-\right)$, system remains in given state ("1" or "–1" respectively). However, if these amplitudes are greater than A_+ or A_-, the system switches to the opposite state [51].

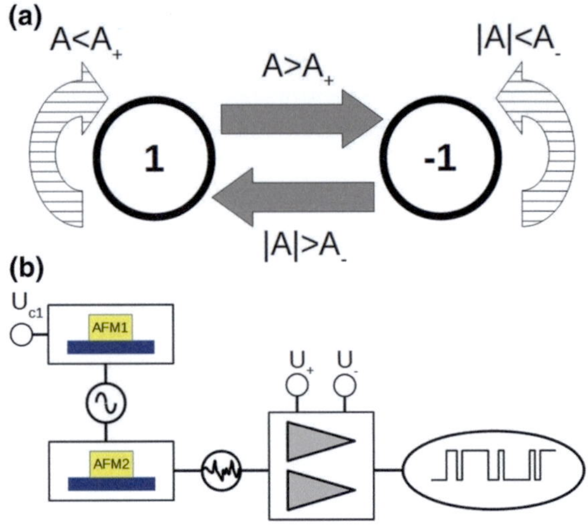

Fig. 25 The Markov-like stochastic signal converter: **a** schematic illustration of the device operation scheme; **b** principal scheme for the p-bit implementation using an AFM SHO

The schematic operation of such a converter is shown in Fig. 25a. Moreover, by changing the values of A_+ and A_- thresholds, we can change the probability of "1" or "–1" state in the system. For example, consider the case where $A_+ \to \infty$. If the system enters the "1" state, it will stay there forever because the switching condition $A > A_+$ will never be fulfilled. So, the overall probabilities of the states will be $p_+ \approx 1$ and $p_- \approx 1$.

Similar to the case of "p-bit" discussed above for $I_i = 1$, the principal scheme of AFM "p-bit" is shown in Fig. 22b. As discussed above, the first AFM SHO driven by DC current produces a THz input signal for the second one, which generates a stochastic output signal. This stochastic signal is then applied to the Markov-like converter (which takes two control voltages U_+, U_-) and the resulting signal is a "p-bit" like.

As discussed above for the proposed converter, we need two control amplitudes A_+ and A_- that should be tuned for the required "p-bit" state probabilities. The results of applying such a converter to the AFM SHO stochastic output signal are shown in Fig. 26. As we can see, choosing equal threshold amplitudes A_+ and A_- leads to a symmetric state distribution in the output signal. This feature can be explained by looking back at the initial Eq. (6). The change ϕ to $-\phi$ (which can be considered as the change of rotation of the Neel vector from clockwise to counterclockwise) due to the symmetry of the equation is equal to the change of phase of the AC signal by $\pi/2$ thus solution does not differ much. This means that the peak distribution in the stochastic AFM SHO output signal is the same for positive and negative amplitudes, so applying equal threshold amplitudes $A_+ = A_-$ will produce an output signal with almost equal distribution of p_+ and p_- states.

Increasing these threshold amplitudes A_+ and A_- one by one will change the "preferred" system state, as can be seen in Fig. 27. It is possible to force the system to go into a "1" or "–1" state by adjusting the appropriate thresholds A_+ and A_-.

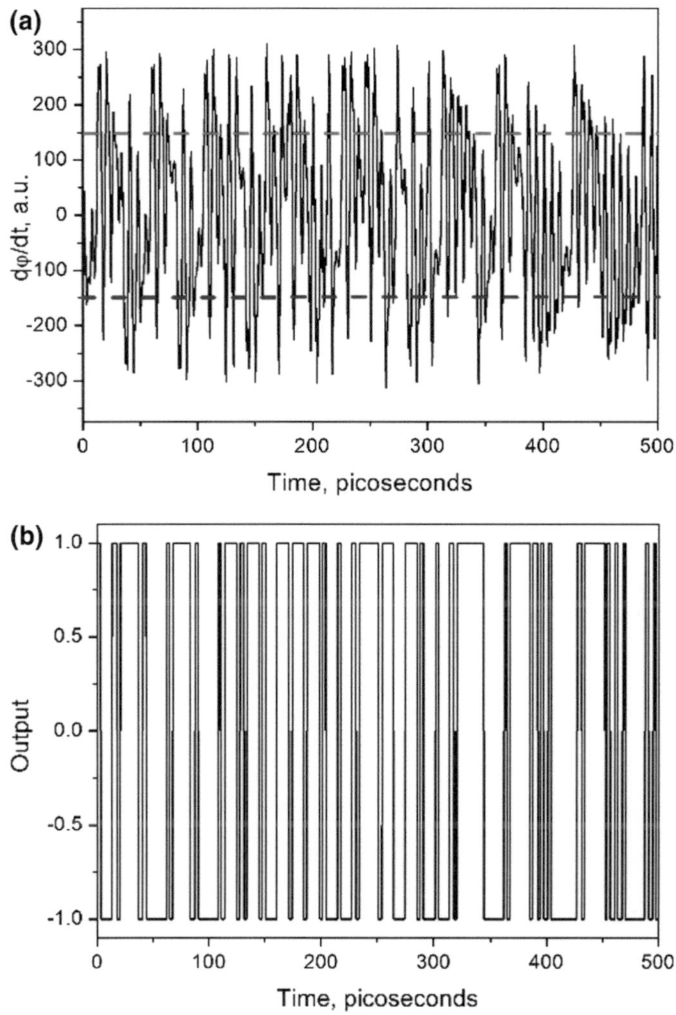

Fig. 26 Input AFM SHO stochastic signal (**a**) and result of this signal conversion (**b**). Red and blue dashed lines indicate the threshold amplitudes $A_+ = A_- = 150$ chosen for the converter so that the output signal has almost equal state distributions: $p_+ = 0.49$, $p_- = 0.51$

This opens up the possibility of creating a "p-bit" element using a Josephson-like AFM SHO. The main advantage of such a device is its operating frequency. Instead of the kHz frequencies typical of MTJ-based devices, in the case of AFM SHO, we have sub-THz or even THz frequencies.

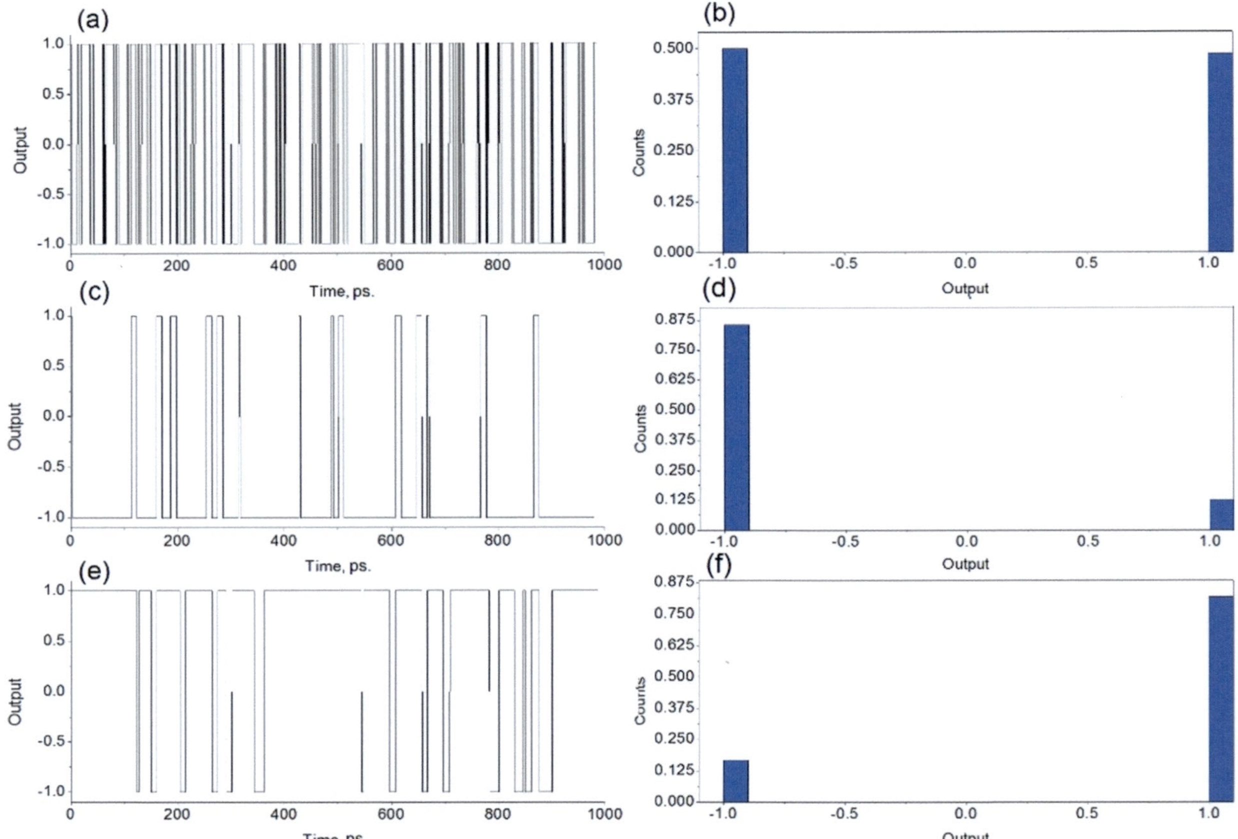

Fig. 27 Output of the Markov-like signal convertor and corresponding distribution of system states for different values of control amplitudes A_+ and A_-: **a–b** $A_+ = A_- = 150$; **c–d** $A_+ = 150$, $A_- = 300$; **e–f** $A_+ = 300$ $A_- = 150$

8 Outlook

In this chapter, the state-of-the-art in the field of antiferromagnetic (AFM) spintronic oscillators is described in detail. As shown, bilayer and multilayer AFM spintronic nanostructures with a "free" magnetic layer made of an insulating or conducting antiferromagnet and biased by a DC and/or AC current can perform as sources of (sub-)terahertz frequency signals.

Signal sources based on AFM spintronic oscillators have several important features. First, they are devices driven by a pure spin current and therefore have better energy efficiency than classical electronic devices that use the transfer of electric charges. Second, the use of AFM-based signal sources does not require any special experimental conditions such as cryogenic temperatures, high voltages, strong magnetic fields, etc., and the corresponding devices can in principle be implemented using existing nanotechnology. Finally, the operating frequency of AFM spintronic systems significantly exceeds the operating frequencies of existing ferromagnetic spintronic devices, which makes signal sources based on AFM nanostructures promising for many applications in medicine, communications, materials science, etc.

The operation of AFM spintronic oscillators is based on several related phenomena, namely the spin transfer torque, the spin Hall and the spin pumping effects, and the tunnelling anisotropic magnetoresistance. These phenomena, which are well studied theoretically and experimentally in ferromagnetic spintronics, have recently begun to be studied for AFM systems, and the first experiments performed have shown good agreement between theory and experiment. Nevertheless, the practical realisation of AFM-based devices nowadays faces several unsolved problems, mainly related to the AFM device fabrication technology. For this reason, the development of antiferromagnetic spintronics today is mainly focused on the theoretical and numerical study of different AFM-based devices, including the aforementioned (sub-)terahertz frequency signal sources.

Several schemes of AFM spin Hall oscillators (SHOs) based on bilayer AFM/heavy metal nanostructures and oscillators based on antiferromagnetic tunnel junctions (ATJs) have been proposed recently. Both devices offer similar capabilities: their AC output power reaches several microwatts in the 0.1–1 THz frequency range, but the mechanism of AC energy extraction from these nanostructures could be completely different. To date, two of the most promising schemes for THz signal sources are the so-called Josephson-like AFM SHO, which was discussed in detail in this chapter, and an ATJ-based SHO.

A Josephson-like AFM SHO is a "classical" bilayer AFM/heavy metal (e.g. NiO/Pt) nanostructure with hard axis and easy plane anisotropy of the AFM layer. Due to this magnetic anisotropy, the rotation of the magnetic vectors of the magnetic sublattices in the AFM layer is non-uniform in time, resulting in magnetisation dynamics similar to the phase dynamics in superconducting Josephson junctions.

In this chapter, we have shown that, depending on the form of an input signal, a current-driven Josephson-like AFM SHO can work as a source of (sub-)terahertz frequency signals with different spectral characteristics. In the simplest case, where the input DC current exceeds a certain generation threshold, it can perform as a source of a harmonic AC output signal. In the subcritical regime, when the input DC current is below the threshold, the SHO can work as a spike generator driven by an additional input signal in the form of short current pulses. Under the action of such a combined current, including DC and pulsed components, the oscillator can generate a short output current spike when the total amplitude of the input signal exceeds the threshold. We have shown that in this operating regime, a Josephson-like AFM SHO can function as an artificial "neuron" or as a logic element capable of implementing AND, OR, and other logic operations. Our previous study suggests that three such elements may be sufficient to realise a FULL ADDER device, which is a key element in computing hardware.

Like the other types of oscillators, a Josephson-like AFM SHO can be synchronised to an external AC signal. Depending on the amplitudes of the DC and AC components of an input current and the current frequency, conventional synchronisation (at the input signal frequency) and fractional synchronisation (when the input signal and generation frequencies are in a rational relationship to each other) can be achieved for such a SHO. In addition, when the oscillator is biased by a pure AC harmonic current, it can exhibit complex behaviour caused by the stochastic magnetisation dynamics excited in a SHO. We have shown that this regime of stochastic generation of the SHO can be used to implement an ultra-fast "p-bit" device, which is promising for probabilistic computing at terahertz frequencies.

The results presented in this chapter open up new possibilities in the physics and potential applications of antiferromagnetic spintronic oscillators, including the emerging fields of ultra-fast neuromorphic and probabilistic computing.

Acknowledgements This work was supported in part by Grants Numbers 22BF07-03 and M/37-2023 from the Ministry of Education and Science of Ukraine, Grant Number 16F-2022 from the National Academy of Sciences of Ukraine, and the NATO SPS.MYP G5792 grant.

References

1. Gutfleisch, O., Willard, M.A., Brück, E., Chen, C.H., Sankar, S.G., Liu, J.P.: Thin Solid Films **23**, 821 (2010)
2. Coey, J.M.: Magnetism and magnetic materials, Cambridge University Press (2000)
3. Zhang, S., Zhao, D.: Advances in magnetic materials: processing, properties, and performance, CRC Press (2017)
4. Han, M.G.: Electromagnetic Materials and Devices, BoD–Books on Demand (2020)
5. Francis, L.A., Poletkin, K.: Magnetic sensors and devices: technologies and applications, CRC Press (2017)
6. Spaldin, N.A.: Magnetic materials: fundamentals and applications, Cambridge University Press (2010)
7. Krishnan, K.M.: Fundamentals and applications of magnetic materials, Oxford University Press (2016)
8. Cherepanov, V., Kolokolov, I., L'vov, V.: Physics Reports **229**, 81 (1993)

9. Demokritov, S.O., Demidov, V.E., Dzyapko, O., Melkov, G.A., Serga, A.A., Hillebrands, B., Slavin, A.N.: Nature **443**, 430 (2006)
10. Ignatiev, A.A.: Magnetoelectronics of Microwaves and Extremely High Frequencies in Ferrite Films, Springer Science & Business Media (2009)
11. Serga, A.A., Chumak, A.V., Hillebrands, B.: J. Phys. D Appl. Phys. **43**, 264002 (2010)
12. Wu, M., Hoffmann, A.: Recent advances in magnetic insulators-from spintronics to microwave applications, Academic Press (2013)
13. Prokopenko, O.V., Bozhko, D.A., Tyberkevych, V.S., Chumak, A.V., Vasyuchka, V.I., Serga, A.A., Malyshev, V.Y.: Ukr. J. Phys. **64**, 888 (2019)
14. Wolf, S.A., Awschalom, D.D., Buhrman, R.A., Daughton, J.M., von Molnár, V.S., Roukes, M.L., Treger, D.M.: Science **294**, 1488 (2001)
15. Žutić, I., Fabian, J., Sarma, S.D.: Rev. Mod. Phys. **76**, 323 (2004)
16. Chumak, A.V., Vasyuchka, V.I., Serga, A.A., Hillebrands, B.: Nat. Phys. **11**, 453 (2015)
17. Barnes, S.E., Maekawa, S.: Concepts in Spin Electronics, Oxford University Press (2006)
18. Tsymbal, E.Y., Zutic, I.: Spin transport and magnetism, CRC Press (2012)
19. Joshi, V.K.: Engineering science and technology, an international journal **19**, 1503 (2016)
20. Hirohata, A., Yamada, K., Nakatani, Y., Prejbeanu, I.L., Diény, B., Pirro, P., Hillebrands, B.: J. Magn. Magn. Mater. **509** (2020)
21. Houssameddine, D., Ebels, U., Delaët, B., Rodmacq, B., Firastrau, I., Ponthenier, F., Dieny, B.: Nat. Mater. **6**, 447 (2007)
22. Fang, B., Carpentieri, M., Louis, S., Tiberkevich, V., Slavin, A., Krivorotov, I.N., Zeng, Z.: Phys. Rev. Appl. **11**, 014022 (2019)
23. Tomasello, R., Fang, B., Artemchuk, P., Carpentieri, M., Fasano, L., Giordano, A., Prokopenko, O., Zeng, Z., Finocchio, G.: Phys. Rev. Appl. **14**, 024043 (2020)
24. Artemchuk, P.Y., Prokopenko, O.V., Bankowski, E.N., Meitzler, T.J., Tyberkevych, V.S., Slavin, A.N.: AIP Adv. **11**, 025234 (2021)
25. Hafarov, A., Prokopenko, O., Sidorenko, S., Makarov, D., Vladymyrskyi, I.: Chapt. 4 in: Modern Magnetic and Spintronic Materials: Properties and Applications. NATO Science for Peace and Security Series B: Physics and Biophysics, p. 73–94, Springer (2020). https://doi.org/10.1007/978-94-024-2034-0_4
26. Gomonay, H.V., Loktev, V.M.: Phys. Rev. B **81**, 144427 (2010)
27. Gomonay, E.V., Loktev, V.M.: Low Temperature Physics **40**, 17 (2014)
28. Jungwirth, T., Marti, X., Wadley, P., Wunderlich, J.: Nat. Nanotechnol. **11**, 231 (2016)
29. Jungfleisch, M.B., Zhang, W., Hoffmann, A.: Phys. Lett. A **382**, 865 (2018)
30. Jungwirth, T., Sinova, J., Manchon, A., Marti, X., Wunderlich, J., Felser, C.: Nat. Phys. **14**, 200 (2018)
31. Baltz, V., Manchon, A., Tsoi, M., Moriyama, T., Ono, T., Tserkovnyak, Y.: Rev. Mod. Phys. **90**, 015005 (2018)
32. Fukami, S., Lorenz, V.O., Gomonay, O.: J. Appl. Phys. **128** (2020)
33. Han, J., Cheng, R., Liu, L., Ohno, H., Fukami, S.: Nat. Mater. **22**, 684 (2023)
34. Marti, X., Fina, I., Frontera, C., Liu, J., Wadley, P., He, Q., Ramesh, R.: Nat. Mater. **13**, 367 (2014)
35. Marrows, C.: Science **351**, 558 (2016)
36. Kosub, T., Kopte, M., Hühne, R., Appel, P., Shields, B., Maletinsky, P., Makarov, D.: Nat. Commun. **8**, 13985 (2017)
37. Olejník, K., Schuler, V., Martí, X., Novák, V., Kašpar, Z., Wadley, P., Jungwirth, T.: Nat. Commun. **8**, 15434 (2017)
38. Olejník, K., Seifert, T., Kašpar, Z., Novák, V., Wadley, P., Campion, R.P., Jungwirth, T.: Sci. Adv. **4**, eaar3566 (2018)
39. Cheng, R., Xiao, D., Brataas, A.: Phys. Rev. Lett. **116**, 207603 (2016)
40. Khymyn, R., Lisenkov, I., Tiberkevich, V., Ivanov, B.A., Slavin, A.: Sci. Rep. **7**, 43705 (2017)
41. Sulymenko, O., Prokopenko, O., Tiberkevich, V., Slavin, A., Ivanov, B., Khymyn, R.: Phys. Rev. Applied **8**, 064007 (2017)

42. Khymyn, R., Lisenkov, I., Voorheis, J., Sulymenko, O., Prokopenko, O., Tiberkevich, V., Slavin, A.: Sci. Rep. **8**, 15727 (2018)
43. Sulymenko, O., Prokopenko, O., Lisenkov, I., Åkerman, J., Tyberkevych, V., Slavin, A., Khymyn, R.: J. Appl. Phys. **124**, 152115 (2018)
44. Sulymenko, O., Prokopenko, O., Tiberkevich, V., Slavin, A.: IEEE Magn. Lett. **9**, 3104605 (2018)
45. Lee, D.K., Park, B.G., Lee, K.J.: Phys. Rev. Appl. **11**, 054048 (2019)
46. Puliafito, V., Khymyn, R., Carpentieri, M., Azzerboni, B., Tiberkevich, V., Slavin, A., Finocchio, G.: Phys. Rev. B **99**, 024405 (2019)
47. Zhong, H., Qiao, S., Yan, S., Liang, L., Zhao, Y., Kang, S.: J. Magn. Magn. Mater. **497**, 166070 (2020)
48. Yan, H., Feng, Z., Qin, P., Zhou, X., Guo, H., Wang, X., Liu, Z.: Adv. Mater. **32**, 1905603 (2020)
49. Ovcharov, R.V., Galkina, E.G., Ivanov, B.A., Khymyn, R.S.: Phys. Rev. Appl. **18**, 024047 (2022)
50. Wolba, B., Gomonay, O., Kravchuk, V.P.: Phys. Rev. B **104**, 024407 (2021)
51. Slobodianiuk, D.V., Prokopenko, O.V.: J. Appl. Phys. **134**, 153903 (2023)
52. Artemchuk, P.Y., Sulymenko, O.R., Louis, S., Li, J., Khymyn, R.S., Bankowski, E., Prokopenko, O.V.: J. Appl. Phys. **127**, 063905 (2020)
53. Lebrun, R., Ross, A., Bender, S.A., Qaiumzadeh, A., Baldrati, L., Cramer, J., Kläui, M.: Nature **561**, 222 (2018)
54. Haley, S.C., Maniv, E., Wu, S., Cookmeyer, T., Torres-Londono, S., Aravinth, M., Analytis, J.G.: Nat. Commun. **14**, 4691 (2023)
55. Gurevich, A.G., Melkov, G.A.: Magnetisation oscillations and waves, CRC Press (2020)
56. Jiles, D.: Introduction to magnetism and magnetic materials, CRC Press (2015)
57. Slonczewski, J.C.: J. Magn. Magn. Mater. **159**, L1 (1996)
58. Berger, L.: J. Appl. Phys. **49**, 2156 (1978)
59. Stiles, M.D., Zangwill, A.: Phys. Rev. B **66**, 014407 (2002)
60. Ralph, D.C., Stiles, M.D.: J. Magn. Magn. Mater. **320**, 1190 (2008)
61. Tsoi, M., Jansen, A.G.M., Bass, J., Chiang, W.C., Seck, M., Tsoi, V., Wyder, P.: Phys. Rev. Lett. **80**, 4281 (1998)
62. Kiselev, S.I., Sankey, J.C., Krivorotov, I.N., Emley, N.C., Schoelkopf, R.J., Buhrman, R.A., Ralph, D.C.: Nature **425**, 380 (2003)
63. Lee, K.J., Deac, A., Redon, O., Nozieres, J.P., Dieny, B.: Nat. Mater. **3**, 877 (2004)
64. Krivorotov, I.N., Emley, N.C., Sankey, J.C., Kiselev, S.I., Ralph, D.C., Buhrman, R.A.: Science **907**, 228 (2005)
65. Sankey, J.C., Cui, Y.T., Sun, J.Z., Slonczewski, J.C., Buhrman, R.A., Ralph, D.C.: Nat. Phys. **4**, 67 (2008)
66. Wang, C., Cui, Y.T., Katine, J.A., Buhrman, R.A., Ralph, D.C.: Nat. Phys. **7**, 496 (2011)
67. Demidov, V.E., Urazhdin, S., Ulrichs, H., Tiberkevich, V., Slavin, A., Baither, D., Demokritov, S.O.: Nat. Mater. **11**, 1028 (2012)
68. Linder, J.: Phys. Rev. B **84**, 094404 (2011)
69. Gomonay, H.V., Kunitsyn, R.V., Loktev, V.M.: Phys. Rev. B **85**, 134446 (2012)
70. Cheng, R., Xiao, J., Niu, Q., Brataas, A.: Phys. Rev. Lett. **113**, 057601 (2014)
71. Železný, J., Wadley, P., Olejník, K., Hoffmann, A., Ohno, H.: Nat. Phys. **14**, 220 (2018)
72. Hirohata, A.: Magnetochemistry **8**, 37 (2022)
73. Sinova, J., Valenzuela, S.O., Wunderlich, J., Back, C., Jungwirth, T.: Rev. Mod. Phys. **87**, 1213 (2015)
74. Demidov, V.E., Ulrichs, H., Gurevich, S.V., Demokritov, S.O., Tiberkevich, V.S., Slavin, A.N., Urazhdin, S.: Nat. Commun. **5**, 3179 (2014)
75. Chen, T., Dumas, R.K., Eklund, A., Muduli, P.K., Houshang, A., Awad, A.A., Åkerman, J.: Proc. IEEE **104**, 1919 (2016)
76. Awad, A.A., Dürrenfeld, P., Houshang, A., Dvornik, M., Iacocca, E., Dumas, R.K., Åkerman, J.: Nat. Phys. **13**, 292 (2017)

77. Awad, A.A., Houshang, A., Zahedinejad, M., Khymyn, R., Åkerman, J.: Appl. Phys. Lett. **116** (2020)
78. Noack, T.B., Vasyuchka, V.I., Bozhko, D.A., Heinz, B., Frey, P., Slobodianiuk, D.V., Serga, A.A.: Physica Status Solidi (b) **256**, 1900121 (2019)
79. Vaidya, P., Morley, S.A., van Tol, J., Liu, Y., Cheng, R., Brataas, A., Del Barco, E.: Science **368**, 160 (2020)
80. Wang, H., Xiao, Y., Guo, M., Lee-Wong, E., Yan, G.Q., Cheng, R., Du, C.R.: Phys. Rev. Lett. **127**, 117202 (2021)
81. Czeschka, F.D., Dreher, L., Brandt, M.S., Weiler, M., Althammer, M., Imort, I.M., Goennenwein, S.T.: Phys. Rev. Lett. **107**, 046601 (2011)
82. Prokopenko, O.V., Krivorotov, I.N., Bankowski, E.N., Meitzler, T.J., Tiberkevich, V.S., Slavin, A.N.: J. Appl. Phys. **114** (2013)
83. Prokopenko, O.V., Tiberkevich, V.S., Slavin, A.N.: Functional materials **21**, 206 (2014)
84. Prokopenko, O.V.: Ukr. J. Phys. **60**, 104 (2015)
85. Park, B., Wunderlich, J., Marti, X., Holy, V., Kurosaki, Y., Yamada, M., Yamamoto, H., Nishide, A., Hayakawa, J., Takahashi, H., others: Nat. Mater. **10**, 347 (2011)
86. Kriegner, D., Výborný, K., Olejník, K., Reichlová, H., Novák, V., Marti, X., Jungwirth, T.: Nat. Commun. **7**, 11623 (2016)
87. Wadley, P., Howells, B., Železný, J., Andrews, C., Hills, V., Campion, R.P., Jungwirth, T.: Science **351**, 587 (2016)
88. Ivanov, B.A.: Low Temp. Phys. **40**, 91 (2014)
89. Darula, M., Doderer, T., Beuven, S.: Supercond. Sci. Technol. **12**, R1 (1999)
90. Prokopenko, O., Bankowski, E., Meitzler, T., Tiberkevich, V., Slavin, A.: IEEE Magn. Lett. **2**, 3000104 (2011)
91. Shtanko, O., Prokopenko, O.: in 2020 IEEE 10th International Conference Nanomaterials: Applications & Properties (NAP), NMM04 (2020); https://doi.org/10.1109/NAP51477.2020.9309712
92. Shtanko, O., Prokopenko, O.: in 2021 IEEE 11th International Conference Nanomaterials: Applications & Properties (NAP), NMM04 (2021); https://doi.org/10.1109/NAP51885.2021.9568573
93. Shtanko, O., Prokopenko, O.: in 2022 IEEE 12th International Conference Nanomaterials: Applications & Properties (NAP), NMM 04 (2022); https://doi.org/10.1109/NAP55339.2022.9933996
94. Pikovsky, A., Rosenblum, M., Kurths, J.: Synchronization: a universal concept in nonlinear sciences, Cambridge University Press (2003)
95. Kaka, S., Pufall, M.R., Rippard, W.H., Silva, T.J., Russek, S.E., Katine, J.A.: Nature **437**, 389 (2005)
96. Mancoff, F.B., Rizzo, N.D., Engel, B.N., Tehrani, S.: Nature **437**, 393 (2005)
97. Slavin, A.N., Tiberkevich, V.S.: Phys. Rev. B **74**, 104401 (2006)
98. Ruotolo, A., Cros, V., Georges, B., Dussaux, A., Grollier, J., Deranlot, C., Fert, A.: Nat. Nanotechnol. **4**, 528 (2009)
99. Slavin, A., Tiberkevich, V.: IEEE Trans. Magn. **47**, 1875 (2009)
100. Tiberkevich, V., Slavin, A., Bankowski, E., Gerhart, G.: Appl. Phys. Lett. **95** (2009)
101. Sanid, C., Murugesh, S.: J. Phys. D Appl. Phys. **47**, 065005 (2014)
102. Erokhin, S., Berkov, D.: Phys. Rev. B **89**, 144421 (2014)
103. Sulymenko, O.R., Prokopenko, O.V.: J. Phys. Stud. **21**, 4801–1 (2017)
104. Slobodianiuk, D.V., Prokopenko, O.V.: J. Nano- Electron. Phys. **12**, 05013 (2020)
105. Slobodianiuk, D., Prokopenko, O.: in 2021 IEEE 11th International Conference Nanomaterials: Applications & Properties (NAP), NMM04 (2021); https://doi.org/10.1109/NAP51885.2021.9568506
106. Oktavio, M.: Phys. Rev. B **29**, 1231 (1984)
107. Oktavio, M.: Phys. A **163**, 248 (1990)
108. Kautz, R., Monaco, R.: J. Appl. Phys. **57**, 875 (1985)
109. Kautz, R.: J. Appl. Phys. **58**, 424 (1985)

110. LeCun, Y., Bengio, Y., Hinton, G.: Nature **521**, 436 (2015)
111. Sangwan, V.K., Hersam, M.C.: Nat. Nanotechnol. **15**, 517 (2020)
112. Torrejon, J., Riou, M., Araujo, F.A., Tsunegi, S., Khalsa, G., Querlioz, D., Grollier, J.: Nature **547**, 428 (2017)
113. Bradley, H., Louis, S., Trevillian, C., Quach, L., Bankowski, E., Slavin, A., Tyberkevych, V.: AIP Adv. **13** (2023)
114. Fukami, S., Ohno, H.: J. Appl. Phys. **124** (2018)
115. Camsari, K., Faria, R., Sutton, B., Datta, S.: Phys. Rev. X **7**, 031014 (2017)
116. Borders, W., Pervaiz, A.Z., Fukami, S., Camsari, K., Ohno, H., Datta, S.: Nature **573**, 390 (2019)
117. Bapna, M., Majetich, S.: Appl. Phys. Lett. **111**, 243107 (2017)

Terahertz Signal Detectors Based on Antiferromagnetic Spintronic Nanostructures

Volodymyr Prokopenko and Oleksandr Prokopenko

Abstract Antiferromagnetic spintronic detectors are nanoscale multilayer devices consisting of a layer of antiferromagnet and capable of generating a low frequency or DC output voltage under the action of a (sub-) terahertz frequency input signal. In a general case, an input AC signal current applied to the detector excites the current-driven (sub-) terahertz frequency dynamics of magnetic sublattices in its antiferromagnetic layer, which then leads to the generation of the detector output voltage across the whole structure due to several possible effects depending on the device materials and geometry. Like the conventional ferromagnet-based spintronic detectors, which are now considered as promising signal detectors for microwave applications, the detectors based on antiferromagnetic nanostructures can be promising for many terahertz applications, including terahertz signal detection, spectrum analysis, and energy harvesting. In this chapter, we review the principles of operation of antiferromagnetic spintronic detectors, discuss their key features compared to the conventional ferromagnet-based detectors, and also review the recent advances, trends, and problems in the physics and applications of antiferromagnetic spintronic detectors.

Keywords Spintronics · Antiferromagnet · Signal detector · Magnetic dynamics · Multilayer structures · Antiferromagnetic tunnel junction

1 Introduction

Antiferromagnetic (AFM) spintronics [1–6] is an emerging field of modern physics of magnetism and magnetic technologies focused on the study and exploitation of dynamic and static magnetic states in antiferromagnets. In contrast to ferromagnetic spintronics, which has been actively developed over the past 30 years [7–9], AFM

V. Prokopenko (✉) · O. Prokopenko
Educational and Scientific Institute of High Technologies, Taras Shevchenko National University of Kyiv, 64/13 Volodymyrs'ka Str, Kyiv 01601, Ukraine
e-mail: v.o.prokopenko@gmail.com

A. D. Pogrebnjak et al. (eds.), *Nanocomposite and Nanocrystalline Materials and Coatings*, Advanced Structured Materials 214,
https://doi.org/10.1007/978-981-97-2667-7_4

spintronics emerged only 10–15 years ago [1–6, 10–27]. Moreover, active studies of spintronic effects in AFM materials and nanostructures actually started only 5–10 years ago, after the first successful experiments in 2018 [28]. Thus, it can be assumed that serious research in the field of AFM spintronics is just beginning, and perhaps the most important (mainly experimental) results will be obtained in the next 5–10 years, which makes this field of spintronics promising from both scientific and practical points of view [1–6, 10–28]. At the same time, it is necessary to take into account the following facts, which objectively hinder the development of AFM spintronics [1–6, 10–28]:

(i) Currently, there is no technology to fabricate high-quality AFM films and film nanostructures (single crystal, with a single domain antiferromagnetic state, with a certain predetermined anisotropy of the antiferromagnetic nanostructure layer, etc.).
(ii) The physics of antiferromagnets is more complex than the physics of ferromagnetic materials since in antiferromagnets it is necessary to consider the dynamics of two or more magnetic sublattices (see Sect. 2 in Chap. 3). This complicates theoretical calculations and numerical modelling for AFM systems;
(iii) The net magnetisation in antiferromagnets is close to zero (zero in a fully compensated antiferromagnet, see Sect. 2 of Chap. 3). As a result, all magnetisation-dependent spintronic effects, in particular the classical magnetoresistance effect (well known for layered ferromagnetic nanosystems [7–9]), are very weak, which makes it difficult to observe these effects experimentally in AFM materials and limits the range of their possible practical applications;
(iv) The characteristic frequencies of the magnetic oscillations in antiferromagnets are determined by a very strong internal exchange field and are typically in the sub-terahertz and terahertz band, which greatly complicates work with AFM systems and requires the use of complex and expensive experimental equipment [18, 29];
(v) Finally, antiferromagnets are insensitive to weak DC magnetic fields, which makes it essentially impossible to control the state of the antiferromagnet by applying such fields to an AFM sample, i.e. to use the traditional methodology for experiments with ferromagnetic materials, in particular experiments with ferromagnetic spintronic devices [7–9].

At the same time, antiferromagnetic spintronic devices have a number of key advantages over lower-frequency spintronic devices based on ferromagnetic materials:

(i) The operating frequencies of AFM spintronic devices are significantly higher than those of ferromagnetic spintronic devices, which can form the basis of new technological solutions in terahertz electronics [1–6, 18];
(ii) The information encoded in a particular magnetic state of an antiferromagnetic nanodevice is protected from external magnetic fields, since the antiferromagnet is insensitive to magnetic fields significantly weaker than its internal exchange field of ~1000 T;

(iii) Since external fields, even relatively strong ones, ~1–10 T, do not really affect the state of the antiferromagnet, and its net magnetisation is typically close to zero (see Sect. 2 of Chap. 3), the AFM-based system appears "invisible" in a magnetic field and thus cannot be detected by existing magnetic sensors;
(iv) AFM nanoelements, due to their magnetic "neutrality", do not magnetically interact with the neighbouring elements, both magnetic and non-magnetic, such as ferromagnetic spintronic or semiconductor elements. This makes it possible to significantly improve the parameters of memory devices using AFM materials, in particular to reduce the size of the memory cell to 10–20 nm and even smaller [13–16].

All of the above, despite the aforementioned difficulties in the realisation of AFM spintronic devices, makes this field one of the most promising for the development of future electronics, in particular for the realisation of ultra-fast neuromorphic electronics (see Chap. 3).

Despite a number of differences between ferromagnets and antiferromagnets, the operating principles of AFM and ferromagnetic spintronic devices are similar in many respects. For example, recent experimental and theoretical work has shown that antiferromagnets (as well as ferromagnets) are sensitive to spin-polarised current [10, 11, 17–28]. The corresponding effects can be used to control the magnetic state and magnetic dynamics of antiferromagnets and composite structures based on them [1–6, 10–28, 30]. However, the peculiarities of the excitation, flow and interaction of spin currents with magnetic sublattices of the AFM system, the influence of spin transfer and spin pumping processes on the magnetisation dynamics in AFM spintronic nanostructures have not yet been thoroughly investigated.

In Chap. 3, the physics and potential applications of antiferromagnetic spintronic nano-oscillators based on bilayer or multilayer nanostructures containing an AFM layer were discussed in detail. In this chapter, we will consider the working principle of other practically important AFM devices, such as detectors and spectrum analysers of electromagnetic signals based on multilayer AFM spintronic nanostructures.

Detectors based on AFM nanostructures actually use the same physical effects that are important for AFM spintronic oscillators (see Sect. 2 of Chap. 3), but AFM signal detectors are potentially easier to implement. This is due to the following:

(i) In an AFM signal detector, an AC input electromagnetic signal of (sub-) terahertz frequency is converted into a low-frequency or DC output signal that can be easily registered and measured using conventional low-frequency electronics. At the same time, the output signal of an AFM spintronic oscillator (see Chap. 3) is a (sub-) terahertz-frequency signal, which means that its measurement requires rather complex and expensive experimental equipment [17–26];
(ii) For the practical implementation of an AFM oscillator capable of producing a quasi-monochromatic output AC signal, it is crucial that the AFM layer of the spintronic nanostructure is in a single-domain state [1–6]. Only under these conditions will the magnetic dynamics be coherent throughout the whole AFM

layer of the nanostructure. Conversely, if the AFM layer has some domain structure, the signals generated in different domains will be out of phase, which will significantly degrade the spectral characteristics of the entire device. However, the presence of a domain structure is less important for spintronic AFM detectors. In such systems, the presence of the domain structure of the AFM layer only affects the summation of the low-frequency signals generated in each AFM domain, and for such low-frequency signals, possible dephasing (caused by incoherence of magnetic dynamics in different AFM domains) seems to be less significant than for AFM nano-oscillators [17–26];

(iii) The practical testing of the AFM spintronic detector is quite simple if there is a sufficiently powerful source of quasi-monochromatic (sub-) terahertz radiation or a source of powerful electromagnetic pulses with a spectral component at the required frequency that is close the antiferromagnetic resonance frequency. Initial studies have shown that antiferromagnets are sensitive to such radiation, making it possible to use AFM spintronic nanostructures to detect signals at a specific frequency [31–34]. At the same time, the problems associated with the relatively low efficiency of converting the energy of an input AC signal into the energy of a low frequency output signal are not critical for AFM detectors because the power of the input AC signal can vary over a fairly wide range, which is sufficient to produce a large detector output signal that is easy to measure. However, for AFM spintronic nano-oscillators, this problem is significant because the output AC signal of the oscillator is quite weak due to the not very high energy conversion efficiency. Although it is possible to increase the output power of such a signal source by increasing the low-frequency drive current, this can only be done as long as this current is below the electrical breakdown threshold of the device, which naturally limits the output power of typical AFM nano-oscillators to ~1 μW [17–26] (see Chap. 3).

Next, in Sect. 2 of this chapter, we briefly review the basic physical effects underlying the operation of AFM spintronic detectors. In Sects. 3 and 4, respectively, the properties of currently known AFM signal detectors based on bilayer and multilayer AFM spintronic nanostructures are considered. The physics and applications of spectrum analysers based on a single current-driven AFM detector and their arrays are also discussed in Sect. 4.

2 Principles of Operation of Antiferromagnetic Signal Detectors

The general principle of operation of an AFM signal detector is similar to that of any other microwave or terahertz-frequency detector. When a sub-terahertz or terahertz AC input signal is applied to an AFM spintronic nanostructure, it excites magnetisation dynamics in an AFM layer of the nanostructure. This dynamics produces a low frequency or DC signal, which is the output signal of the detector. By measuring

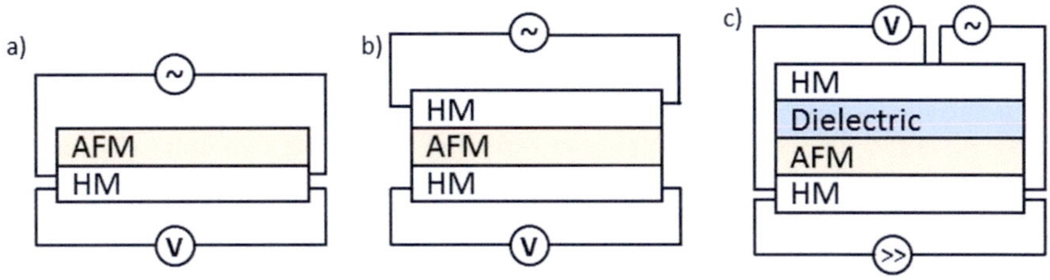

Fig. 1 Schematics of AFM nanostructures consisting of antiferromagnetic (AFM), dielectric and heavy metal (HM) layers used to create terahertz signal detectors, and connection of these nanostructures to external electrical AC and DC circuits: **a** bilayer HM/AFM nanostructure; **b** three-layer HM/AFM/HM nanostructure; **c** four-layer HM/AFM/D/HM antiferromagnetic tunnel junction structure. When an input AC signal is applied to the structure, the AFM nanostructure generates a DC voltage, which is used as the output signal of the detector

such a low frequency or DC output signal, one can determine the properties of the AC input signal acting on the detector.

Although the general operating principle of all AFM signal detectors is the same, the specific mechanism for converting the detector's AC input signal into its low-frequency or DC output signal is different for detectors based on various types of AFM spintronic nanostructures. In this context, the physics and technology of AFM signal detectors appear to be more "diverse" than for signal detectors based on ferromagnetic spintronic nanostructures [35–38]. The fundamental differences in the operation of various AFM signal detectors are mainly due to the diversity of the AFM spintronic nanostructures on which the detectors are based, as well as to the different mechanisms of forming the low-frequency output signal of the detector. Such diversity is a characteristic feature of AFM systems, in particular, it is inherent to AFM spintronic nano-oscillators [17–26], discussed in Chap. 3.

There are currently three basic designs of AFM signal detectors, shown in Fig. 1.

In the simplest case, the AFM signal detector is a bilayer film nanostructure containing a layer of anisotropic antiferromagnet and a layer of heavy metal (HM) characterised by a strong spin–orbit interaction [39] (Fig. 1a). In fact, such a nanostructure is identical to the one used for AFM spin Hall oscillators [17, 19, 20, 26] (see also Sect. 2 in Chap. 3). In such a nanostructure, the input AC current flowing along the HM layer excites a perpendicular AC spin current due to the Hall spin effect [40] (this process is fundamentally no different from that for AFM spintronic nano-oscillators). The AC spin current, in turn, excites the magnetic dynamics in the adjacent AFM layer, and this, in antiferromagnets with biaxial anisotropy (typical for Josephson-like AFM nano-oscillators [17, 19, 20, 26], see Chap. 3), can cause the appearance of a DC spin current flowing back into the HM layer [39]. As a result of the inverse spin Hall effect (ISHE) [40], this outgoing spin current in the HM layer is converted into an electric DC current perpendicular to the direction of the input AC current. Since the heavy metal has a certain finite resistance, the output signal of the detector is a transverse output DC voltage measured across the HM layer. In this detector configuration, the HM layer is used simultaneously as an input and output electrode, which simplifies the detector design but can sometimes cause unwanted parasitic coupling between electric and spin currents in the HM layer.

To avoid the parasitic couplings between different electric and spin currents, a more advanced detector design is a nanostructure containing two outer HM layers, between which the AFM layer is located [41] (Fig. 1b). In such a detector, an AC signal is applied to only one of the HM layers, e.g. the top HM layer in Fig. 1b. At the same time, due to the spin Hall effect [40] and the spin-transfer torque effect [42, 43], the input AC signal results to the excitation of magnetic dynamics in the AFM layer. Subsequently, due to the spin pumping effect [42], the dynamics of the Neel vector in the AFM layer leads to the appearance of an inverse DC spin current that penetrates another HM layer (bottom HM layer in Fig. 1b), which finally leads to the generation of a transverse DC output voltage, which is the output signal of the AFM detector [41]. In this detector design, the AC and DC circuits are effectively separated, which makes it easier to control the state of the device.

The most complex structure is the AFM signal detector based on an antiferromagnetic tunnel junction (ATJ), which is a four-layer HM/AFM/D/HM nanostructure containing two outer layers of heavy metal (HM), an inner AFM layer and a thin dielectric (D) layer serving as a tunnelling barrier for charge carriers [21, 27] (Fig. 1c). The physical principle of operation of such a signal detector differs significantly from the previous two AFM detectors and is more reminiscent of the principle of operation of spin diodes based on ferromagnetic materials [35–38]. When a drive current is passed through the bottom HM layer of such a structure, a transverse spin current is generated, which excites the magnetic dynamics in the adjacent AFM layer. As a result, due to the phenomenon of tunnelling anisotropic magnetoresistance (TAMR) [44], the magnetoresistance of the entire structure starts to oscillate at a frequency determined by the magnitude of the applied drive current. If an AC current created by an input AC signal is passed through a nanostructure with an oscillating TAMR, the oscillations of the input current and the TAMR oscillations are mixed inside the structure. Provided that the frequencies of these oscillations are close, this will result in the generation of an output DC voltage across the whole nanostructure [27]. Obviously, this design of the AFM signal detector is based on the most complex four-layer AFM nanostructure, and for the detector to work, up to three independent circuits have to be implemented: a drive current circuit, an input circuit for an AC input signal, and an output circuit for measuring the detector's output signal—a low-frequency or DC voltage. However, in spite of the above, such a design turns out to be one of the most promising, since it allows (i) to use isotropic AFM layers without a certain type of anisotropy; (ii) to better control the state of the detector (by selecting the desired drive current), and (iii) to significantly improve the magnitude of the detector's output signal (in fact, the detector sensitivity) by varying the geometric parameters of the detector (for example, the thickness of the dielectric layer serving as a tunnelling barrier). Therefore, the characteristics of the latter type of AFM signal detector will be discussed in more detail below in Sect. 4.

In conclusion, one should note that the physical phenomena crucial for the operation of AFM signal detectors (in particular, the spin-transfer torque effect [42, 43], the spin Hall effect [40], and the spin pumping effect [42]) are identical to those discussed in Sect. 2 of Chap. 3 for AFM spintronic oscillators. In addition to these effects, the inverse spin Hall effect (ISHE) [40] in heavy metals, which is the inverse effect to the conventional (direct) spin Hall effect (see Sect. 2.2 of Chap. 3), plays an important role in AFM signal detectors. Due to the ISHE, charge carriers with opposite spin directions (different projections of the spin magnetic moment on any chosen axis) redistributing (i.e. forming a spin current) in a heavy metal with a strong spin–orbit interaction, acquire a velocity component perpendicular to the spin current due to the spin–orbit interaction. So, when a spin current flows in a heavy metal, an electric current perpendicular to the spin current is formed in a HM layer.

3 Signal Detectors Based on Bilayer and Three-Layer Antiferromagnetic Nanostructures

3.1 Detectors Based on HM/AFM Structures

One of the first types of an AFM signal detector based on an HM/AFM nanostructure with an AFM layer having biaxial magnetic anisotropy was proposed and theoretically investigated in [39]. Such a detector is a conventional bilayer system in which a HM layer (platinum in [39]) is used as an electrode through which an input AC signal $\mathbf{j}_{AC}$ enters the system and simultaneously as an electrode from which an output DC signal V_{DC} is registered (Fig. 2).

In [39], the Pt/NiO structure is considered, where the antiferromagnetic NiO layer has biaxial anisotropy, which is a critical feature for the detector operation. When a linearly-polarised AC current with a density of $\mathbf{j}_{AC}$ flows through the Pt layer, a transverse AC spin current is generated due to the spin Hall effect [40]. This AC spin

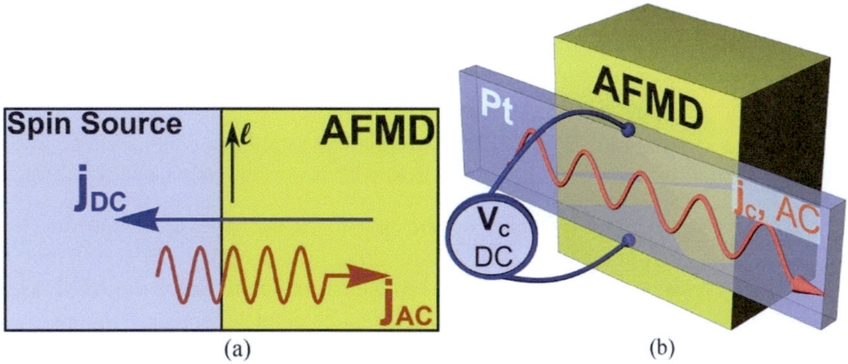

Fig. 2 Detector of (sub-) terahertz signals based on a HM/AFM nanostructure: **a** the scheme of the AC-DC current conversion in the detector; **b** the proposed scheme for the experimental observation of the spin current rectification

current excites two evanescent modes in the antiferromagnet layer, which leads to the formation of a DC spin current with density $\mathbf{j}_{DC}$ flowing back to the HM layer via the AFM/HM interface [39]. In the platinum layer, this DC spin current is converted into an electric DC current due to the ISHE [40], resulting in the generation of a transverse DC voltage V_{DC}.

If one introduces a coordinate system $\mathbf{e}_1, \mathbf{e}_2, \mathbf{e}_3$, where the easy axis of the antiferromagnet corresponds to the unit vector $\mathbf{e}_3$, the hard axis corresponds to the unit vector $\mathbf{e}_2$, and the intermediate axis corresponds to the unit vector $\mathbf{e}_1$, the input AC current with density $\mathbf{j}_{AC}$ can be written as

$$\mathbf{j}_{AC} = j_{AC}(\mathbf{e}_1 \cos\theta + \mathbf{e}_2 \sin\theta)\sin(\omega t). \tag{1}$$

Here, we have taken into account that the current is linearly polarised, with the polarisation vector lying in the plane $(\mathbf{e}_1, \mathbf{e}_2)$ perpendicular to the easy axis $(\mathbf{e}_3)$. The angle θ is the angle of the current polarisation, and ω is the angular frequency of the harmonic AC current. Under the action of the current density (1), as noted above, magnetic dynamics is excited in the adjacent antiferromagnetic layer, which results in the generation of a transverse output DC spin current with density [39]

$$\mathbf{j}_{DC} = \mathbf{e}_3 j_{AC}^2 B, \tag{2}$$

where the coefficient B that characterises the efficiency of converting the input AC power ($\sim j_{AC}^2$) into an output DC signal ($\sim j_{DC}$) can be calculated as [39]:

$$B = \frac{\gamma\hbar\omega_{ex}}{ec^2 M_S}\sin(2\theta)\left[\lambda_2 \tanh\left(\frac{d}{\lambda_1}\right) - \lambda_1 \tanh\left(\frac{d}{\lambda_2}\right)\right]. \tag{3}$$

Here γ is the modulus of the gyromagnetic ratio, $\hbar$ is the reduced Planck constant, ω_{ex} is the exchange angular frequency, e is the modulus of the electron change, c is the speed of AFM magnons, M_S is the saturation magnetisation of the AFM sublattice, $\lambda_{1,2} = c/\sqrt{\omega_{1,2}^2 - \omega^2}$ is the spatial decay of spin excitations along $(\mathbf{e}_1, \mathbf{e}_2)$ axes, $\omega_{1,2}$ are the angular frequencies of the AFM resonances, and d is the thickness of an AFM layer.

Analysing expressions (2) and (3), a number of important conclusions can be drawn:

(1) At $\lambda_1 = \lambda_2$, when the biaxial anisotropy of the AFM layer disappears, the antiferromagnet becomes a system with uniaxial anisotropy, the coefficient B in (3) becomes zero. Thus, the HM/AFM nanostructure in which the AFM layer has no biaxial anisotropy cannot be used as a detector of terahertz signals.
(2) The effect of rectifying an input AC signal in the HM/AFM nanostructure is quadratic, since $V_{DC} \sim j_{DC} \sim j_{AC}^2$. Accordingly, just as it is done for spin-torque microwave detectors [35–38], it is possible and appropriate to introduce the concept of volt-watt sensitivity R for AFM signal detectors: $R = V_{DC}/P_{AC} = B/S$, where P_{AC} is the power of the input AC signal, S is the cross-sectional area

of the device (the contact area between the HM and AFM layers in Fig. 2). It is obvious that the sensitivity R increases with decreasing device size. However, when considering such behaviour in nanoscale samples, it is necessary to take into account the conditions for the existence of the desired magnetic state in the AFM layer and the existence of fluctuations of different nature in the system.

(3) As the frequency of the input AC signal approaches one of the AFM resonant frequencies, the coefficient B increases significantly, resulting in a large increase in the volt-watt sensitivity R of the detector in the frequency range close to the AFM resonant frequency.

(4) Even for HM/AFM structures where the thickness of the AFM layer $d \to \infty$, the coefficient B decreases but does not reach zero. This means that, unlike many other AFM spintronic devices, the considered AFM signal detector can operate even in the case of HM/AFM structures with very thick antiferromagnetic layers.

Thus, the proposed detector based on the HM/AFM nanostructure with biaxial anisotropy of the AFM layer can operate as a non-resonant quadratic detector of linearly-polarised input AC signals with a sufficiently high volt-watt sensitivity near the AFM resonance frequency of the antiferromagnetic layer.

Figure 3 shows the characteristic dependence of the volt-watt sensitivity R of the considered detector on the thickness d of the AFM layer (Fig. 3a) and the frequency of the input AC signal (for the detector with $d \to \infty$, when $R \to R_\infty$; Fig. 3b). The curves in Fig. 3 are calculated using the following parameters [39]: spin polarisation angle $\theta = \pi/4$, cross-sectional area of the device $S = 30 \times 30$ nm^2, exchange angular frequency for NiO is $\omega_{\text{ex}} \approx 2\pi \cdot 27,5$ THz, $c \approx 38$ km/s is the speed of the AFM magnons, $M_S \simeq 350$ kA/m is the saturation magnetisation of the AFM sublattice, $\omega_1 \simeq 2\pi \cdot 220$ GHz and $\omega_2 \simeq 2\pi \cdot 1,1$ THz are the angular frequencies of the AFM resonances. An estimate of the output DC voltage of the device under typical experimental conditions made in [39] gives a value of $V_{\text{DC}} \sim 40\ \mu V$, which is sufficient for some practical applications.

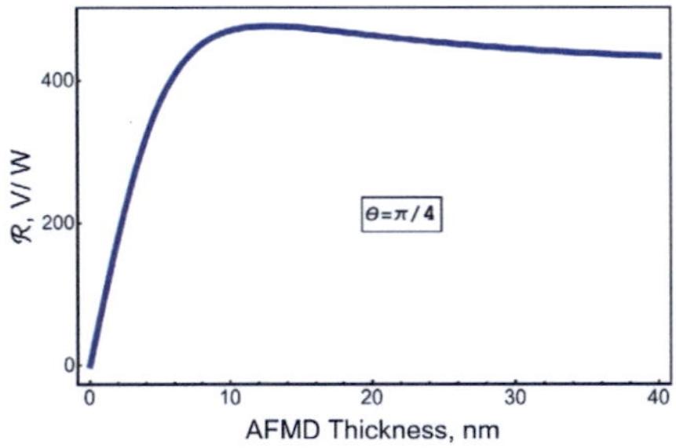

(a)

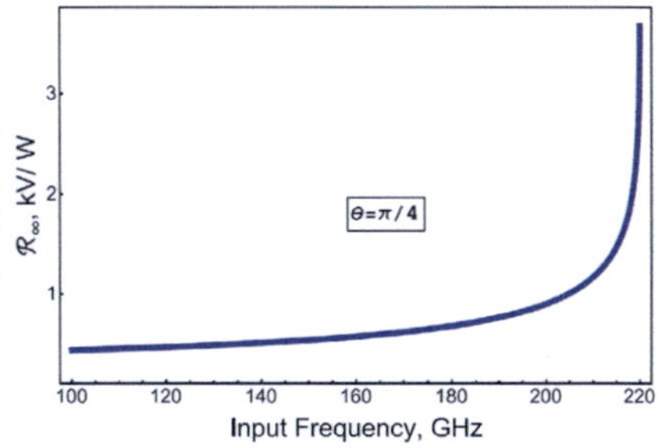

(b)

Fig. 3 Dependence of volt-watt sensitivity R of the AFM signal detector on **a** thickness of AFM layer d for input signal with frequency $\omega = 2\pi \cdot 10$ GHz; **b** on input signal frequency $\omega/2\pi$ for HM/AFM structure with infinite thickness of AFM layer ($d \to \infty$, $R \to R_\infty$). Parameters used for calculations are indicated in the text and in [39]

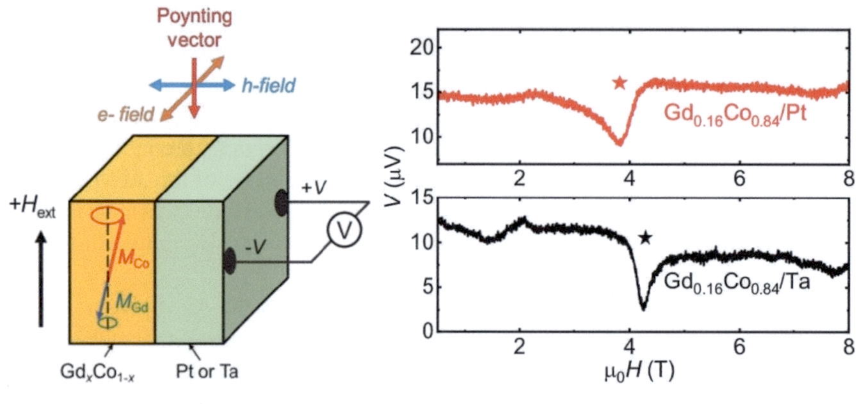

(a) (b)

Fig. 4 **a** The Gd_xCo_{1-x}/HM nanostructure studied in [45]. **b** Experimentally measured field dependencies of the output DC voltage of the signal detectors based on the Gd_xCo_{1-x}/HM nanostructures with antiferromagnetic resonance mode

Significant progress in the use of bilayer spintronic nanostructures has recently been made in [45], where the effect of quasi-resonant detection of a strong AC signal has been experimentally demonstrated. In [45], the possibility of detecting AFM dynamics in Gd_xCo_{1-x}/Ta and Gd_xCo_{1-x}/Pt nanostructures excited by a strong input AC signal generated in a gyrotron operating at 154 GHz was investigated (Fig. 4a). Using a weak bias DC magnetic field H_{ext} parallel to the sample's plane as a control parameter, the authors of [45] were able to obtain the field dependence of the initial transverse DC voltage $\sim 10 - 15\ \mu V$ across the HM layer of the nanostructure. The result obtained in [45] confirms the possibility of *practical* implementation of signal detectors using AFM resonance mode with acceptable performance characteristics.

3.2 Detector Based on a HM/AFM/HM Structure

Another type of AFM signal detector based on a three-layer HM/AFM/HM nanostructure (Fig. 5a) has recently been proposed in [46]. In the structure considered, the HM layers were made of platinum (Pt) and the antiferromagnetic layer was made of Ir_xMn_{1-x}. The principle of operation of such a detector is partially similar to that of AFM signal detectors based on bilayer HM/AFM nanostructures (see Sect. 3.1). Similar to the detectors based on HM/AFM nanostructures, in the three-layer HM/AFM/HM system, magnetic oscillations in the AFM layer are excited by the input AC signal and the DC drive current applied to one of the HM layers (the bottom Pt layer in Fig. 5a). The excited magnetic oscillations of the AFM sublattices due to the spin pumping effect [42] generate a spin current $\mathbf{j}_{SP}$ that flows through the AFM/HM interface into the other HM layer (the top Pt layer in Fig. 5a). This spin current is, in turn, converted into a transverse electric current, producing the detector's output signal, a DC voltage V_{OUT} applied across the top Pt layer of the nanostructure.

It is shown in [46] that the proposed detector is a resonance-type quadratic detector of terahertz frequency electromagnetic signals, and in this sense, it is a higher-frequency analogue of the known resonance-type spin-torque microwave detectors [35–38, 47]. The output voltage of such a detector can be calculated as [46]

$$V_{\mathrm{OUT}}(\omega) = V_0 \frac{\omega_{\mathrm{AC}}^2 \omega_{\mathrm{AFMR}} \omega}{\left(\omega_{\mathrm{AFMR}}^2 - \omega^2\right)^2 + \left(\alpha_{\mathrm{eff}} \omega_{\mathrm{ex}} \omega\right)^2}, \quad (4)$$

where V_0 is the characteristic maximum voltage generated by the detector (see Eq. (7) in paper [46]), $\omega_{\mathrm{AC}} = \sigma I_{\mathrm{AC}}$ is the characteristic value of the spin-transfer torque acting on the AFM sublattices, σ is the spin-torque coefficient described in Sect. 2.2

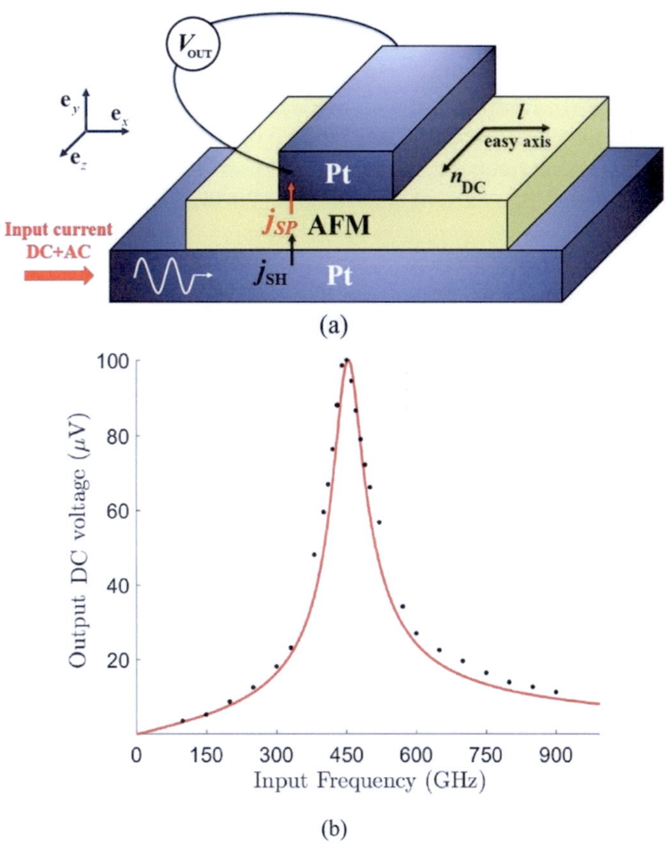

Fig. 5 **a** Schematic of the signal detector based on a HM/AFM/HM nanostructure. **b** Amplitude-frequency curve of the detector calculated for experimentally acceptable system parameters (solid line—theory, points—micromagnetic simulations). For details see [46]

of Chap. 3, I_{AC} is the amplitude of the input AC signal, $\omega_{AFMR} = \sqrt{\omega_{ex}\omega_e}$ is the frequency of the antiferromagnetic resonance, ω_{ex} and ω_e are the frequencies corresponding to the exchange field and the uniaxial anisotropy field in the AFM layer, α_{eff} is the effective Gilbert damping parameter that takes into account the spin pumping [42] from an AFM layer. An example of the calculation of the detector's output DC voltage according to Eq. (4) for experimentally possible values of the system's parameters (see [46] for details) is shown in Fig. 5b.

Although the proposed detector has a more complex structure than the detectors based on bilayer HM/AFM nanostructures, it has a number of key advantages over the detectors discussed in Sect. 3.1, as well as a number of specific features, namely:

(i) The considered AFM detector is controlled by a drive DC current applied to the bottom layer of the HM/AFM/HM nanostructure. This allows the detector's work point to be chosen by adjusting the value of the drive DC current, thereby varying the detector's operating frequency (in fact, the frequency ω_{AFMR}, which becomes a function of the drive current) and its volt-watt sensitivity R;

(ii) Under certain conditions, this detector can have a very high volt-watt sensitivity $R \sim 10^5$ V/W comparable to that of a spin-torque microwave detector [36–38, 48, 49];

(iii) The detector's amplitude-frequency curve has a Lorentzian shape, which can be used to create frequency detectors and spectrum analysers, similar to those implemented on the base of spin-torque microwave oscillators and detectors [47, 50];

(iv) The detector can only be implemented if its AFM layer has uniaxial anisotropy, otherwise $\omega_{AFMR} \to 0$, and the detector output signal $V_{OUT} \sim \omega_{AFMR}$ disappears.

4 Signal Detectors Based on Antiferromagnetic Tunnel Junctions

In Sect. 3, it was pointed out that the principles of operation of AFM signal detectors based on HM/AFM and HM/AFM/HM nanostructures are quite similar and, in fact, consist in converting the output spin current, which is dependent on the input AC signal, into a DC voltage V_{DC}. At the same time, the principle of operation of AFM signal detectors based on antiferromagnetic tunnel junction (ATJ) is quite different due to the specific features of this nanostructure.

The ATJ is a four-layer nanostructure consisting of an AFM layer, a dielectric layer acting as a tunnel barrier and two outer HM layers. An example of such a nanostructure is a Pt/IrMn/MgO/Pt junction with a layer of the conductive antiferromagnet Ir_xMn_{1-x} (IrMn), a dielectric MgO layer and current-carrying layers of Pt (Fig. 6a) considered in our papers [21, 27, 51, 52].

The principle of operation of a detector based on a Pt/IrMn/MgO/Pt contact is as follows. An electric drive current I_{drive} flowing in the bottom Pt layer (see Fig. 6a) causes a transverse spin current I_{SH} due to the Hall spin effect [40]. This spin current

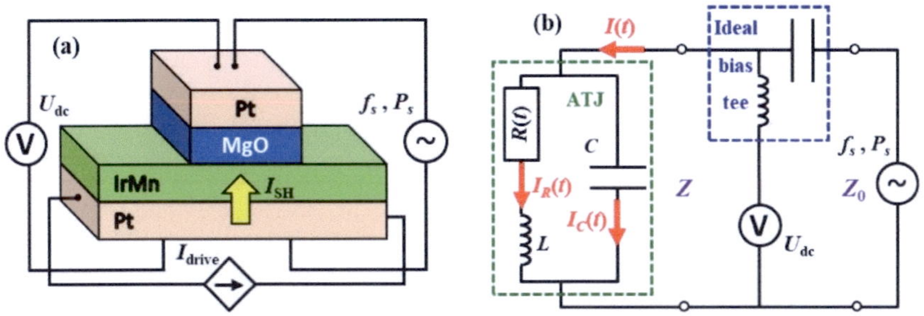

Fig. 6 **a** Pt/IrMn/MgO/Pt antiferromagnetic tunnel junction (ATJ) connected to external circuits. The drive current I_{drive} generates the transverse spin current I_{SH}, which excites the rotation of the magnetisation of the AFM sublattices and thereby AC variations of the junction resistance $R(t)$. When an input AC signal of power P_S and frequency f_S is applied to the junction, an output DC voltage V_{DC} is generated across the structure. **b** An electrical model of an ATJ-based signal detector connected to external AC and DC circuits

penetrates into the IrMn layer and excites the rotation of its magnetic sublattices, which causes oscillations of the tunnel anisotropic magnetoresistance $R(t)$ of the junction at the terahertz frequency. By changing the value of the drive current I_{drive}, it is possible to change the rotation frequency of the magnetic sublattices of the antiferromagnet, thereby controlling the frequency of oscillations of the magnetoresistance $R(t)$ of the entire nanostructure. The effect of tunnelling anisotropic magnetoresistance (TAMR) [44] associated with the anisotropy of IrMn, which leads to a change in the nanostructure resistance when the magnetic sublattices of the antiferromagnet are rotated, plays a fundamental role in these processes. If we now pass an input AC signal current $I(t)$ through the Pt/IrMn/MgO/Pt junction so that it flows through all layers of the nanostructure, the voltage between the Pt layers of the junction will be defined as $V(t) = I(t)R(t)$. Assuming that the input AC signal is harmonic, $I(t) = I_S \cos(2\pi f_S t)$, and the tunnel magnetoresistance of the junction is $R(t) = R_0 + \Delta R \cos(2\pi f_R t + \Delta\varphi_R)$, and the frequencies f_S and $f_R = f_R(I_{\text{drive}})$ are almost the same, mixing of the signals $I(t)$ and $R(t)$ can lead to the appearance of a DC voltage component $V_{\text{DC}} = 0,5 I_S \Delta R \cos(\Delta\varphi_R)$, where I_S is the amplitude of the input AC current, ΔR is the amplitude of the ATJ resistance oscillations, $\Delta\varphi_R$ is the phase difference between the oscillations $I(t)$ and $R(t)$. The DC voltage V_{DC} is the output signal of the ATJ-based signal detector.

The physics of AFM nanostructures, especially such multilayers as ATJs, is quite complex, so to determine the output DC voltage of an ATJ-based detector, it is convenient to use a simplified electrical model of an ATJ, first proposed for an ATJ-based signal source [21] and later used for a signal detector [27]. According to this model, an ATJ can be represented as an equivalent circuit consisting of an oscillating magnetoresistance $R(t)$, intrinsic junction inductance L and its capacitance C (Fig. 6b). This electrical scheme is connected through an ideal bias tee to an external circuit, which has an impedance Z_S at frequency f_S.

In [27, 51, 52], it is shown that the expression for the output DC voltage V_{DC} of an ATJ-based signal detector can be written as

$$V_{DC} = \frac{I_S \Delta R}{2} \frac{1-\xi}{(1-\xi)^2+\beta^2} = \left(\frac{\eta}{2+\eta}\right)\sqrt{\frac{P_S R_0}{2}} \frac{1-\xi}{(1-\xi)^2+\beta^2}, \qquad (4.5)$$

where I_S is the amplitude of the input AC current associated with the input AC signal with power P_S, ΔR is the amplitude of oscillations of the junction magnetoresistance, η is the TAMR coefficient, R_0 is the equilibrium ATJ resistance, $\xi = (2\pi f_S)^2 LC$ and $\beta = 2\pi f_S R_0 C$ are the dimensionless parameters characterising the resonant and inertial properties of the junction, respectively.

The electrical characteristics of a junction, R_0, ΔR, L and C, are functions of its electrical and geometrical parameters. In the simplest case, these dependencies have the form [21, 27, 51, 52]:

$$R_0 = \frac{\mathrm{RA}\exp[\kappa d]}{a^2}, \quad \Delta R = \frac{\eta}{2+\eta} R_0, \quad C = \frac{\varepsilon\varepsilon_0 a^2}{d}, \quad L = \mu_0 d, \qquad (4.6)$$

where RA is the resistance-area product characterising the junction fabrication technology, κ is the tunnelling barrier parameter, d in the thickness of the MgO layer, a is the effective lateral size of the junction (its cross-sectional area is a^2), ε is the relative permittivity of MgO, and ε_0 and μ_0 are the vacuum permittivity and permeability, respectively.

Taking into account, Eqs. (5) and (6), the output DC voltage of the detector V_{DC} appears to be essentially dependent on the junction geometry (Fig. 7). The strongest dependence of V_{DC} is observed on the thickness of the tunnel barrier d, which is associated with the exponential dependence of R_0 on d (Fig. 7a). It can be seen that depending on the frequency f_S of an input signal, there is an optimum value of $d \sim 1$ nm. The dependence $V_{DC}(a)$, although monotonic, is weaker (Fig. 7b). It can be seen that beyond a certain value $a = a_0$, further reduction of the lateral junction size a does not really affect the output DC voltage of the detector.

Since the frequency of the Pt/IrMn/MgO/Pt magnetoresistance oscillations can change with the drive current I_{drive}, this opens up the possibility of using such an ATJ as a floating-frequency signal detector or spectrum analyser like the ferromagnetic systems considered in [50]. Our studies in [27] showed that the ATJ-based spectrum analyser with optimised parameters can provide a frequency measurement error of ~5 MHz in the frequency range from 0.5 to 0.75 THz (i.e. the frequency error of the device is about 10^{-4} %).

At the same time, a characteristic feature of an ATJ-based signal detector is the monotonically decreasing dependence of the detector output DC voltage $V_{DC}(f_S)$ on the input signal frequency f_S, as shown in Fig. 8. This dependence shows that such a detector is only effective in the range of sufficiently low subterahertz frequencies. In the frequency range ~ 1 THz, the detector output DC voltage can be below the thermal noise level.

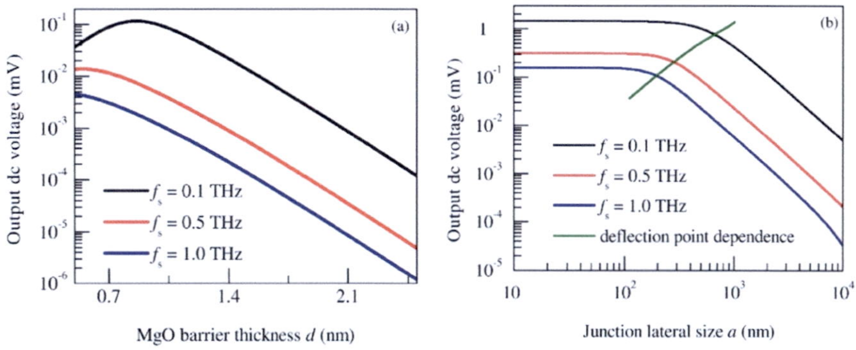

Fig. 7 Output voltage V_{DC} of an ATJ-based signal detector with typical parameters [27] calculated for different input signal frequencies as a function of **a** MgO barrier thickness d and **b** the lateral size of an ATJ a (MgO barrier thickness $d = 1$ nm is used)

Fig. 8 Frequency dependence of the output DC voltage V_{DC} (blue line) and the Johnson–Nyquist thermal noise voltage (red line) calculated for an ATJ with typical parameters at a temperature of 293 K [27]

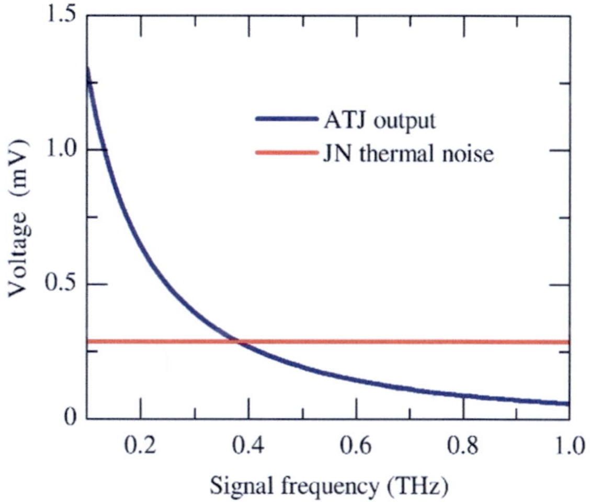

To analyse the effect of temperature T on the noise-handling and detection properties of an ATJ-based signal detector, it was assumed in [51, 52] that some detector parameters are functions of temperature, namely:

$$d(T) = d_0(1 + \alpha T),\ \mathrm{RA}(T) = \mathrm{RA}_0(1 - \chi T),\ \eta(T) = \eta_0\left(1 - \left(\frac{T}{T_\mathrm{N}}\right)^{3/2}\right)^7, \tag{7}$$

where d_0, RA_0 and η_0 are the effective values given at temperature $T = 0$ K, α is the thermal expansion coefficient for MgO, χ is the thermal coefficient of resistance of an ATJ, and T_N is the Neel temperature for IrMn. Equation (7) has been obtained under

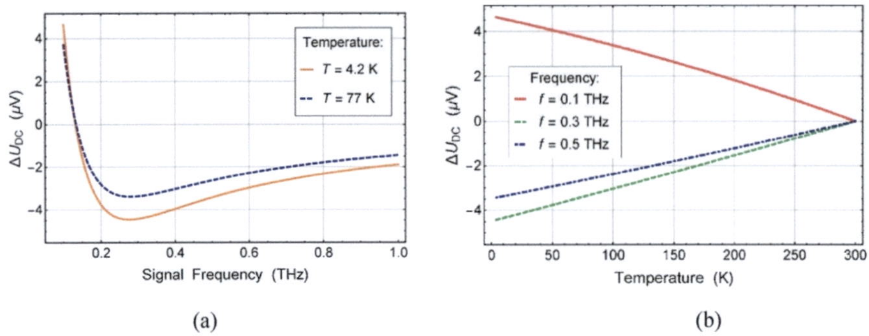

Fig. 9 Deviation of the detector output DC voltage $\Delta U_{\mathrm{DC}}(T) = V_{\mathrm{DC}}(T) - V_{\mathrm{DC}}(300\ \mathrm{K})$ as a function of **a** input signal frequency and **b** temperature [51, 52]

the assumptions that have been proved for ferromagnetic systems [36–38, 53, 54]. Experimental data from papers [44, 55] have been used to derive these expressions.

After substituting (7) into (6), numerical modelling revealed that the output DC voltage V_{DC} of an ATJ-based detector changes slightly with temperature T. It is therefore convenient to introduce a temperature-dependent voltage deviation $\Delta U_{\mathrm{DC}}(T) = V_{\mathrm{DC}}(T) - V_{\mathrm{DC}}(300\ \mathrm{K})$ and study the behaviour of this function. Examples of such calculations are shown in Fig. 9. It can be seen that over a wide range of temperatures the changes $\Delta U_{\mathrm{DC}}(T)$ are quite small, so it may seem that the effect of temperature on the characteristics of an ATJ-based signal detector is not significant. However, two important points must be noted:

(i) Although the value $\Delta U_{\mathrm{DC}}(T)$ itself changes very little with temperature T (the case of weak temperature dependence of the TAMR coefficient η is considered), the signal-to-noise ratio can change very much. Consequently, the noise-handling properties of the detector will improve significantly at cryogenic temperatures;

(ii) As can be seen from Fig. 9, there is a certain frequency at which the deviation of the detector's output DC voltage ceases to be temperature dependent, $\Delta U_{\mathrm{DC}}(T) = 0$. This may be important in the design of thermally-stable detectors and spectrum analysers.

5 Outlook

This chapter describes in detail the state-of-the-art in terahertz signal detection based on antiferromagnetic (AFM) spintronic nanostructures consisting of an AFM layer and one or more heavy metal (HM) layers characterised by a strong spin–orbit interaction. As shown, there are three main types of such detectors based on the bilayer (HM/AFM), three-layer (HM/AFM/HM) and more complex four-layer (HM/AFM/D/HM) nanostructures. All of these detectors are capable of converting an input AC signal at THz into an output DC voltage using different physical mechanisms of energy conversion with fairly good conversion efficiency.

The operation of the AFM signal detector is based on the spin-transfer torque and spin Hall effects, which are used to excite the magnetisation dynamics in the AFM layer of the detector nanostructure. To form a detector's DC output, the spin pumping and inverse spin Hall effect can be used in detectors based on HM/AFM and HM/AFM/HM nanostructures. In the antiferromagnetic tunnel junction HM/AFM/D/HM structures, an output DC signal is formed by the mixing between the input AC signal current traversing the whole structure and its DC current-driven oscillating tunnelling anisotropic magnetoresistance.

Although the recent experiment [45] has demonstrated the possibility of experimental realisation of AFM-based signal detectors, this field is currently at the initial stage of development, and many important results for different types of terahertz signal detectors would only be obtained in the future. Nevertheless, the existing theoretical and numerical models of such devices predict that AFM-based spintronic signal detectors can operate in different modes (resonant and non-resonant detection modes) and have a volt-watt sensitivity $\sim 10^5$ V/W to an input AC signal (introduced for a quadratic detector) comparable to that of the best ferromagnetic spintronic detectors.

In addition, AFM-based detectors can be used to develop frequency detectors using current-driven HM/AFM/HM and HM/AFM/D/HM nanostructures. A similar approach based on the use of a single current-driven AFM-based detector or an array of independent detectors with different detection frequencies could be used for the implementation of spectrum analysers including the thermally stable devices suitable for precise measurements.

The results presented in this chapter open up new possibilities in the physics and potential applications of antiferromagnetic spintronic detectors, which can be promising from both scientific and practical points of view and can represent a new stage in the development of terahertz electronics.

Acknowledgements This work was supported in part by Grants Numbers 22BF07-03 and M/37-2023 from the Ministry of Education and Science of Ukraine, Grant Number 16F-2022 from the National Academy of Sciences of Ukraine, the NATO SPS.MYP G5792 grant, and the STCU grant No. 9918 under the "2023 IEEE Magnetism in Ukraine" initiative.

The authors thank all brave defenders of Ukraine who made finalising this publication possible.

References

1. Jungwirth, T., Marti, X., Wadley, P., Wunderlich, J.: Nat. Nanotechnol. **11**, 231 (2016)
2. Jungfleisch, M.B., Zhang, W., Hoffmann, A.: Phys. Lett. A **382**, 865 (2018)
3. Jungwirth, T., Sinova, J., Manchon, A., Marti, X., Wunderlich, J., Felser, C.: Nat. Phys. **14**, 200 (2018)
4. Baltz, V., Manchon, A., Tsoi, M., Moriyama, T., Ono, T., Tserkovnyak, Y.: Rev. Mod. Phys. **90**, 015005 (2018)

5. Fukami, S., Lorenz, V.O., Gomonay, O.: J. Appl. Phys. **128**, 070401 (2020)
6. Han, J., Cheng, R., Liu, L., Ohno, H., Fukami, S.: Nat. Mater. **22**, 684 (2023)
7. Žutić, I., Fabian, J., Sarma, S.D.: Rev. Mod. Phys. **76**, 323 (2004)
8. Tsymbal, E.Y., Zutic, I., Handbook of Spin Transport and Magnetism: CRC press (2012)
9. Hirohata, A., Yamada, K., Nakatani, Y., Prejbeanu, I.L., Diény, B., Pirro, P., Hillebrands, B.: J. Magn. Magn. Mater. **509**, 166711 (2020)
10. Gomonay, H.V., Loktev, V.M.: Phys. Rev. B **81**, 144427 (2010)
11. Gomonay, E.V., Loktev, V.M.: Low Temp. Phys. **40**, 17 (2014)
12. Marti, X., Fina, I., Frontera, C., Liu, J., Wadley, P., He, Q., Ramesh, R.: Nat. Mater. **13**, 367 (2014)
13. Wadley, P., Howells, B., Železný, J., Andrews, C., Hills, V., Campion, R.P., Jungwirth, T.: Science **351**, 587 (2016)
14. Kosub, T., Kopte, M., Hühne, R., Appel, P., Shields, B., Maletinsky, P., Makarov, D.: Nat. Commun. **8**, 13985 (2017)
15. Olejník, K., Schuler, V., Martí, X., Novák, V., Kašpar, Z., Wadley, P., Jungwirth, T.: Nat. Commun. **8**, 15434 (2017)
16. Olejník, K., Seifert, T., Kašpar, Z., Novák, V., Wadley, P., Campion, R.P., Jungwirth, T.: Sci. Adv. **4**, 3566 (2018)
17. Khymyn, R., Lisenkov, I., Tiberkevich, V., Ivanov, B.A., Slavin, A.: Sci. Rep. **7**, 43705 (2017)
18. Sulymenko, O., Prokopenko, O., Tiberkevich, V., Slavin, A., Ivanov, B., Khymyn, R.: Phys. Rev. Appl. **8**, 064007 (2017)
19. Khymyn, R., Lisenkov, I., Voorheis, J., Sulymenko, O., Prokopenko, O., Tiberkevich, V., Slavin, A.: Sci. Rep. **8**, 15727 (2018)
20. Sulymenko, O., Prokopenko, O., Lisenkov, I., Åkerman, J., Tyberkevych, V., Slavin, A., Khymyn, R.: J. Appl. Phys. **124**, 152115 (2018)
21. Sulymenko, O., Prokopenko, O., Tiberkevich, V., Slavin, A.: IEEE Magn. Lett. **9**, 064007 (2018)
22. Lee, D.K., Park, B.G., Lee, K.J.: Phys. Rev. Appl. **11**, 054048 (2019)
23. Puliafito, V., Khymyn, R., Carpentieri, M., Azzerboni, B., Tiberkevich, V., Slavin, A., Finocchio, G.: Phys. Rev. B **99**, 024405 (2019)
24. Ovcharov, R.V., Galkina, E.G., Ivanov, B.A., Khymyn, R.S.: Phys. Rev. Appl. **18**, 024047 (2022)
25. Wolba, B., Gomonay, O., Kravchuk, V.P.: Phys. Rev. B **104**, 024407 (2021)
26. Slobodianiuk, D.V., Prokopenko, O.V.: J. Appl. Phys. **134**, 153903 (2023)
27. Artemchuk, P.Y., Sulymenko, O.R., Louis, S., Li, J., Khymyn, R.S., Bankowski, E., Prokopenko, O.V.: J. Appl. Phys. **127**, 063905 (2020)
28. Lebrun, R., Ross, A., Bender, S.A., Qaiumzadeh, A., Baldrati, L., Cramer, J., Kläui, M.: Nature **561**, 222 (2018)
29. Das, S., Anveshkumar, N., Dutta, J., Biswas, A. (eds.): Advances in Terahertz Technology and Its Applications. Springer (2021)
30. Gomonay, E., Loktev, V.: Fiz. Nizk. Temp. **41**, 898 (2015)
31. Afanasiev, D., Hortensius, J.R., Ivanov, B.A., Sasani, A., Bousquet, E., Blanter, Y.M., Mikhaylovskiy, R.V., Kimel, A.V., Caviglia, A.D.: Nat. Mater. **20**, 607 (2021)
32. Hortensius, J.R., Afanasiev, D., Matthiesen, M., Leenders, R., Citro, R., Kimel, A.V., Mikhaylovskiy, R.V., Ivanov, B.A., Caviglia, A.D.: Nat. Phys. **17**, 1001 (2021)
33. Matthiesen, M., Hortensius, J.R., Mañas-Valero, S., Kapon, I., Dumcenco, D., Giannini, E., Šiškins, M., Ivanov, B.A., van der Zant, H.S.J., Coronado, E., Kuzmenko, A.B., Afanasiev, D., Caviglia, A.D.: Phys. Rev. Lett. **130**, 076702 (2023)
34. Blank, T.G.H., Grishunin, K.A., Ivanov, B.A., Mashkovich, E.A., Afanasiev, D., Kimel, A.V.: Phys. Rev. Lett. **131**, 096701 (2023)
35. Tulapurkar, A.A., Suzuki, Y., Fukushima, A., Kubota, H., Maehara, H., Tsunekawa, K., Djayaprawira, D.D., Watanabe, N., Yuasa, S.: Nature **438**, 339 (2005)

36. Prokopenko, O.V., Krivorotov, I.N., Meitzler, T.J., Bankowski, E., Tiberkevich, V.S., Slavin, A.N.: Spin-torque microwave detectors. In: Demokritov, S.O., Slavin, A.N. (eds.) Magnonics: From Fundamentals to Applications, Topics in Applied Physics, vol. 125, pp. 143–161. Springer (2013)
37. Prokopenko, O.V., Slavin, A.N.: Low Temp. Phys. **41**, 353 (2015)
38. Artemchuk, PYu., Prokopenko, O.V.: Detection of Microwave and Terahertz-Frequency Signals in Spintronic Nanostructures. In: Kaidatzis, A., Sidorenko, S., Vladymyrskyi, I., Niarchos, D. (eds.) Modern Magnetic and Spintronic Materials: Properties and Applications (NATO Science for Peace and Security Series B: Physics and Biophysics), pp. 1–26. Springer (2020)
39. Khymyn, R., Tiberkevich, V., Slavin, A.: AIP Adv. **7**, 055931 (2017)
40. Sinova, J., Valenzuela, S.O., Wunderlich, J., Back, C., Jungwirth, T.: Rev. Mod. Phys. **87**, 1213 (2015)
41. Safin, A., Puliafito, V., Carpentieri, M., Finocchio, G., Nikitov, S., Stremoukhov, P., Kirilyuk, A., Tyberkevych, V., Slavin, A.: Appl. Phys. Lett. **30**, 222411 (2020)
42. Cheng, R., Xiao, J., Niu, Q., Brataas, A.: Phys. Rev. Lett. **113**, 157601 (2014)
43. Železný, J., Wadley, P., Olejník, K., Hoffmann, A., Ohno, H.: Nat. Phys. **14**, 220 (2018)
44. Park, B.G., Wunderlich, J., Martí, X., Holý, V., et al.: Nat. Mater. **10**, 347 (2011)
45. Funada, S., Ishikawa, Y., Kimata, M., Hayashi, K., Sano, T., Sugi, K., Fujii, Y., Mitsudo, S., Shiota, Y., Ono, T., Moriyama, T.: Phys. Rev. Appl. **19**, L031003 (2023)
46. Safin, A., Puliafito, V., Carpentieri, M., Finocchio, G., Nikitov, S., Stremoukhov, P., Kirilyuk, A., Tyberkevych, V., Slavin, A.: Appl. Phys. Lett. **117**, 222411 (2020)
47. Artemchuk, P.Yu., Zhang, J., Prokopenko, O.V., Bankowski, E.N., Meitzler, T.J., Krivorotov, I.N., Katine, J.A., Tyberkevych, V.S., Slavin, A.N.: IEEE Magn. Lett. **12**, 4502205 (2021)
48. Cheng, X., Boone, C.T., Zhu, J., Krivorotov, I.N.: Phys. Rev. Lett. **105**, 047202 (2010)
49. Zhang, L., Fang, B., Cai, J., Carpentieri, M., Puliafito, V., Garescì, F., Khalili Amiri, P., Finocchio, G., Zeng, Z.: Appl. Phys. Lett. **113**, 102401 (2018)
50. Louis, S., Sulymenko, O., Tiberkevich, V., Li, J., Aloi, D., Prokopenko, O., Krivorotov, I., Bankowski, E., Meitzler, T., Slavin, A.: Appl. Phys. Lett. **113**, 112401 (2018)
51. Prokopenko, V., Prokopenko, O.: 2022 IEEE 12th International Conference "Nanomaterials: Applications & Properties" (NAP-2022), NMM03 (2022). https://doi.org/10.1109/NAP55339.2022.9934516
52. Prokopenko, V., Vlasenko, A., Prokopenko, O.: IEEE 41st International Conference on Electronics and Nanotechnology (ELNANO-2022), vol. 219 (2022). https://doi.org/10.1109/ELNANO54667.2022.9927047
53. Prokopenko, O., Melkov, G., Bankowski, E., Meitzler, T., Tiberkevich, V., Slavin, A.: Appl. Phys. Lett. **99**, 032507 (2011)
54. Prokopenko, O.V., Bankowski, E., Meitzler, T., Tiberkevich, V.S., Slavin, A.N.: IEEE Trans. Magn. **48**, 3807 (2012)
55. Wang, Y.Y., Song, C., Cui, B., Wang, G.Y., Zeng, F., Pan, F.: Phys. Rev. Lett. **109**, 137201 (2012)

Layer-by-Layer Chitosan/PCL Electrospun Membrane Loaded with Copper Nanoparticles as Antibacterial Wound Healing Dressing

Viktoriia Korniienko, Yuliia Varava, Rafal Banasiuk, Valeriia Korniienko, Kateryna Diedkova, Oksana Petricenko, Disha Arora, Anastasiia Denysenko, Roman Moskalenko, and Maksym Pogorielov

Abstract Wound healing is a multifaceted process influenced by numerous factors, including bacterial infections that can lead to complications. Chronic wounds, often burdened by bacterial colonization, require antimicrobial agents to prevent contamination and infection. The prevalence of multi-drug-resistant bacteria complicates wound healing, highlighting the need for new solutions. In our research, we developed electrospun Chitosan/PCL patches embedded with copper nanoparticles (CuNPs), conducting a thorough analysis of their properties and effectiveness. CuNPs have previously demonstrated effectiveness in disrupting bacterial cell walls, affecting signaling pathways vital for bacterial survival and colony expansion. In addition, chitosan also exhibited antibacterial activities and demonstrated the ability to facilitate wound healing. Our findings reveal that CuNPs possess antibacterial activity within specific concentration ranges and are highly biocompatible. These Ch/PCL-CuNP materials significantly inhibit bacterial growth in vitro, showcasing their potential as medical composites and offering a promising alternative in combating antimicrobial resistance.

Keywords Antibacterial components · Copper nanoparticles · Chitosan · Electrospinning · Wound healing

V. Korniienko (✉) · V. Korniienko · K. Diedkova · O. Petricenko · D. Arora · M. Pogorielov
Institute of Atomic Physics and Spectroscopy, University of Latvia, Jelgavas street, 3, Riga 1004, Latvia
e-mail: maksym.pogorielov@lu.lv

Y. Varava · V. Korniienko · K. Diedkova · A. Denysenko · R. Moskalenko
Biomedical Research Centre, Sumy State University, R-Korsakova street, 2, Sumy 40007, Ukraine

R. Banasiuk
Nanowave sp. z o.o., Postepu 14B, 02-676 Warsaw, Poland

A. D. Pogrebnjak et al. (eds.), *Nanocomposite and Nanocrystalline Materials and Coatings*, Advanced Structured Materials 214,
https://doi.org/10.1007/978-981-97-2667-7_5

1 Introduction

Wound healing constitutes a complex, finely regulated physiological process encompassing sequential stages of hemostasis, inflammation, proliferation, and remodeling [1]. Some patients contend with concurrent medical conditions, such as diabetes and immunosuppression, along with exposure to adverse factors, including wound contamination in combat or hospital-acquired infections [2]. These factors can obstruct the normal course of wound healing, fostering the progression to a chronic pathological state.

Recent developments, notably the emergence of drug-resistant strains among common wound bacteria, have heightened the risks associated with prolonged hospitalization, repeated surgeries, and even amputation [3]. In general, multi-drug-resistant bacteria represent a significant global health concern, contributing to a wide spectrum of infections, both in healthcare facilities and the community [4]. These infections often manifest as cases of bacteremia, upper and lower respiratory tract infections, skin and soft-tissue infections, osteomyelitis, and septic arthritis, making Methicillin-resistant *Staphylococcus aureus* (MRSA) or other resistant bacteria a prominent pathogen in the realm of infectious diseases [5].

Chronic wounds, frequently infiltrated and overwhelmed by bacterial colonization, necessitate the use of antimicrobial agents to avert contamination, colonization, and bacterial infection. Commonly, wound management employs antimicrobial and antibiotic treatments to eliminate bacteria at the wound site [6]. However, the application of these treatments to wounds devoid of bacterial biofilms can result in a toxic wound environment, tissue necrosis, and ultimately hindered tissue repair [7]. Although antibiotics can be topically administered along with dressings, this approach faces challenges due to the increasing prevalence of antibiotic-resistant bacteria, rendering it an ineffective option [8].

Consequently, a novel therapeutic approach involving antimicrobial nanomaterials and delivery mechanisms is actively under investigation, as evidenced by a burgeoning volume of scientific literature and patent activity [9]. Within the realm of wound dressings with antimicrobial properties, two categories are notable. Antibiotic-based dressings offer selective, specific antibacterial effects with low toxicity, yet they often prove ineffective against multi-resistant bacterial infections, particularly those found in healthcare settings [10]. In contrast, antiseptic dressings wield broad-spectrum antibacterial capabilities with a lower likelihood of microbial resistance [11]. However, their use can entail dose-dependent cytotoxicity toward tissues and skin cells.

The global antibiotics market is substantial, reaching USD 40.7 billion in 2020, with a projected compound annual growth rate of 4.5% from 2021 to 2028. Key drivers include the rising prevalence of infectious diseases, including antibiotic-resistant infections, and supportive government legislation [12]. Methicillin-resistant Staphylococcus aureus (MRSA) has emerged as a major contributor to persistent human infections, with case fatality rates for Staphylococcus aureus bacteremia remaining alarmingly high, particularly for MRSA [4]. Its ability to form biofilms on various surfaces is a primary factor in MRSA's antibiotic resistance. Given the lack of approved treatments specifically effective against MRSA biofilms, further efforts are imperative to identify more efficient therapeutic approaches.

Nanomedicine has emerged as a promising avenue for addressing infections involving resistant pathogens, with metal nanoparticles exhibiting notable antimicrobial properties [13]. While silver dressings are frequently utilized by healthcare professionals, their effectiveness in significantly reducing bacterial infection remains inconclusive [14]. Moreover, in vitro studies have indicated that silver may have adverse effects on critical wound-healing cells, including keratinocytes and fibroblasts. The consensus regarding the efficacy of silver-releasing dressings remains uncertain, as some studies have suggested an acceleration in wound healing [15]. Nevertheless, given the increasing threat of antibiotic-resistant bacterial strains, it is imperative to prevent infections at wound sites. This underscores the need for continued research and the exploration of more efficient therapeutic approaches.

Copper, with a long history in medicine, is gaining renewed attention for its potential to combat antibiotic resistance [16]. Copper nanoparticles (CuNPs) have demonstrated effectiveness in disrupting bacterial cell walls, affecting signaling pathways vital for bacterial survival and colony expansion [17]. They have proven effective against gram-positive and gram-negative bacteria, including methicillin-resistant staphylococcus, as well as fungal and viral pathogens.

Moreover, copper plays a pivotal role in wound healing and skin regeneration by inducing vascular endothelial growth factor (VEGF) and angiogenesis [19]. Research is ongoing to explore various combinations of copper nanoparticles with other components to create highly effective and safe antibacterial drugs. While research on the clinical use of metal nanoparticles is underway, it is essential to address concerns about their potential toxicity, oxidative stress-inducing properties, and non-biodegradability. Striking a balance in nanoparticle concentration and controlling their size and shape is a widely explored strategy to mitigate these issues.

Exploratory research is being conducted into various amalgamations of copper nanoparticles with other compounds, aiming to develop potent and safe antibacterial medications. For instance, a complex comprising copper and palladium nanoparticles exhibited a robust bactericidal effect against *Pseudomonas aeruginosa* in a mouse model of skin wounds, attributed to the generation of intracellular reactive oxygen species and more favorable DNA fragmentation compared to silver nanoparticles [19].

In a separate investigation, a combination of sodium alginate, deferoxamine, and copper nanoparticles significantly accelerated the healing of diabetic wounds, enhanced angiogenesis, and mitigated long-term inflammation in a diabetic mouse wound model. Cytotoxicity analyses underscored its remarkable biocompatibility [20]. There is also research highlighting the integration of copper nanoparticles into a cellulose acetate framework, with potential applications in various biomedical studies, including wound treatment [21].

Another study illustrated that copper nanoparticles, when incorporated into a complex involving chitosan hydrogel and black phosphorus nanosheets, exhibited substantial potential for enhancing postoperative tumor therapy. Not only did they display antibacterial properties, but they also hindered local regeneration of residual tumor cells [22]. However, it's essential to acknowledge that clinical use of treatments reliant on precious metals remains restricted. One drawback of nanoparticles is their propensity to induce oxidative stress, a critical public health concern with potentially life-threatening consequences.

Moreover, the stability of metal nanoparticles is a notable consideration as they do not biodegrade and may accumulate, posing potential risks to certain organs [23]. Despite these concerns, research into the clinical application of metal nanoparticles is advancing in several countries across Europe and the world. Strategies such as optimizing nanoparticle concentration and controlling their size and shape are widely employed to address potential toxicity concerns [24].

Development of would patches with antibacterial agents, including copper nanoparticles, still remains a challenge in term of active substance release and effectiveness in all stages of wound healing process. Electrospinning technique has risen as a promising tool for development of novel wound care materials [25]. This technology offers several distinct advantages for the development of wound patches. This innovative technique enables the creation of nanofibrous structures, which mimic the extracellular matrix of human tissues. Electrospun fibers have an incredibly high surface area, allowing for enhanced absorption of wound exudate and efficient delivery of therapeutic agents. Also, this technique allows for precise control over fiber composition, size, and alignment, enabling the design of wound patches tailored to specific applications and wound types. Electrospun wound patches can be engineered to provide sustained release of medications, promoting optimal wound healing and reducing the need for frequent dressing changes [26]. It is important that electrospinning can be applied to a wide range of materials, including biodegradable polymers, antimicrobial agents, and growth factors, expanding the versatility of wound patch development. There are a few research that demonstrated the application of electrospinning for the development of copper-loaded wound patches, and we aimed to demonstrate a novel technique of Chitosan/PCL electrospun patches, loaded with copper nanoparticles and provide a comprehensive assessment of their properties and effectiveness.

2 Materials and Methods

2.1 *Copper Nanoparticle Synthesis*

Nanoparticles were synthesized using copper nitrate as a precursor. Water solutions of $Cu(NO_3)_2$ [1M], l-ascorbic acid [1M], $NaHCO_3$ [1M], $NaCO_3$ [1M], TRIS HCl pH = 8.5 [0.5M], polyvinyl pyrrolidone—100 mg/ml, polyvinyl alcohol—50 mg/ml and NaClO [5% w/w] were made beforehand. NaClO was stored in 5 °C, the rest in room temperature. The standard procedure went as follows: To a 250 ml clear

Table 1 Tested combinations for CuNP synthesis, amounts of solutions added in milliliters

Sample	Polyvinyl pyrrolidone	Polyvinyl alcohol	$Cu(NO_3)_2$	L-ascorbic acid	TRIS HCl	NaClO	$NaCO_3$	$NaHCO_3$
CuNP-1	10	8	10	8	0	1	0	2
CuNP-2	8	8	4	8	1	0	0	3
CuNP-3	8	8	4	8	1	0	0	2.5
CuNP-4	8	8	4	8	1	0	2	0

borosilicate glass bottle 200 ml of deionized water was added, next under mechanical string both polyvinyl alcohol and polyvinyl pyrrolidone. After the endcapping agents, the precursor was added followed by l-ascorbic acid. To start the formation of nanoparticles, different concentrations of compounds were added (Table 1). After combining all of the reagents, the solutions were irradiated for 15 min under 5W 420 nm light source.

2.2 Characterization of the CuNPs by Dynamic Light Scattering

DLS measurements of the particles' hydrodynamic size distributions; particles' average hydrodynamic diameter (DH), polydispersity index (PDI), and zeta-potential in the aqueous medium were performed by a Zetasizer Nano ZS instrument (Malvern Instruments Ltd., Malvern, UK) with Malvern Instruments Ltd. Software 7.12. Cu nanoparticles were analyzed with the following specifications: medium, water; refractive index: 1.330; viscosity: 0.8872 cP; temperature, 25 °C; dielectric constant, 78.5. Cu nanoparticles: refractive index of materials: 0.25. Detection angle was 173°, with a wavelength of 633 nm. The data were analyzed using the multimodal number distribution software equipped with the instrument. The measurements were repeated three times in order to check their reproducibility.

2.3 Production of Electrospun Membrane

Polycaprolactone (PCL) with Mn = 80.000 g/mol, Chitosan powder with low molecular weight and Poly(ethylene oxide) (PEO) were obtained from Sigma Aldrich (Saint Louis, MI, USA), Chloroform (purity—99%) was purchased from Chempur (Poland), *N*, *N*-Dimethylformamide (DMF) and Acetic Acid (AA) (purity—99.8–100.5%) were derived from Honeywell (Charlotte, North Carolina, US). To create a Ch/PEO solution, 0.04 g Ch powder and 0.18g PEO were dissolved in 8 mL of 50% of aqueous AA by 24 h stirring with 300 rpm on a magnetic stirrer at room temperature until a uniform solution was achieved. Before electrospinning, the solutions were allowed to rest at room temperature for 30 min to eliminate any trapped air bubbles. To prepare the electrospinning solution, a total of 0.96 g. of PCL was completely dissolved in 8 mL of a solvent mixture consisting of chloroform and DMF, in a ratio of 3:1. The dissolution process involved stirring the mixture at a speed of 300 rpm on a magnetic stirrer for 3 h, at room temperature.

To obtain Ch/PCL, electrospun membrane was used a layer-by-layer electrospinning method, where the PCL nanofiber membrane was taken as a basis. For the production of PCL, electrospun fiber mats utilizing the following parameters: a power of 17 kv, a distance of 170 mm from the tip to the collector, a feeding rate of 1 mL per hour, and a rotation speed of the drum (with a diameter of 7 centimeters) set to 350 rpm.

A Ch/PEO membrane was spun on top of the PCL electrospun scaffold using the following parameters: a feeding rate of 200 μl per hour, a power of 17 kv, a distance of 13 cm from the tip to the collector, a needle inner diameter of 0.8 millimeters and a rotation speed of the drum set to 400 rpm.

CuNPs were added to Ch/PCL membrane by drop-coating: 10 μl with the following concentrations (254, 127, 63.5, 31.8 15.8 and 7.9 μg/ml) were dropped per sample 1 × 1 cm. After procedure, all samples were drying 24 h.

The structure of as-spun membrane was assessed using SEO-SEM Inspect S50-B (SELMI, Sumy, Ukraine) microscope.

2.4 Determination of the Minimum Inhibitory Concentration (MIC) of CuNPs

The antimicrobial efficacy of the copper nanoparticles (CuNPs-1, CuNPs-2, CuNPs-3, CuNPs-4) was determined by the standard methods of Clinical and Laboratory Standards Institute (CLSI) against *Staphylococcus aureus (S. aureus), Enterococcus faecalis (E. faecalis), Staphylococcus epidermidis (S. epidermidis), Pseudomonas aeruginosa (P. aeruginosa), Escherichia coli (E. coli).* Different concentrations of nanoparticles were prepared, and MIC was calculated by microtiter plate dilution method. The MBC was determined by the lowest concentration that kills 99.9% of the initial bacterial strain.

2.5 Determination of the Antimicrobial Action of Ch/PCL-CuNPS

The bacteria used for this experiment were *Escherichia coli* and *Staphylococcus aureus*. The strains were inoculated in 10 mL of Mueller Hinton liquid media broth. The bacteria were grown overnight at 37 °C. 50 μL of a stationary culture of bacteria was inoculated to a final concentration of 1×10^5 CFU/mL for *E. coli* and *S. aureus* and applied in tests. All experiments were performed in triplicate.

Disk Diffusion Method

The antimicrobial activity (contact inhibition) was determined using the disk diffusion method. Then, 50 μL of bacterial broth was spread on a plate of solid Mueller Hinton (MHA) medium. After that, disks of the electrospun membrane (5 mm of diameter, sterilized with UV-light radiation) impregnated with a suspension of the CuNPs (10 μl per disk) were placed in each plate. Different concentrations (254, 127, 63.5, 31.8, 15.8, and 7.9 μg/ml) were tested in order to evaluate the dose-dependent antibacterial effect of the samples. The plates were incubated upside down at 37 °C. The diameter of the inhibition zone was measured after 24 h, including the diameter of the disk. Controls were performed with a non-impregnated membrane ("0" μg/ml). A positive control was initially made with Fox30 Cefoxitin.

Colony-Forming Units

The bacterial viability in the presence of the samples was tested against *E. coli* and *S. aureus* bacterial models at different time points of the experiment. Membrane samples of 0.5 cm^2 size were sterilized with UV-light radiation and placed in a sterile 24-well plastic plate with 2 ml of pre-prepared bacterial suspension. After the exposure time, the samples were serially diluted and cultured on Muller-Hinton agar plates, and final bacterial concentrations (CFU/mL) were determined by counting the colonies. Controls (as-spun non-impregnated samples) were done in parallel to determine the bacterial death over time (2, 4, 6, 8, and 24 h).

2.6 Biocompatibility Assessment

HaCaT cell line was utilized to evaluate the cytotoxicity and biocompatibility of the test samples. The materials were subjected to sterilization by immersing them in 70% ethanol for 1 h. Subsequently, the membranes underwent three 5 min rinses with phosphate-buffered saline (PBS) to remove any residual ethanol. Before the experiment, HaCaT were cultured in 75 cm^2 cell culture flasks under standard conditions, which included a humidified atmosphere with 5% CO_2 at 37 °C, with a medium change every 2–3 days. The culture medium used was Dulbecco's modified Eagle medium/nutrient mixture F-12 (DMEM/F-12) with L-glutamine, supplemented with 100 units of penicillin, 100 units of streptomycin, 2.5 units of amphotericin B per mL, and 10% fetal bovine serum, all of which were sourced from Sigma-Aldrich, USA.

Circular samples with a diameter of 5 mm were positioned on a 96-well plate (Sarstedt), and HaCaT was seeded onto the membranes at a density of 10^4 cells/cm^2, with complete cell culture media added. Wells without membranes, with cells, were employed as a positive control, while wells without cells and membranes served as the negative control. To evaluate cell adhesion at 24 h and cell proliferation on the 3rd and 7th days, the Alamar blue colorimetric assay was applied. Alamar blue solution from Sigma-Aldrich, USA, equivalent to 10% of the medium volume, was introduced into each well, including the negative and positive controls. The plates were incubated at 37 °C for 8 h. The absorbance was measured using a microplate reader Tecan Infinite M200 Pro (Tecan Trading AG, Switzerland) at wavelengths of 570 and 600 nm.

3 Results and Discussion

3.1 Material Characterization

The characteristics of copper nanoparticles were dependent on the synthesis parameters. The hydrodynamic size (Z-average) measured 219.7 ± 8.5 nm for CuNP-1, 1475 ± 28.2 nm for CuNP-2, 1038 ± 19.0 nm for CuNP-3, and 88.93 ± 3.4 nm for CuNP-4. The Z potential also varied among different types of nanoparticles, ranging from −2.46 ± 0.73 for CuNP-1 to −11.4 ± 0.57 for CuNP-4. Several authors declare that the antibacterial properties of copper nanoparticles are influenced by their size, with smaller nanoparticles generally exhibiting greater efficacy in this regard. Smaller nanoparticles have a higher surface area per unit mass. This increased surface area provides more sites for interactions with bacterial cells, facilitating greater contact and disruption of cellular processes as well as due to their ability to release more copper ions into the bacterial environment [27]. Additionally, they can penetrate bacterial cells more easily. They can enter the cells through various mechanisms, disrupting vital processes like DNA replication and damaging cellular structures. Several research already demonstrated that CuNP with size below 100 nm tends to produce higher levels of ROS when they interact with the surrounding environment. It is also important that smaller nanoparticles are less likely to induce resistance in bacteria. The smaller size allows them to interact with bacteria in multiple ways, making it harder for bacteria to develop specific resistance mechanisms [28].

The as-spun PCL membrane consists of randomly oriented fibers with an average diameter of 1.76 ± 0.65 μm, which enables the formation of interconnected pores. PCL is a well-known material with a low degradation rate, making it suitable as mechanical support for wound patches (Fig. 1a). Its high porosity facilitates gas exchange and the absorption of exudate from the wound bed. The chitosan membrane (Fig. 1b) is characterized by a spider-like network of thin fibers with an average diameter of 159.8 ± 25.5 nm and small-diameter pores. Chitosan is recognized for its antibacterial properties against a wide range of microorganisms and its ability to stimulate wound healing [29].

The layer-by-layer Chitosan/PCL membrane, as shown in Fig. 1c, combines the highly porous PCL base with the chitosan network on top. These materials comprise a flexible, highly porous PCL layer with advanced sorption capabilities and a thin chitosan layer that adds antibacterial and wound-regeneration properties. Notably, the drop casting of CuNPs does not alter the morphology of the Chitosan/PCL membrane.

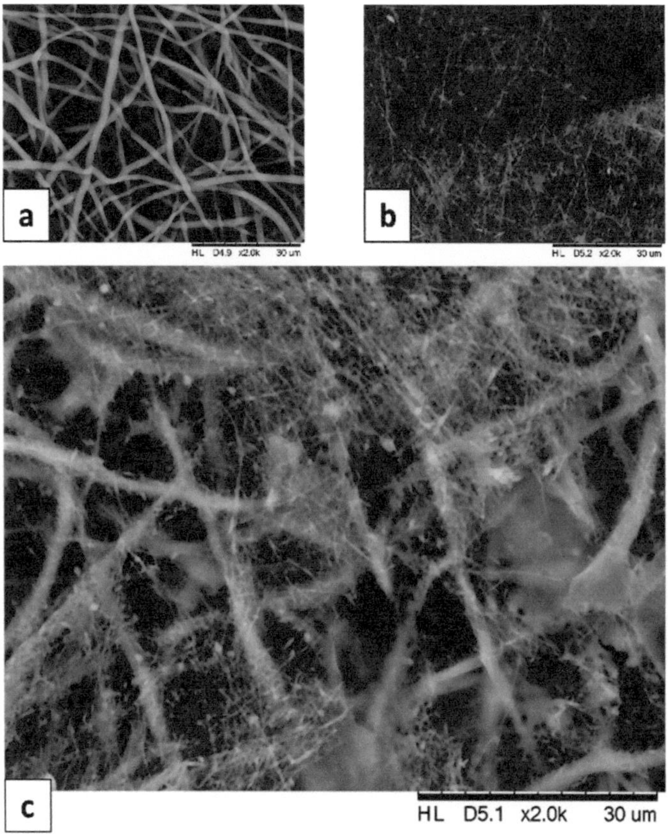

Fig. 1. Scanning electron microscopy of PCL (**a**), chitosan (**b**), and layer-by-layer Ch/PCL (**c**) electrospun membranes (magnification ×2000)

3.2 Minimal Inhibitory Concentration of CuNPs

The MIC of four types of CuNPs against selected bacterial strains was found to be in range 127–254 μg/mL (Table 2) that is consistent with that previously obtained by Clara Ortega-Nieto et al. [30].

Determination of the Antimicrobial Action of Ch/PCL-CuNPs

Several studies have indicated that nanoparticles can display antimicrobial properties in dry conditions, implying that direct contact might serve as a potential mechanism for their antimicrobial effects [31]. In this particular investigation, the antimicrobial potential of Chitosan/PCL nanofibers infused with varying concentrations of CuNPs was examined by culturing *E. coli* and *S. aureus* colonies on MHA plates. The results obtained are depicted in Fig. 2. Unfortunately, no inhibition zones were observed, signifying that the prepared test sample did not exhibit sufficient antibacterial activity. Across all tested concentrations (which corresponded to the minimum inhibitory concentration (MIC) evaluation experiment), no antimicrobial activity was evident.

Cu NPs are recognized for their susceptibility to oxidation in the presence of air and water. Interestingly, both their oxidized intermediate (cuprous oxide Cu_2O) and the resulting product (copper oxide CuO) are acknowledged for exhibiting antimicrobial activities, effectively targeting a variety of pathogens [32]. The in vitro inhibitory effect of tested samples was similar against both Gram-negative than Gram-positive (Fig. 3), and stronger at higher concentration of incorporated nanoparticles as antimicrobial actions of metal-based nanoparticles can be revealed by ROS generation.

The impact of Chitosan/PCL-CuNPs on inhibiting bacterial growth during a 6 h incubation in a bacterial suspension is evident when compared to control Chitosan/PCL samples containing 0 μg/mL of nanoparticles. The experiments did not exhibit significant inhibition in the growth of *S. aureus* and *E. coli* in comparison to the control group. In contrast, samples impregnated with 127–254 μg/mL of CuNPs showed a marked reduction in bacterial growth. CuNP-containing samples demonstrated the ability to inhibit the growth of both Gram-positive and Gram-negative

Table 2. MIC of four types of CuNPs against different bacterial strains

Samples	Bacterial strains				
	S. epidermidis (μg/mL)	*S. epidermidis* (μg/mL)	*E. faecalis* (μg/mL)	*P. aeruginosa* (μg/mL)	*E. coli* (μg/mL)
CuNPs-1	127	127	254	127	127
CuNPs-2	127	254	254	127	127
CuNPs-3	127	254	127	127	127
CuNPs-4	127	127	127	127	127

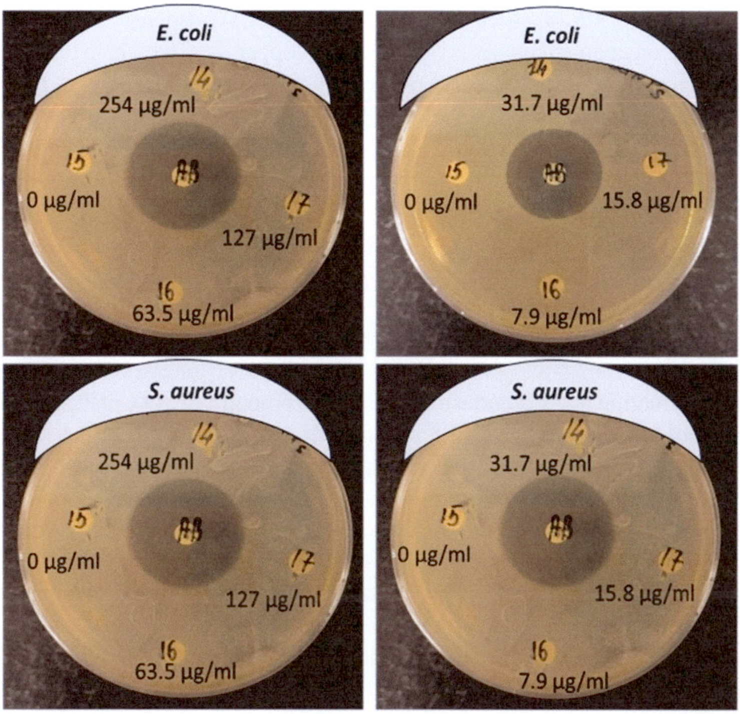

Fig. 2. Evaluation of the antibacterial activity of Ch/PLA-CuNPs membrane (concentration of nanoparticles incorporated to disks is indicated in µg/mL, "0" means Ch/PCL non-impregnated sample)

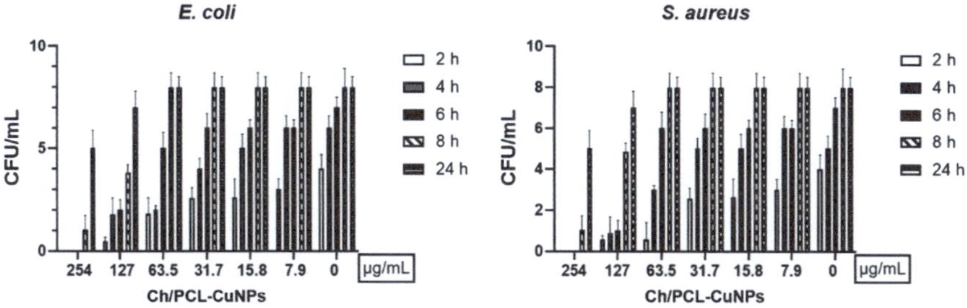

Fig. 3. Ch/PCL-CuNPs nanocomposite time-dependent antimicrobial activity (concentration of nanoparticles incorporated to disks is indicated in µg/mL, "0" means Ch/PCL non-impregnated sample)

bacteria for up to 8 h of co-cultivation in the bacterial suspension. The proposed antimicrobial mechanism involves the release of copper ions, which, in turn, supports the Fenton reaction, ultimately leading to the production of hydroxyl radicals [33]. The dynamics of reactive oxygen species (ROS) production and the subsequent reduction in the concentration of antioxidant compounds unveil a time-dependent mode of action for CuNP-containing samples, resulting in progressive oxidative damage to cellular components.

3.3 Biocompatibility Assessment

Resazurin reduction assay demonstrated high biocompatibility of Ch/PCL membrane with appropriate HaCaT cell proliferation up to day 7 (Fig. 4). At the same time, the addition of CuNP in MIC concentration and above showed low cell survival (below 60%) up to day 5 with reduction of cell number by day 7. Concentration of CuNPs 15.78 μg/mL and below demonstrated an appropriate biocompatibility profile. It should be noted, that infected wound bad represent by mixture of bacteria and immune cells (macrophages, leucocytes, etc.) in exudate that significantly dilute any compounds we used for treatment. Despite the high toxicity of CuNPs in MIC concentrations, their amount will be diluted and do not provide toxic effect to skin cells.

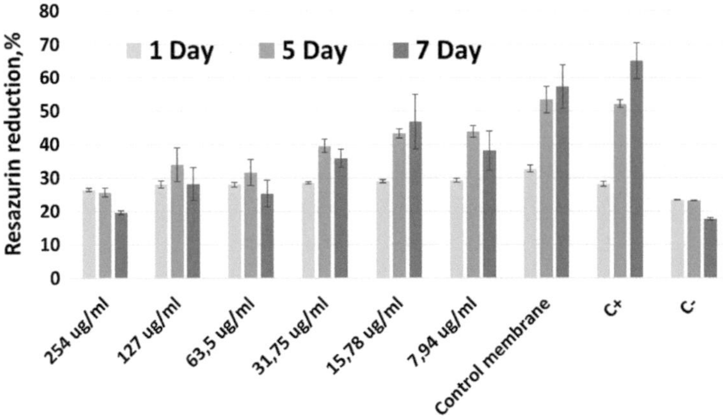

Fig. 4. Data of resazurin reduction assay in different time points after HaCaT cells seeding on Ch/PCL membrane with CuNPs in different concentrations. C+—positive control (cells only), C-—negative control (DMEM media only)

4 Conclusion

Antimicrobial tests revealed that CuNPs had highly bactericidal effect on the Gram-positive and Gram-negative bacteria with the MIC range of 127–254 μg/mL. It seems that the disk diffusion method based on minimum bactericidal concentration results cannot be used as a test for materials impregnated by CuNPs, while colony-forming units (time-dependent kill assay) showed to be useful methods. The result obtained from this study shows that CuNPs-4 has potential bactericidal effects against all tested at a concentration of 127 μg/mL. These nanoparticles can be incorporated in medicaments, as it possesses potent antimicrobial efficacy. Further results confirmed the enhancement of Ch/PCL-CuNP nanomaterials antimicrobial activity due to copper nanoparticle incorporation. In addition, particle size plays an important role in the antimicrobial properties of nanoparticles. The bactericidal impact of copper nanoparticles is contingent on both their size and the concentration over time. These characteristics open avenues for producing nanostructured antimicrobials effective against human pathogens, especially in addressing skin and soft tissue infections. The integration of CuNPs with nanofibrous electrospun membranes presents an opportunity to create novel antimicrobial composites, positioning them as a promising alternative in the fight against antimicrobial resistance.

Acknowledgments This work was supported by the Latvian-Ukrainian Joint Programme of Scientific and Technological Cooperation "Novel copper-loaded wound healing patches with advanced regeneration capacity and antibacterial effect".

References

1. Rodrigues, M., Kosaric, N., Bonham, C.A., Gurtner, G.C.: Cells **11**(19), 2953 (2022)
2. Aitcheson, S.M., Frentiu, F.D., Hurn, S.E., Edwards, K., Murray, R.Z.: Molecules **26**(16), 4917 (2021)
3. White, R.J., Cooper, R., Kingsley, A.: Br. J. Nurs. **10**(9), 563 (2001)
4. Giurazza, R., Mazza, M.C., Andini, R., Sansone, P., Pace, M.C., Durante-Mangoni, E.: Life (Basel) **11**(6), 519 (2021)
5. Satoskar, A.A., Parikh, S.V., Nadasdy, T.: Nat. Rev. Nephrol. **16**(1), 32 (2020)
6. Bowers, S., Franco, E.: Am. Fam. Phys. **101**(3), 159 (2020)
7. Powers, J.G., Higham, C., Broussard, K., Phillips, T.J.: J. Am. Acad. Dermatol. **74**(4), 607 (2016)
8. Haalboom, M.: Curr. Med. Chem. **25**(41), 5772 (2018)
9. Mishra, A., Pradhan, D., Halder, J., Biswasroy, P., Rai, V.K., Dubey, D., Kar, B., Ghosh, G., Rath, G.J.: Inorg. Biochem. **237**, 111938 (2022)
10. Obagi, Z., Damiani, G., Grada, A., Falanga, V.: Surg. Technol. Int. **35**, 50 (2019)
11. Everett, E., Mathioudakis, N.: Ann. N. Y. Acad. Sci. **1411**(1), 153–165 (2018)
12. Parmanik, A., Das, S., Kar, B., Bose, A., Dwivedi, G.R., Pandey, M.M.: Curr. Microbiol. **79**(12), 388 (2022)
13. Mofazzal Jahromi, M.A., Sahandi Zangabad, P., Moosavi Basri, S.M., Sahandi Zangabad, K., Ghamarypour, A., Aref, A.R., Karimi, M., Hamblin, M.R.: Adv. Drug Deliv. Rev. **123**(33) (2018)

14. Oleshko, O., Liubchak, I., Husak, Y., Korniienko, V., Yusupova, A., Oleshko, T., Banasiuk, R., Szkodo, M., Matros-Taranets, I., Kazek-Kęsik, A., Simka, W., Pogorielov, M.: Materials (Basel) **13**(19), 4359 (2020)
15. Lo, S.F., Chang, C.J., Hu, W.Y., Hayter, M., Chang, Y.T.: J. Clin. Nurs. **18**(5), 716 (2009)
16. Rowland, J.L., Niederweis, M.: J. Bacteriol. **195**(16), 3724 (2013)
17. Ingle, A.P., Duran, N., Rai, M.: Appl. Microbiol. Biotechnol. **98**(3), 1001 (2014)
18. Li, S., Xie H., Li, S., Kang, Y.J.: Exp. Biol. Med. (Maywood) **237**(1), 77 (2012)
19. Huang, X., Li, T., Zhang, X., Deng, J., Yin, X.: Mater. Sci. Eng. C **129**, 112392 (2021)
20. Li, S., Wang, X., Chen, J., Guo, J., Yung, M., Wan, G., Yan, C., Li, W., Machens, H.-G., Rinkevich, Y., Yang, X., Song, H., Chen, Z.: Int. J. Biol. Macromol. **202**, 657 (2022)
21. Woźniak-Budych, M.J., Staszak, K., Staszak, M.: Molecules **28**(18), 6687 (2023)
22. Su, Y., Zhang, X., Ren, G., Zhang, Z., Liang, Y., Wu, S., Shen, J.: Chem. Eng. J. **400**, 125949 (2020)
23. Alavi, M., Kamarasu, P., McClements, D.J., Moore, M.D.: Adv. Colloid Interface Sci. **306**, 102726 (2022)
24. Alphandéry, E.: Int. J. Mol. Sci. **21**(12), 4412 (2020)
25. Liu, M., Duan, X.P., Li, Y.M., Yang, D.P., Long, Y.Z.: Mater. Sci. Eng. C Mater. Biol. Appl. **76**, 1413 (2017)
26. Deineka, V., Sulaieva, O., Pernakov, N., Radwan-Pragłowska, J., Janus, L., Korniienko, V., Husak, Y., Yanovska, A., Liubchak, I., Yusupova, A., Piątkowski, M., Zlatska, A., Pogorielov, M.: Mater. Sci. Eng. C Mater. Biol. Appl. **120**, 111740 (2021)
27. Tang, S., Zheng, J.: Adv. Healthc. Mater. **7**(13), e1701503 (2018)
28. Yin, I.X., Zhang, J., Zhao, I.S., Mei, M.L., Li, Q., Chu, C.H.: Int. J. Nanomed. **15**, 2555 (2020)
29. Alven, S., Aderibigbe, B.A.: Int. J. Mol. Sci. **21**(24), 9656 (2020)
30. Ortega-Nieto, C., Losada-Garcia, N., Pessela, B.C., Domingo-Calap, P., Palomo, J.M.: ACS Bio. Med. Chem. Au. **3**(4), 349 (2023)
31. Yang, X., Chung, E., Johnson, I., Ren, G., Cheong, Y.-K.: Appl. Sci **11**(10), 4520 (2021)
32. Sharma, D., Shandilya, P., Saini, N.K., Singh, P., Thakur, V.K., Saini, R.V., Mittal, D., Chandan, G., Saini, V., Saini, A.K.: Mat. Today. Chem. **19**, 100391 (2021)
33. Salah, I., Parkin, I.P., Allan, E.: RSC Adv. **11**, 18179 (2021)

A Comparative Study of Microstructure and Properties of TiZrN/NbN and TiSiN/NbN Nanolaminate Coatings

Olga Maksakova, Vyacheslav Beresnev, Serhii Lytovchenko, and Diana Kaynts

Abstract In this work, we present a comparative analysis of the structure and properties of cathodic-arc multilayer TiZrN/NbN and TiSiN/NbN nanolaminates. It has been found that both nanolaminates are reasonable for tribological applications inasmuch demonstrate improved mechanical properties like super hardness of 36–44 GPa and Young's modulus of 396–435 GPa. The multilayer structure of both nanolaminates reduces the interfacial shear stress of the coating; however, the TiSiN/NbN nanolaminate possesses higher fracture toughness and bonding strength. The complete cohesive-adhesive failure of the TiSiN/NbN coating from the substrate occurs at a load of 47 N. Phase segregation, microstructure evolution, and grain refinement strengthening are the alternative backgrounds for enhanced mechanical and adhesive properties of TiSiN/NbN nanolaminate, while for TiZrN/NbN nanolaminate it is the solid-solution hardening. In summary, the presence of thin amorphous-crystalline TiSiN layers enhances the strength of nanolaminates better than the solid-solution TiZrN crystalline layers.

Keywords Nanolaminates · Nanocomposites · Cathodic-arc deposition · Amorphous-crystalline · Solid-solution · Hardness · Adhesion

O. Maksakova (✉)
Faculty of Materials Science and Technology, Institute of Materials Science, Slovak University of Technology in Bratislava, 25, Jána Bottu Str., 917 24 Trnava, Slovakia
e-mail: maksakova.tereshenko@gmail.com

O. Maksakova · V. Beresnev · S. Lytovchenko
V.N. Karazin Kharkiv National University, 4, Svoboby Sq., Kharkiv 61000, Ukraine

D. Kaynts
Uzhhorod National University, Uzhhorod, Ukraine

A. D. Pogrebnjak et al. (eds.), *Nanocomposite and Nanocrystalline Materials and Coatings*, Advanced Structured Materials 214,
https://doi.org/10.1007/978-981-97-2667-7_6

1 Introduction

Over the past few decades, hard coating materials and technologies have been widely used in the cutting tools industry because they can significantly improve cutting efficiency and extend tool life [1]. Many studies report that the excellent properties of transition metal nitride hard coatings (TMNs) can be obtained by adjusting component materials and process conditions of the coatings [2–4]. Generally speaking, TMN exhibits superior properties when bonded with functional elements that play a crucial role in achieving better properties [5].

For example, TiZrN is one of the promising ternary compounds for hard and protective coatings. Abadias et al. [6] comprehensively studied the structure evolution of $Ti_{1-x}Zr_xN$ films produced by dc magnetron sputtering for a wide compositional range. Their results indicated that all $Ti_{1-x}Zr_xN$ films are single-phase solid solutions with rock-salt crystal structures and stable in the entire compositional range. Generally, TiZrN films possess a higher hardness compared to binary TiN and ZrN films due to solid-solution hardening originating from the difference in atomic size between Ti and Zr atoms [7]. Furthermore, the hardness of the $Ti_{1-x}Zr_xN$ hard coatings is found to be related to the ratio between Ti and Zr, and the maximum hardness is expected to be 1:1 due to the largest inhomogeneous strain [8]. For practical applications, Knotek et al. [9] reported that TiZrN coated inserts greatly improved the wear performance compared to TiN and ZrN coated, and the TiZrN coating offered outstanding protection against abrasive wear loads at elevated temperatures. Purushotham et al. [10] also noted that the wear resistance and adhesion strength of the TiN coating can be enhanced by implanting zirconium to form the $Ti_{1-x}Zr_xN$ phase in the coating. Furthermore, $Ti_{1-x}Zr_xN$ coatings have high thermal stability [11, 12]. The $Ti_{1-x}Zr_xN$ materials still maintain their crystal structure and hardness up to ~30 GPa after annealing in a vacuum up to 1200° C for 2 h. An early work by Knotek and Barimani [13] claimed the occurrence of slightly compositional fluctuations and upward diffusion from the X-ray diffraction (XRD) results of the annealed $Ti_{30}Zr_{70}N$ films. On the contrary, Hoerling et al. [8] reported that no spinodal decomposition was observed on the annealed TiZrN samples with similar composition. They explained that the discrepancies may be due to a limited driving force for transformation in the high-temperature end and the kinetics limit of atomic diffusion in the low-temperature end. The recent study by Tung et al. [12] on the vacuum annealed TiZrN with different compositions also indicated that the spinodal decomposition was not observed even after heat treatment at 1000 °C for 3 h. The results further confirmed the high thermal stability of TiZrN.

Another well-acknowledged ternary system that shows interesting properties for many industrial machining tools, is TiSiN. Adding Si to TiN leads to a hardening of $TiN–Si_3N_4$ nanocomposites of TiN/SiN_x multilayers and metastable solid solutions [14–16]. However, the system does not form a thermodynamically stable ternary phase. The solid solutions of the cubic structure $(Ti_{1-x}Si_x)N$ are the least explored of the TiSiN structures mentioned previously. Outstanding questions are for the occupational sites of Si in the TiN lattice and the corresponding saturation limit.

The mechanisms of phase separation including element segregation and the nature of metastable SiN_x phases forming coherent interfaces to TiN and their eventual transformation to a-Si_3N_4 also need to be clarified.

The expected site for Si in TiN for the solid solutions is the substitution for Ti. This supposition is supported by density functional theory (DFT) calculation tests assuming a nitrogen atmosphere [17]. According to ab initio calculations, solid solutions, where $x = 0.125$, $x = 0.25$, $x = 0.5$, $x = 0.75$, and $x = 1$, exhibit a lattice parameter that deviates less than 0.5% from that calculated for TiN [17, 18]. This makes it difficult to differentiate between phase segregation and solid-solution alloying in the cubic state by diffraction or electron microscopy. For the eventuality of a Si_3N_4 tissue phase forming, which can be amorphous, high-resolution transition electron microscopy is needed. However, the strong segregation tendency of the species in a solid nitride solution will eventually lead to phase separation into TiN and SiN_x. The analysis challenge then becomes one of detecting the smallest volumes of SiN_x in a crystalline matrix as an initial phase-separation product. The phase of SiN_x closest to the equilibrium expected for the deposition conditions in this work is a-Si_3N_4, which has a substantially larger molar volume compared to c-TiN. However, instead, and as is known for the case of TiN–SiN_x nanocomposites and TiN/SiN_x multilayers, a (semi) coherent metastable SiN_x tissue phase with a thickness of a few monolayers can be formed and stabilized by pseudomorphic forces [19]. For larger thicknesses, this tissue phase transforms into a-Si_3N_4.

Furthermore, the research of new hard coatings has focused on the layer and the whole structure design of the coatings. In particular, the introduction of multilayered architectures to transiting metal nitride coatings has attracted extensive interest in improving their mechanical properties and performances. The multilayer structure will act as a crack inhibitor and, therefore, the resistance to coating fractures will increase. This effect can be the result of various mechanisms such as the crack deflection at weak interfaces [20], crack tip shielding by plastic deformation in combination with strong interfaces [21], a favorable distribution of residual stress [22], a crack deflection due to large differences in the cohesive strength between the adjacent layers [23], and differences in the elastic and/or morphological properties between the adjacent layers [24]. In addition, the properties of the surface can be improved by the deposition of layers that have separately different types of effects on the surface, such as protection against corrosion and wear, heat insulation, electrical conductivity, diffusion barrier, and adhesion to the substrate.

In several investigations, it has been stated that multilayer nanolaminates can exhibit better fracture resistance compared to conventional multilayer coatings with a modulation period of $\lambda \leq 100$ nm. The properties of nanolaminates are not determined by the rule of mixtures [25, 26]. For example, ultra-high hardness can be achieved in nanolaminates when their individual layer thickness is less than the slip-plane dislocation length [27, 28] and extremely low thermal conductivity arises when the individual layer thickness of the nanolaminate is less than the mean free path [29]. The presence of high-density interfaces in nanolaminates affects their deformation behavior and strength; they offer superior strength [30, 31], improved ductility and fracture toughness [32, 33], radiation damage resistance [34], excellent wear resistance [31, 35], thermal stability [36], corrosion resistance [37], improved biocompatibility [38], and good optical properties [39].

To date, research efforts on nitride multilayer nanolaminates have started to appear. Wong et al. [40] investigated the microstructure and properties of AlN/ZrN and AlN/TiN nanolaminate coatings with a thickness of approximately 2 μm and the period thickness is 1–160 nm. It was found that with a critical thickness of ~2 nm for the AlN layer, the AlN/TiN nanolaminates exhibit a highly textured [111] oriented superlattice structure and an improvement in film hardness up to 30 GPa. In the case of AlN/ZrN, no highly textured structure, no cubic AlN phase, and no hardness enhancement were observed. The results indicate that the lattice structure and lattice matching in AlN-containing nanolaminates are important for the formation of the metastable cubic AlN phase and the other properties associated with the formation of the ordered microstructure. Kravchenko et al. [41] surveyed the mechanical behavir of nanolaminate $(Al_{50}Ti_{50})N/ZrN$ coatings, where they focused on basic concepts of structural phase analysis of nanolaminate systems (interface and grain morphology, chemical bonding state, residual stresses, etc.). The isostructurality of fcc-(Ti, Al)N and B1-ZrN lattices was found to contribute to the growth of local epitaxial grains in multilayer nanolaminate. The highest mechanical parameters were obtained for the nanolaminate with the smallest modulation period $\lambda \sim 15$ nm. Beltrami et al. [42] reported on magnetron sputtered nanolaminate coatings consisting of alternated CrN coupled with cubic tungsten nitride (β-W_2N) or molybdenum nitride (γ-Mo_2N) with an individual layer thickness of 5–100 nm. Multilayer CrN/Mo_2N showed synergistic improvements in microhardness and corrosion resistance, while CrN/W_2N combined the high microhardness of W_2N and the low corrosion current of CrN.

To the best of our knowledge, no report exists on the development and experimental analysis of TiZrN/NbN and TiSiN/NbN nanolaminates. In an attempt to fill this vacancy, here we report on the comparison of the microstructure and mechanical properties of cathodic-arc deposited TiZrN/NbN and TiSiN/NbN nanolaminates.

2 Material and Methods

2.1 Deposition

The design concept of multilayer nanolaminates in this study is shown schematically in Fig. 1. It can be expected that both nanolaminates could be utilized for tribological applications with improved mechanical properties affected by the nanocomposite structure in the case of TiSiN/NbN and solid-solution hardening in the case of TiZrN/NbN.

After the structure of the materials, filtered cathodic vacuum arc physical vapor deposition (FCVA-PVD) was used to deposit nanolaminates composed of different types of coating material types. The schematic layout of the cathodic-arc deposition system is depicted in Fig. 2. This technology is one of the most widely used due to

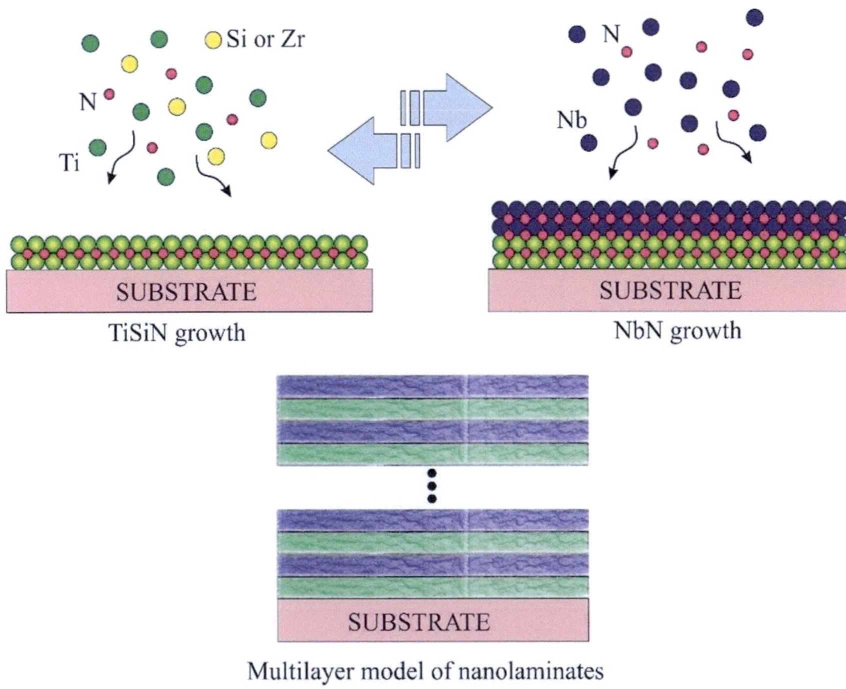

Fig. 1 Schematic diagrams showing the multilayer growth mode of nanolaminates

its high deposition rate with table rotation and high energy of ions. In many cases, the ions generated by the cathodic arc are multiply-charged and have a near-optimal kinetic energy for forming dense and adherent coatings on complex shapes [43, 44]. In addition, this method has the advantages of good adhesion between the coating and substrate, high deposition rate, simple pretreatment, and room temperature operation. The characteristics of this technique are the combined effects that can take place when the pulsed plasma bombards the substrate surface: deposition, implantation, shock pressure, and thermal phenomena. These effects play a dominant role in the surface processing, depending on the operation plasma parameters and the properties of the substrate [45].

In this investigation, nanolaminates were deposited on stainless steel (12X18H9T or an analogue of SUS321 and 321S51) with 10 mm × 10 mm × 3 mm dimensions. They were sanded to 2000 grit and further polished with finer grit diamond paste (the particle size 1 μm) to a mirror-like surface. The Ti and Si elements for TiSiN/NbN were provided by the sintered TiSi target (chemical specification in at. % is $Ti_{85}Si_{15}$) and the solid Nb target (purity specification is 99.8%). The Ti and Zr elements for TiZrN/NbN were provided by the sintered TiZr target (chemical specification in at. % is $Ti_{75}Zr_{25}$) and the Nb target (the purity specification is 99.8%). Nitrogen was supplied by injecting N2 gas with a purity of 99.6% into the deposition chamber. Before deposition, the substrates were ultrasonically cleaned with acetone, alcohol,

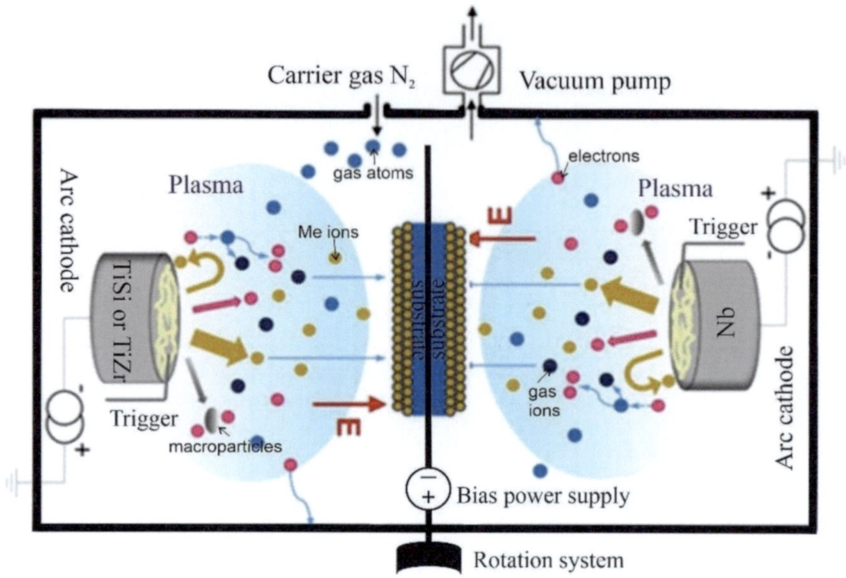

Fig. 2 Scheme of the CAE-PVD system used for the deposition of the nanolaminates

and ultrapure water for 25 min and then fixed to the vertical substrate holder inside the chamber. At first, we evacuated the chamber to less than 4×10^{-3} Pa with the aim of eliminating as much as possible residual oxygen impurity pollution and improving the quality of nanolaminates. Afterward, the synthesis of the multilayer nanolaminates started. The deposition time was 40 min, the working gas pressure was 0.5 Pa, and the negative bias voltage was -200 V. In the process of coating preparation, the cathode target current and the filter system current were set to 95 A and 2.0 A, respectively. The multilayer architecture was achieved due to the continuous rotation of the substrate holder at a speed of 6 rpm. More clearly, the specific deposition conditions and parameters of multilayer nanolaminates are summarized in Table 1.

Table 1 Deposition parameter of multilayer TiZrN/NbN and TiSiN NbN nanolaminates

Parameter	Value
TiSi target, at. %	85/15
TiZr target, at. %	75/25
Nb target, %	99.8
Arc current, A	95
Negative bias, V	-200
Working gas pressure, Pa	0.5
Substrate temperature, °C	300
Deposition time, min	40

Eventually, multilayer TiZrN/NbN and TiSiN/NbN nanolaminates with a total thickness of approximately 2 microns and a total number of layers of 120 were deposited under the same deposition parameters.

2.2 Characterization

The surface morphology was characterized using field emission scanning electron microscopy (FE-SEM) on a FEI Nova NanoSEM 450. The microscope was operated in secondary electron imaging (SEI) mode. The chemical composition was analyzed by X-ray photoelectron spectroscopy (XPSThermo Scientific K-Alpha) with an X-ray source of Al Kα (hv = 1486.6 eV) operating at an acceleration voltage of 12 kV. Before executing the analysis, the specimen was pre-sputtered by 2 keV Ar + ions for 2 min to remove surface contamination. For the fitting of the XPS spectra, XPSPEAK 4.1 was used. The background was subtracted by Shirley's method [46]. The microstructure was analyzed by high-resolution transmission electron microscopy (HRTEM, Tecnai G2 F20). TEM samples were prepared using the standard lift-out method in a focused ion beam (FIB) with the FEI Scios DualBeam. To protect the microstructure of the samples from FIB-induced damage during TEM sample preparation, we placed protective platinum (Pt) layers on the surface of nanolaminates before starting the milling operation. The nanohardness and elastic modulus were measured using a nanoindentation instrument (TTX-NHT2, Anton Paar) with a Berkovich diamond indenter. In the test, in order to avoid the effect of the substrate, the indentation depth was controlled within 10% of the coating thickness. The normal load was 50 mN and ten indentation tests were conducted. The data were averaged to ensure the accuracy of the results. The scratch test was conducted on a Multifunctional friction tester (Bruker UMT-3) to assess the adhesion strength. A Rockwell diamond indenter with a radius of 200 μm was displaced with respect to the nanolaminate surface for a defined distance (12 mm), with constant speed (1 mm/min) and with normal forces that increased linearly from 0.9 to 50 N. In addition to the acoustic emission sensor, the SEM system was carried out to identify the points and to obtain the failure mode based on the surface morphologies.

3 Results

3.1 Microstructural Analysis

The surface morphology of multilayer TiZrN/NbN and the TiSiN/NbN nanolaminates and corresponding macroparticle size statistics are shown in Fig. 3. It can be observed that all nanolaminates prepared by the FCVA technique exhibit a rather smooth and compact microstructure without cracks. Only a handful quantity of microparticles and pits (caused by macroparticles that fell away) appear on the coating surface, which is always unavoidable for the FCVA technique. The unstable cathode arc and the irregular movement of the arc spots are important reasons for the formation of microparticles. Since we applied the special filtration system, the number of microparticles and pits on the surface of prepared nanolaminates decreased significantly compared with the coatings prepared by conventional arc ion plating [47] or the cathodic-arc plasma evaporation technique [48, 49]. It is obvious that the number of microparticles and pits for the TiSiN/NbN nanolaminate is less than for the TiZrN/NbN.

Figure 4 shows an XPS wide spectrum corresponding to the top layers of the multilayer TiZrN/NbN and TiSiN/NbN nanolaminates. Titanium (22.9 at. %), zirconium (20.3 at. %), niobium (14.2 at. %), and nitrogen (42.6 at. %) are identified as elements of TiZrN/NbN nanolaminate, while titanium (27.2), silicon (9.5 at%), niobium (18.0 at. %), and nitrogen (45.3 at. %) are identified as elements of TiSiN/NbN nanolaminate. In addition to the constituent elements, oxygen and carbon are identified, which may be due to the small amount of these elements still in the deposition chamber and their adsorption into the deposited nanolaminates during synthesis.

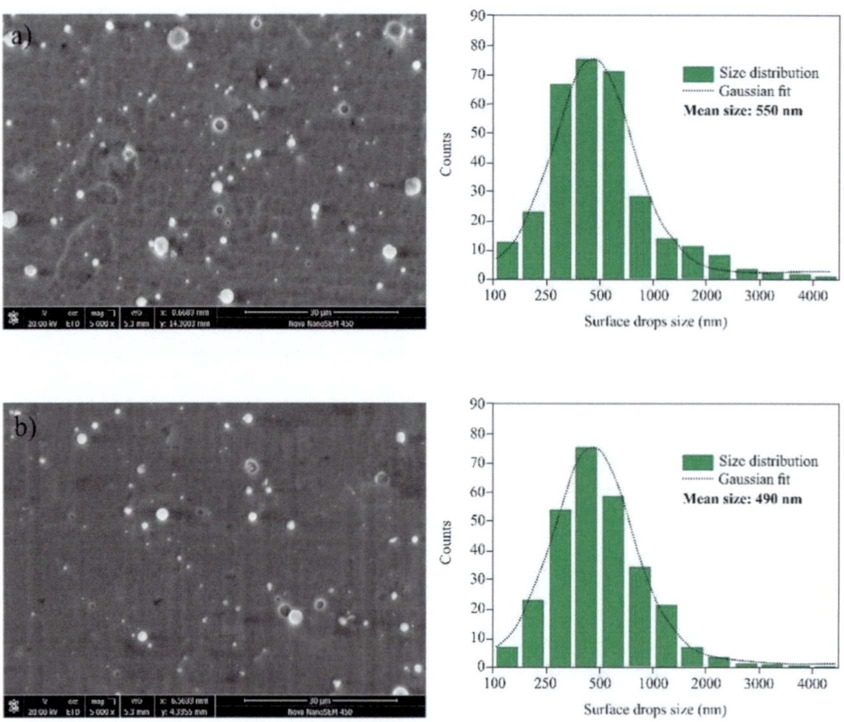

Fig. 3 SEM images of surface morphologies of multilayer TiZrN/NbN (**a**) and TiSiN/NbN (**b**) nanolaminates (the right panel represents the graph of the surface drop size of the corresponding nanolaminate obtained from analysis of multiple SEM images)

The content of the Si element in TiSiN/NbN nanolaminate is really high. Since the energy for Si_3N_4 formation (△G298 = −665.4 kJ/mol) is much lower than that of TiN (△G298 = −308.3 kJ/mol) [50], more and more N atoms react with Si atoms to form a stable Si–N bond at high N_2 pressure. Therefore, the secondary sputtering effect of Si ions on the substrate is weakened, and the Si content in the coating is increased.

TEM analysis was used to further investigate the phase composition and microstructure of the experimental nanolaminates. Figure 5 shows high-resolution TEM (HRTEM) images and corresponding fast Fourier transform (FFT) patterns of TiZrN and NbN layers of TiZrN/NbN nanolaminate. As can be seen in Fig. 5a, a large crystalline grain is formed within the TiZrN layer. Most of its interface has a clear.

border, while the area of the planes' misfit is also evident (white and pale orange lines). This grain is surrounded by some disordered region creating a tissue. The d-value measured from the selected grain in the TiZrN layer is 0.2533 nm and is between

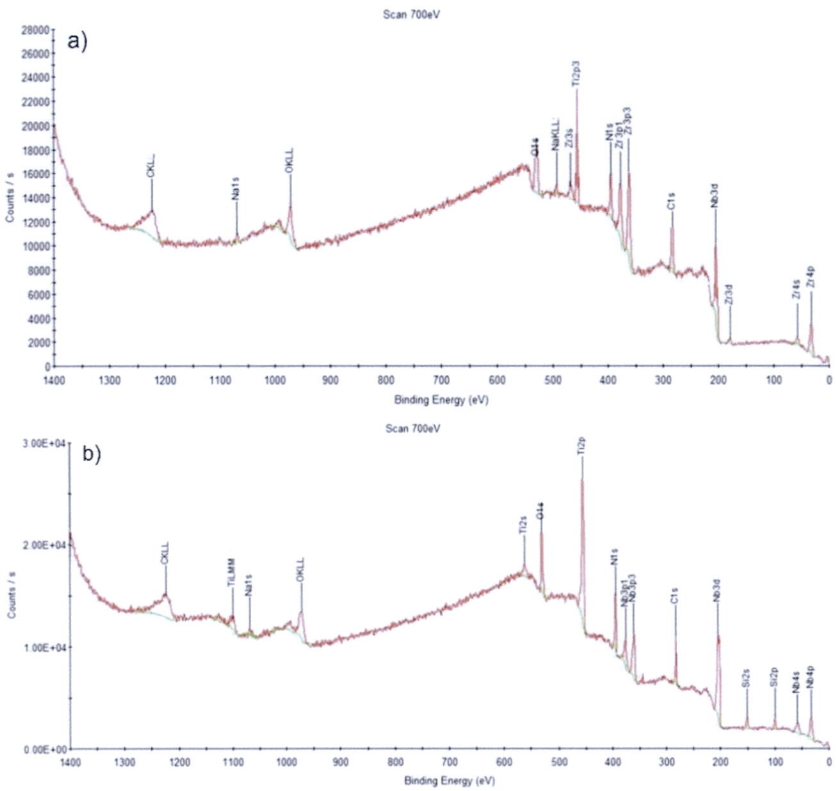

Fig. 4 XPS wide spectrum of the surface of multilayer TiZrN/NbN (**a**) and TiSiN/NbN (**b**) nanolaminates

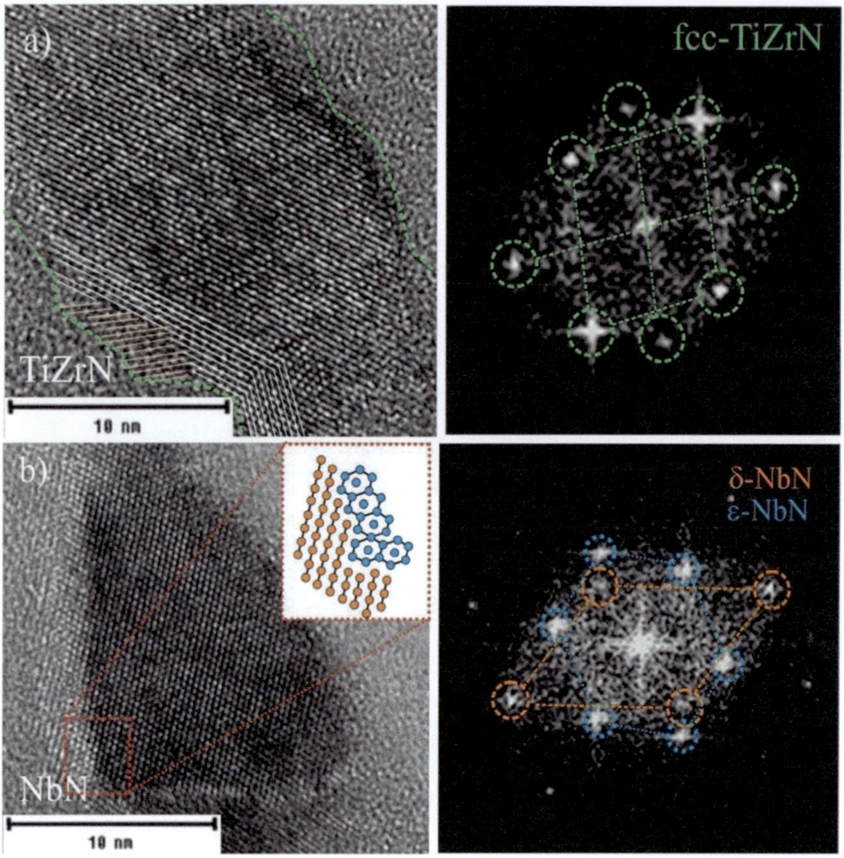

Fig. 5 HRTEM images and corresponding FFT patterns of TiZrN (**a**) and NbN (**b**) layers of multilayer TiZrN/NbN nanolaminate

fcc-TiN (111) (d(111) = 0.2449) and fcc-ZrN (d(111) = 0.2649) that specify the formation of a (Ti,Zr)N solid-solution phase in TiZrN nanolayers. One or more of the following conditions affect the formation of the TiZrN solid-solution phase, namely: (i) a synergistic effect of mutual diffusion and solubility of Zr in the TiN lattice; (ii) stored energy and stress relaxation as a driving force; and (iii) a combination of several driving forces, such as temperature, stored energy, and stress. The FTT pattern in Fig. 5a indicates that the TiZrN layer exhibits an fcc structure with (111) preferred orientation. For a cubic crystal with face centered cubic structure, there are three basal planes, namely, (100), (110), and (111). Planar densities of (110), (100), and (111) are calculated to be about $1.41/a^2$, $2/2a^2$, and $2.31/a^2$ atoms/unit area, where a is the lattice constant. Therefore, for fcc crystals, the (110) plane has the lowest planar density and is the most open basal plane, while (111) has the highest planar density and is the closest basal plane. When the high deposition parameters at which the TiZrN/NbN nanolaminate was synthesized (-200 V and 0.5 Pa) are taken into account, the formation of the plane with the highest planar density is the most favorable. Figure 5b shows the crystalline morphology of the NbN nanolayer. The formed nanograin has a border with numerous plane misfits (marked by color lines) that indicate the different ordered grains and/or coexistence of different crystalline structures. This grain is also surrounded by completely disordered regions. Measured from the selected grain d-spacings of 0.192 nm and while 0.220 nm corresponds to the ε-NbN (10–12) and plane and δ-NbN (200) plane, respectively. These data suggest that the crystal structure of NbN layers is composed of the phases hcp ε-NbN (P63/mmc) and fcc δ-NbN (Fm3m) phases. The FFT diagram taken from the NbN layer (see Fig. 5b) clearly shows a superposition of spots originating from these two crystalline systems.

Figure 6 shows the HRTEM images and the corresponding FFT patterns of TiSiN and NbN layers of multilayer TiSiN/NbN nanolaminate. As can be seen in Fig. 6a, the green areas are TiN nanocrystalline, while the yellow areas are intercrystalline amorphous SiN_x. Meanwhile, the TiSiN layer presents an obvious nanocomposite structure with nanosized crystalline grains encapsulated by the amorphous SiN_x phase coated. In addition, relatively diffused diffraction rings can be observed from the FFT shown in the Fig. 6 insert. These diffraction rings can be referred to as the fcc-TiN (111) and fcc-TiN (200) planes by measuring the characteristic crystal plane spacing of electron diffraction. It is worth noting that the (200) plane exhibits a stronger diffraction intensity compared to the (111) plane, thus proving the formation of the preferred orientation of (200). The formation of preferred orientation in nanoscale coatings has been explained by some models in terms of strain energy, surface energy, and kinetic factors [51, 52]. At the same time, in the FTT pattern, the diffraction peaks of other crystal planes are not identified because of the small grain size and crystallinity of the layer. No clear diffraction spots of SiN_x were obtained, which is attributed to the fact that SiN_x always exists in an amorphous form in the TiSiN layer. As can be seen from Fig. 6b, the NbN layers exhibit a nanocrystalline morphology. One can see the cluster of tiny grains of 1–3 nm in width, which are denoted by blue circles. According to the FFT pattern, the crystal structure of the NbN layers is a mixture of hcp ε-NbN and fcc δ-NbN phases.

From the medium magnification image of TiSiN and NbN layers (not presented here), no columnar structure is observed inside the nanolaminate due to the delay of epitaxial growth provided by the mixed structure [53, 54]. The non-epitaxial growth is quite different from the multilayered TiSiN-based coatings reported in [55]. According to the reported references, the epitaxial growth of multilayered coatings results from the minimization of the interfacial energy during the competition between strain and interfacial energies. In this work, the observed mixed amorphous-crystal structure blocked the columnar structure of TiSiN and NbN. Thus, the NbN layer and TiSiN grains incline to divide into many subgrains to decrease the strain energy and reveal dispersive orientations. This should be responsible for the distinctive amorphous-featured halo observed in the selected area diffraction in the Fig. 6a insert.

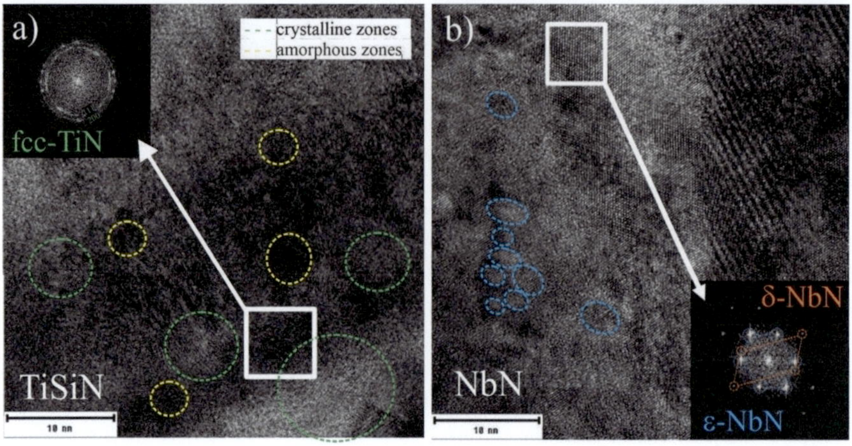

Fig. 6 HRTEM images and corresponding FFT patterns of the TiSiN (**a**) and NbN (**b**) layers of the multilayer TiSiN/NbN nanolaminate

3.2 *Mechanical Properties*

To explore the mechanical properties of nanolaminates, nanoindentation tests were conducted. The rule of thumb of 7–15% of coating thickness has been applied to minimize substrate artifacts [56]. The load–displacement curves of the measured nanolaminates (not presented here) predict the absence of shear band formation, as no 'pop-in' was found during the loading in the load–displacement curves. The hardness (H), elastic modulus (E), measured during nanoindentation, as well as elastic strain to failure (H/E), resistance to plastic deformation (H^3/E^2), and yield strength (σ_{ys}), calculated based on H and E, are summarized in Table 2. The hardness values of the TiZrN/NbN and TiSiN/NbN nanolaminates are 36.6 ± 2.9 and 44.5 ± 3.5 GPa, respectively. The elastic moduli of the TiZrN/NbN and TiSiN/NbN nanolaminates are 396.6 ± 30.2 and 435.2 ± 22.3 GPa, respectively. Both coatings show hardness higher than 30 GPa, which is conducive to tribological properties. The H/E and H^3/E^2 ratios are important parameters that predict the tribological properties. The ratio H/E reflects the elastic strain of the coating, while the ratio H^3/E^2 represents the ability of the coating to resist plastic deformation [57, 58]. The high ratios of H/E and H^3/E^2 are beneficial in improving fracture toughness and inhibiting crack propagation [59]. As shown in Table 2, the H/E and H^3/E^2 ratios of TiZrN/NbN and TiSiN/NbN are fairly high.

Table 2 Mechanical properties of multilayer TiZrN/NbN and TiSiN NbN nanolaminates

Parameter	H, GPa	E, GPa	H/E	H^3/E^2	σ_{ys}, GPa
TiZrN/NbN	36.6 ± 2.9	396.6 ± 30.2	0.092 ± 0.002	0.312 ± 0.028	13.5 ± 0.8
TiSiN/NbN	44.5 ± 3.5	435.2 ± 22.3	0.102 ± 0.002	0.465 ± 0.038	16.5 ± 1.2

Yield strength (σ_{ys}) was estimated from Tabor's slip line field theory [60], and the estimated values of the experimental nanolaminates are also summarized in Table 2. The results suggest that the strengths of TiZrN/NbN and TiSiN/NbN nanolaminates are 13.5 ± 0.8 and 16.5 ± 1.2 GPa, respectively. These incremental trends of H, E, H/E, H^3/E^2, and σ_{ys} for TiSiN/NbN nanolaminate are consistent with the other results on TiSiN-based nanomaterials suggesting a transition in the deformation mechanism [14, 61]. In addition, the competing of two different effects, as the compressed out-of-plane interplanar spacing of TiSi and Nb and the amorphous/disordered TiSiN layers and interfaces, also contribute to the mechanical properties' evolution.

3.3 Adhesive Strength

The cohesive and bonding strength are important factors considered in the study of the mechanical properties of coatings and related to the failure of coatings. Figure 7 shows this scratched surface and acoustic emission signals of the experimental nanolaminates. According to the data collected from their scratch test, there is not a lot of noise involved in the acoustic emission signal, which imposes large certainties in judging the critic load. Thus, the failure position of the coatings through the transition point of the acoustic signal was clearly identified. Besides, optical microscopy has been applied to verify the failure position and reveal the beak morphologies of the scratch surface. From the scratch analysis of TiZrN/NbN in Fig. 7a, the start of the peeling initiates at the load of 12.5 N. The acoustic signal fluctuations become compact and leveled. For TiSiN/NbN coatings, the start of the peeling could be observed at a load of 15.2 N, where acoustic signal fluctuations become growth rapidly and densely. The critical failure load of the scratch refers to the bearing capacity of the scratch under external forces [62]. In the scratch test, the minimum critical loads of the nanolaminates as the active peeling continues were 25 N for TiZrN/NbN and 22 N for TiSiN/NbN. Minimum loads of complete peeling of the nanolaminates were 33 N for TiZrN/NbN and 35.9 N for TiSiN/NbN. The results show that both multilayer nanolaminates have a strong cohesive strength.

To establish the stages of failure, the morphologies of the scratched surfaces were studied. The following stages of destruction of the TiZrN/NbN coating can be distinguished. At a load of less than 5 N, the microdroplet phase plastically deforms in the scratch zone. With increasing load to 8 N the first thin cracks appear. A further increase in the load initiates the propagation of the thin cracks throughout the scratch zone. Their location forms a net of cracks that are parallel to each other. Coming close to the 12.5 N load, the degree of deformation of the material in the scratch increases, its depth increases, cracks open, and the peering of the layers of the coatings starts. The cohesive strength of the nanolaminate corresponds to a load of 25 N. At 33 N, the coating was partially lifted from the substrate. The cohesive-adhesive destruction of the coating from the substrate occurs. The microdroplet phase in the TiZrN/NbN nanolaminate structure has a positive effect on the inhibition of destruction. The microdroplet phase of one of the layers, for example, TiZrN, retains the other top

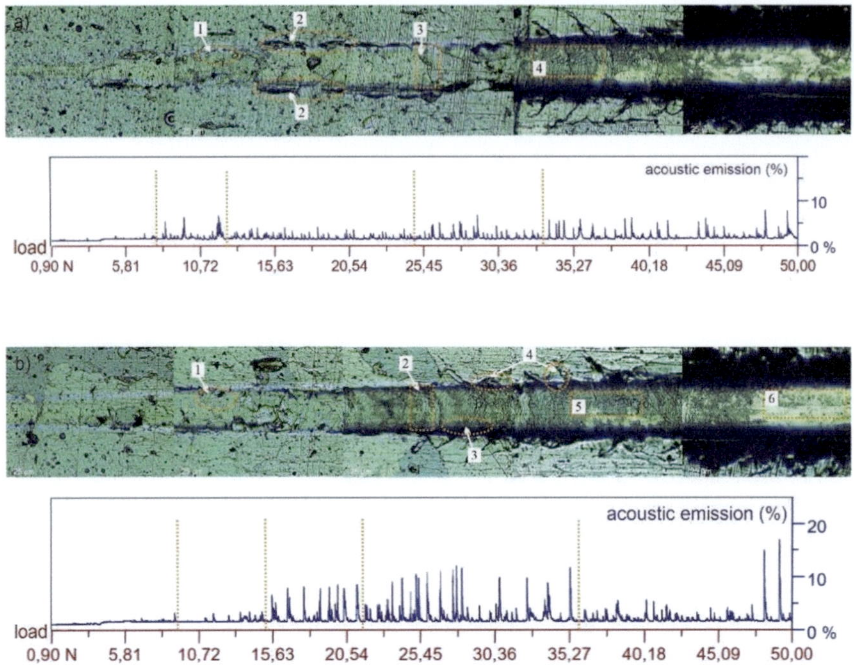

Fig. 7 Surface morphology and acoustic emission signals of scratch tracks from multilayer TiZrN/NbN (**a**) and TiSiN/NbN (**b**) nanolaminates

layer, NbN, of the coating when it is chipped. Also, it slows down the spread of chips at some distance from the scratch. The positive effect of the microdroplet phase on the destruction of combined coatings is consistent with classical ideas about the possibility of breaking cracks on inclusions and particles [63, 64]. The zones marked in Fig. 7a show the defects that were initiated in the scratched track of the TiZrN/NbN nanolaminate. As one can see, zone 1 shows the thin linear cracks; zone 2 demonstrates the coating aggregation at the edges of the scratch; zone 3 depicts the fracture ridges; zone 4 shows the opening of the cracks into the depth.

The nature of damage of the TiSiN/NbN nanolaminate is different. The initial stage of destruction with a load of 9.5 N is the appearance of the first thin cracks in the scratch zone. With a further increase in the load to 11 N, the cracks in the scratch are oriented as counter arcs, forming a circular cluster. Such damage to a scratch is characteristic of an elliptical distribution of stress and strain in an elastic material under normal load [65]. The stress distribution causes circular cracks to appear. As the load increases, their density per unit of scratch length increases. When the load approaches 14 N, the chevron cracks appear, the edges of which extend beyond the scratch boundary. At a load of 15.2 N, the peeling or interfacial spallation becomes evident. An increase in load to 20 N leads to the formation of depressions, apparently due to the release of dislocations in that part of the circular cluster of cracks that are located along the indenter and are formed by internal stresses. Next, agglomeration of the depressions and their merge occurs. The stage of accumulation of depressions due to the release of dislocations corresponds to cohesive peeling at the scratch boundary. Thus, the cohesive strength of the nanolaminate corresponds to a load of 22 N. With a further increase in the load, the peeling into the depth slows down. Under a load of 35.9 N, the cohesive peeling of the coating reaches the substrate, forming foci of coating destruction 10 ÷ 15 μm in size with the bulging of the substrate. Complete cohesive-adhesive failure of the coating from the substrate occurs at a load of 47 N. A distinctive feature of TiSiN/NbN nanolaminate compared to that of TiZrN/NbN is the greater presence of cohesive chips, which are located in different planes and lead to the formation of grooves and steps. The transition of chips into several planes can be explained by a reorientation of the direction of their propagation because of the intersection with nanograin and interphase boundaries. The defects that were initiated in the scratched track of the TiSiN/NbN nanolaminate are shown in Fig. 7b. It is seen that zone 1 demonstrates thin arc cracks; zone 2 shows the net of parallel cracks; zone 3 displays chevron cracks; zone 4 shows cohesive chipping; zone 5 is grooves; zone 6 presents substrate buckling.

4 Discussion

For multilayer TiSiN/NbN nanolaminate, phase segregation, microstructure evolution, and grain refinement strengthening are responsible for the enhanced mechanical and adhesive properties. In particular, the constituent amorphous phase in the TiSiN layer hinders the movement and proliferation of dislocations and cracks at the grain boundary and further strengthens the boundary of the nanocrystalline TiN grains. The existence of the fine nanocrystalline structure in NbN layers also restricts the microcracks in a small area and controls the length of the cracks. In summary, the amorphous phase and a large number of grain boundaries hinder the transgranular propagation of the cracks [61]. These constitute the nanoscale hardening and toughening of the TiSiN/NbN nanolaminate. Furthermore, the strong interfaces between the nanocrystalline NbN layers and the amorphous and nanocrystalline TiSiN layers prevent the sliding of the grain boundary [66]. These constitute the interlayered hardening and toughening of the TiSiN/NbN nanolaminate.

For multilayer TiZrN/NbN nanolaminate, the high mechanical and adhesive parameters are attributed to the solid-solution strengthening and the nanograin strengthening. According to the empirical rule, a solid solution can only be formed if the difference in atomic size of the two elements is <15% [67]. In the case of Zr and Ti, the atomic radii are 0.155 and 0.140 nm, respectively, allowing the formation of a solid solution possible. In addition, nanoscale modulation of nanolaminate enables the formation of a single solid solution with Zr-rich (Ti,Zr)N phase (composition of zirconium in TiZrN layers is 20.3 at. %). Only when the modulation period is low and the activated Zr atoms substitute Ti atoms at the interface, does the stored energy due to stress increase to a critical value and promote the diffusion of the Zr atoms into TiN. This critical energy is the driving force for the formation of solid-solution phase in nanolaminates.

5 Conclusions

In this study, the microstructure, chemical composition, phase state, mechanical properties, and adhesion strength of multilayer TiZrN/NbN and TiSiN/NbN nanolaminates were investigated and comparative analysis was carried out. On the basis of the results obtained, the main conclusions of the study can be summarized as follows:

1. Both multilayer nanolaminates show distinct cell surface morphologies with low droplet components. There are no droplets or defects, such as holes and cracks, that damage the uniformity and continuity of the coatings. The droplet part of TiSiN/NbN nanolaminate is slightly lower compared to that of TiZrN/NbN and is related to the top TiSiN layer.
2. Both multilayer nanolaminates have multiphase structures with nanocrystalline grains. The TiZrN/NbN nanolaminate exhibits the fcc crystallographic structure of solid solution (Ti,Zr)N in TiZrN layers and the mixture of the hcp ε-NbN and fcc δ-NbN phases in the NbN layers. The TiSiN/NbN nanolaminate presents an obvious nanocomposite structure of intercrystalline amorphous SiNx with fcc-TiN nanocrystalline phases in TiSiN layers and the coexistence of hcp ε-NbN and fcc δ-NbN phases in NbN layers. The nanocomposite structure of TiSiN layers has the template effect on NbN layers and results in the formation of zones with ultrafine NbN nanograins that are observed during TEM analysis. This effect is not identified for NbN layers in the TiZrN/NbN nanolaminate.
3. The nanoindentation results indicate that TiSiN/NbN nanolaminate possesses much higher hardness (44.5 ± 3.5 GPa) and Young's modulus (435.2 ± 22.3 GPa) than TiZrN/NbN nanolaminate (36.6 ± 2.9 and 396.6 ± 30.2 GPa, respectively). The other important mechanical values (H/E, H^3/E^2, and σ_{ys}) that predict the tribological performance of the material follow a similar trend.
4. The multilayer structure of both nanolaminates reduces the interfacial shear stress of the coating. However, the TiSiN/NbN nanolaminate possesses higher fracture toughness and bonding strength. The complete cohesive-adhesive failure of the TiSiN/NbN from the steel substrate occurs at a load of 47 N. Phase segregation, microstructure evolution, and grain refinement strengthening correspond to the enhanced mechanical and adhesive properties of TiSiN/NbN nanolaminate.

In summary, the increase of mechanical and adhesive parameters for multilayer TiSiN/NbN nanolaminate is higher than that for TiZrN/NbN suggesting that the presence of thin amorphous-crystalline layers enhances the strength of nanolaminates better than the solid solution with crystalline layers.

Acknowledgements This research is funded by the EU NextGenerationEU through the Recovery and Resilience Plan for Slovakia under project No. 09I03-03-V01-00028. This work is partially supported by the Ukrainian state budget program under No 0124U001127. O. Maksakova and V. Beresnev are grateful for the backing of the Christian Doppler Laboratory for Nanoscale Phase Transformations of the Center for Surface and Nanoanalytics (ZONA) of Johannes Kepler University Linz of Austria.

References

1. Podgornik, B., Hogmark, S., Sandberg, O.: Wear **261**, 15 (2006)
2. Liu, A., Deng, J., Cui, H., Chen, Y., Zhao, J.: Int. J. Refract. Met. Hard. Mater. **31**, 82 (2012)
3. Chawla, V., Jayaganthan, R., Chandra, R.: Surf. Coating. Technol. **204**, 1582 (2010)
4. Hu, Y., Huang, J.-H., Zuo, J.-M.: J. Mater. Res. **31**, 370 (2016)
5. Veprek, S., Reiprich, S.: Thin Solid Films **268**, 64 (1995)
6. Abadias, G., Koutsokeras, L., Siozios, A., Patsalas, P.: Thin Solid Films **538**, 56 (2013)
7. Donohue, L.A., Cawley, J., Brooks, J.S.: Surf. Coat. Technol. **72**, 128 (1995)
8. Hörling, A., Sjölén, J., Willmann, H., Larsson, T., Odén, M., Hultman, L.: Thin Solid Films **516**, 6421 (2008)
9. Knotek, O., Löffler, F., Krämer, G.: Int. J. Refract. Met. Hard Mater. **14**, 195 (1996)
10. Purushotham, K., Ward, L., Brack, N., Pigram, P.J., Evans, P., Noorman, H., Manory, R.R.: Corros. Sci. **50**, 8 (2008)
11. Abadias, G., Koutsokeras, L., Dub, S., Tolmachova, G., Debelle, A., Sauvage, T., Villechaise, P.: J. Vac. Sci. Technol. **28**, 541 (2010)
12. Tung, H.-M., Wu, P.-H., Yu, G.-P., Huang, J.-H.: Vacuum **115**, 12 (2015)
13. Knotek, O., Barimani, A.: Thin Solid Films **174**, 51 (1989)
14. Veprek, S., Veprek-Heijman, M.G.J., Karvankova, P., Prochazka, J.: Thin Solid Films **476**, 1 (2005)
15. Söderberg, H., Molina-Aldereguia, J.M., Hultman, L., Odén, M.: J. Appl. Phys. **97**, 114327 (2005)
16. Flink, A., Larsson, T., Sjölén, J., Karlsson, L., Hultman, L.: Surf. Coat. Technol. **200**, 1535 (2005)
17. Alling, B., Karimi, A., Hultman, L., Abrikosov, I.A.: Appl. Phys. Lett. **92**, 071903 (2008)
18. Söderberg, H., Molina-Aldereguia, J.M., Larsson, T., Hultman, L., Odén, M.: Appl. Phys. Lett. **88**, 191902 (2006)
19. Hultman, L., Bareno, J., Flink, A., Söderberg, H., Larsson, K., Petrova, V., Odén, M., Greene, J.E., Petrov, I.: Phys. Rev. B **75**, 155437 (2007)
20. Chan, K., He, M.Y., Hutchinson, J.W.: Mater. Sci. Eng. A **167**, 57 (1993)
21. Huang, Y., Zhang, H.W.: Acta Metall. Mater. **43**, 1523 (1995)
22. He, M.Y., Heridia, F.E., Wissuchek, D.J., Shaw, M.C., Evans, A.G.: Acta Metall. Mater. **41**, 1223 (1993)
23. Wäppling, D., Gunnars, J., Stahle, P.: Int. J. Fracture. **89**, 223 (1998)
24. Wiklund, U., Hedenqvist, P., Hogmark, S.: Surf. Coat. Technol. **97**, 773 (1997)
25. Gu, C., Wang, F., Huang, P., Xu, K., Lu, T.: Mater. Sci. Eng. A **658**, 381 (2016)
26. Cui, Y., Huang, P., Wang, F., Lu, T., Xu, K.: Thin Solid Films **584**, 270 (2015)
27. Veprek, S.: J. Vac. Sci. Technol. A **17**, 2401 (1999)
28. Yashar, P.C., Sproul, W.D.: Vacuum **55**, 179 (1999)
29. Chen, G.: Phys. Rev. B **57**, 14958 (1998)
30. Bufford, D., Bi, Z., Jia, Q., Wang, H., Zhang, X.: Appl. Phys. Lett. **101**, 223112 (2012)
31. Nasim, M., Li, Y., Wen, M., Wen, C.: Materialia **6**, 100263 (2019)
32. Zhang, J., Zhang, X., Wang, R., Lei, S., Zhang, P., Niu, J., Liu, G., Zhang, G., Sun, J.: Acta Mater. **59**, 7368 (2011)
33. Mara, N.A., Bhattacharyya, D., Dickerson, P.O., Hoagland, R., Misra, A.: Mater. Sci. Forum Trans Tech Publ. **633–634**, 647 (2010)

34. Han, W., Demkowicz, M.J., Mara, N.A., Fu, E., Sinha, S., Rollett, A.D., Wang, Y., Carpenter, J.S., Beyerlein, I.J., Misra, A.: Adv. Mater. **25**, 6975 (2013)
35. Singh, D.R., Chawla, N.: J. Mater. Res. **27**, 278 (2012)
36. Lotfian, S., Mayer, C., Chawla, N., Llorca, J., Misra, A., Baldwin, J., Molina-Aldareguía, J.: Thin Solid Films **571**, 260 (2014)
37. Yogesha, S., Bhat, R.S., Venkatakrishna, K., Pavithra, G., Ullal, Y., Hegde, A.C.: Synth. React. Inorg. Met.-Org. Chem. **41**(1), 65 (2011)
38. Li, W., Kabius, B., Auciello, O.: Annual international conference of the IEEE engineering in medicine and biology, 6237 (2010)
39. Nunes, C., Teixeira, V., Prates, M., Barradas, N., Sequeira, A.: Thin Solid Films **442**(1–2), 173 (2003)
40. Wong, M.-S., Hsiao, G.-Y., Yang, S.-Y.: Surf. Coat. Technol. **133–134**, 160 (2000)
41. Kravchenko, Y.O., Coy, E., Peplińska, B., Iatsunskyi, I., Załęski, K., Kempiǹski, M., Beresnev, V.M., Pshyk, A.V., Pogrebnjak, A.D.: Appl. Surf. Sci. **534**, 147573 (2020)
42. Beltrami, M., Mavrič, A., Zilio, S.D., Fanetti, M., Kapun, G., Lazzarino, M., Sbaizero, O., Čekada, M.: Surf. Coat. Technol. **455**, 129209 (2023)
43. Sanders, D.M., Anders, A.: Surf. Coat. Technol. **78**, 133 (2000)
44. Kravchenko, Y., Maksakova, O., Drodziel, P., Loboda, V.: High Temp. Mater. Process. **20**(1), 85 (2016)
45. Chun, S.Y., Chayahara, A.: Surf. Coat. Technol. **127**, 282 (2000)
46. Shirley, D.A.: Phys. Rev. B **5**, 4709 (1972)
47. Wan, Q., Wu, Z.Y., Liu, Y., Yang, B., Liu, H.D., Ren, F., Wang, P., Xiao, Y.Y., Zhang, J., Zhang, G.D.: Corrosion Sci. **183**, 109264 (2021)
48. Chang, C.L., Lin, C.T., Tsai, P.C., Ho, W.Y., Wang, D.Y.: Thin Solid Films **516**, 5324 (2008)
49. Maksakova, O.V., Zhanyssov, S., Plotnikov, S.V., Konarski, P., Budzynski, P., Pogrebnjak, A.D., Beresnev, V.M., Mazilin, B.O., Makhmudov, N.A., Kupchishin, A.I., Kravchenko, Ya.O.: J. Mater. Sci. **56**, 5067 (2021)
50. Huang, B., Liu, L.T., Du, H.M., Chen, Q., Liang, D.D., Zhang, E.G., Zhou, Q.: Vacuum **204**, 111336 (2022)
51. Greene, J.E., Sundgren, J.-E., Hultman, L., Petrov, I., Bergstrom, D.B.: Appl. Phys. Lett. **67**, 2928 (1995)
52. Pelleg, J., Zevin, L.Z., Lungo, S., Croitoru, N.: Thin Solid Films **197**, 117 (1991)
53. Li, W., Liu, P., Zhu, X., Zhang, K., Ma, F., Liu, X., Chen, X., He, D.: Mater. Sci. Eng. A **610**, 28 (2014)
54. Wo, P.C., Munroe, P.R., Zhou, Z.F., Li, K.Y., Xie, Z.H.: Mater. Sci. Eng. A **527**, 4447 (2010)
55. Xu, Y.X., Chen, L., Liu, Z.Q., Pei, F., Du, Y.: Surf. Coat. Technol. **321**, 180 (2017)
56. Lee, K.W., Chung, Y.W., Chan, C.Y., Bello, I., Lee, S.T., Karimi, A., Patscheider, J., Delplancke-Ogletree, M.P., Yang, D., Boyce, B., Buchheit, T.: Surf. Coat. Technol. **168**(1), 57 (2003)
57. Jirout, M., Musil, J.: Surf. Coating. Technol. **200**, 6792 (2006)
58. Leyland, A., Matthews, A.: Wear **246**, 1 (2000)
59. Ou, Y.X., Chen, H., Li, Z.Y., Lin, J., Pan, W., Lei, M.K.: J. Am. Ceram. Soc. **101**, 5166 (2018)
60. Tabor, D.: J. Colloid Interface Sci. **58**(1), 2 (1977)
61. Veprek, S., Veprek-Heijman, M.G.J.: Surf. Coat. Technol. **201**, 6064 (2007)
62. Zechao, L., Yongxin, W., Xiaoying, C., Zhixiang, Z., Jinlong, L., Xia, L., Liping, W., Qunji, X., Appl, A.C.S.: Mater. Interfaces **10**, 2965 (2018)
63. Pogrebnjak, A.D., Kobzev, A.P., Gritsenko, B.P., Sokolov, S., Bazyl, E., Sviridenko, N.V., Valyaev, A.N., Ivanov, Y.F.: J. Appl. Phys. **87**, 2142 (2000)
64. Xie, Z.H., Hoffman, M., Munroe, P., Bendavid, A., Martin, P.J.: Acta Mater. **56**(4), 852 (2008)
65. Pogrebnjak, A.D., Shablya, V.T., Sviridenko, N.V., Valyaev, A.N., Plotnikov, S.V., Kylyshkanov, M.K.: Surf. Coat. Technol. **111**(1), 46 (1999)
66. Sun, S.Q., Ye, Y.W., Wang, Y.X., Liu, M.Q., Liu, X., Li, J.L., Wang, L.P.: Tribol. Int. **115**, 591 (2017)
67. Oh, D., Kim, H., Chung, C.: J. Korean Acad. Prosthodont. **44**, 740 (2006)

The Development of Nanostructuring Method Metal Surfaces by Electrospark Alloying

O. Haponova, V. Tarelnyk, S. Marchenko, N. Tarelnyk, and I. Konoplianchenko

Abstract The paper analyzes the works dedicated to solving surface nanostructuring methods. A method of obtaining nanostructured coatings by electrospark alloying (ESA) using special technological media (STM) created between the anode and the cathode (detail) and containing nanocrystalline particles in a dosed amount is proposed. The characteristics of the coating structure formation of Armco iron by the Mo electrode used of the STM with carbon nanotubes have been considered. The influence of the ESA regimes and the composition of the STM on the microstructure and hardness of the coatings has been studied. The microstructures after the ESA consist of zones: the "white" layer, the diffusion zone, and the substrate. In the microstructures, nanoscale phases of 40 to 60 nm are detected and are uniformly distributed in the coatings. The addition of Tuball Ocsial nanotubes (0.01 wt%) increases the hardness from 446 to 1438 HV. The coatings are formed with a uniform distribution of molybdenum. Carbon, apparently in the form of carbon nanotubes, is concentrated on the surfaces of the samples being processed. Their introduction into the STM helps to increase hardness and continuity. The addition of nanotubes has a positive effect on the quality of the coatings.

Keywords Electrospark alloying · Coatings · Nanotubes · Special technological media · Chemical composition · Molybdenum

O. Haponova (✉) · S. Marchenko
Sumy State University, Sumy, Ukraine
e-mail: gaponova@pmtkm.sumdu.edu.ua

O. Haponova
Institute of Fundamental Technological Research Polish Academy of Sciences Warsaw, Warsaw, Poland

V. Tarelnyk (✉) · N. Tarelnyk · I. Konoplianchenko
Sumy National Agrarian University, Sumy, Ukraine
e-mail: tarelnyk@ukr.net

A. D. Pogrebnjak et al. (eds.), *Nanocomposite and Nanocrystalline Materials and Coatings*, Advanced Structured Materials 214,
https://doi.org/10.1007/978-981-97-2667-7_7

1 Introduction

In the last years, to improve the quality of surface layers of machine parts, the increasing essentiality is obtained by using the methods of surface treatment with concentrated flows of energy and matter, one of which is the method of electrospark alloying (ESA)—the process of transferring material to the surface of the product by a spark electric discharge [1]. Its specific features, which attract technologists, are [2, 3] the locality of the action, low energy consumption, no volumetric heating of the material, strong connection of the applied material to the base, ease of automation, and possibility of combining operations. Using different electrode materials and the environment, the ESA method can be implemented to carry out the processes that are alternative to the CTT processes, but with significantly lower expenses.

However, challenges for technologists are the development of technologies for obtaining unique coatings having such high-quality properties as increased wear resistance, durability, corrosion resistance, etc. An alternative to the above is a process for nanostructuring a surface by various methods. The technologies of surface micro- and nanostructuring of materials are important for various fields of science and technology. The obtained nanostructures have unique properties. The creation of special nanometer-scale structures on the surfaces of solid bodies results in the improvement of both the physical and mechanical properties of the material. Therefore, there is a need in developing new processes for surface strengthening, which are environmentally friendly and energy efficient, namely, the processes based on the electrospark alloying (ESA) methods. The use of such processes for surface nanostructuring would increase the efficiency of highly responsive equipment.

2 Analysis of Major Achievements and Publications

Successfully obtaining a nanocoating is the basis for its application. Traditional technologies for obtaining nanostructured coatings are chemical vapor deposition (CVD) and physical vapor deposition (PVD). They regulate the thicknesses and grain sizes of the coatings with the help of the technological parameters. On the other hand, such methods as the sol–gel method, laser deposition, thermal spraying, and others have been developed [4–7]. These methods have been widely used and sufficiently described in the literature. It should be noted that the traditional processes for obtaining nanocoatings require special expensive equipment, and they do not always meet the requirements. Therefore, there is a need in developing new methods for surface strengthening, which should be environmentally friendly and energy efficient, and provided with the possibility of application at machine-building enterprises.

Nowadays, the methods of surface nanostructuring based on the use of concentrated flows of energy and matter have been developed. One of those is the method of electrospark alloying (ESA). The ESA technology is used in various industries due to the high adhesion of the formed coatings, the possibility of local surface treatment of the responsible parts of assemblies and mechanisms, including large-sized parts, the relative simplicity of the alloying process, and the absence of strict requirements for surface preparation before the use, and high reliability of the equipment. The technological process of coating is environmentally friendly and characterized by low energy consumption, high profitability, and quick payback of investments [8, 9].

The analysis of the literature on the issue of improving the physical and mechanical properties of working surfaces under the effect of the creation of nanostructured electrospark coatings has shown that the ESA method is promising. The papers of [10, 11] showed that in some ESA modes, it was possible to form a nanostructured layer on a surface. The authors identified possible mechanisms for the formation of nanostructures in ESA coatings. At the same time, regardless of the electrode materials, there appeared a modified layer with a subfine polycrystalline structure on the surface being treated. Thus, in [12], it was shown that after the ESA process with bronze, the "white layer" of the coating on 41Cr4 steel consisted of grains of 10 to 60 nm in size.

In [13], having based on the obtained qualitative characteristics, the author established that the ESA method allowed controlling the dispersion of the crystalline structure of coatings. Moreover, the main principles of the application of the ESA process, which contributed to the formation of coatings with nanostructures and ultra-fine grain structures, were there formulated. Those occurred as follows: the use of electrical modes with low energy and pulse duration, an increase in specific duration of processing, the use of cathode materials with low thermal conductivity, and that of the anode materials with low heat capacity, and the transferring of the anode elements to the cathode in the solid state.

The studies showed that, in addition to grinding the fine structure blocks and increasing the density of dislocations due to the application of the ESA method, the process of nanostructuring the coatings can occur due to the use of the anode materials made of nanocomponents. Therefore, the results of the studies indicate that the ESA method really makes it possible to obtain nanostructured layers; however, the technological features for obtaining such coatings have not been presented in the literature, the choice of the mode parameters of the ESA process, as well as of the anode and cathode materials, and others have not been justified. Therefore, the creation of the technological foundations for the development of the processes for surface nanostructuring by the ESA method is an urgent task.

In addition, most of the new technologies based on the ESA method are application-oriented on the use of either powder materials with finely dispersed particles, or alloys in the form of composite materials containing the main matrix and various inclusions of nanometer-sized particles therein [14–18]. Taking into account the works [17–19], devoted to the study of the structure and properties of the ESA coatings, it should be noticed that the improvement of the physical and mechanical properties, and in particular, the wear resistance thereof, are facilitated by the use of the anode electrodes with inclusions of components having nanostructural structures. At the same time, the amount of the nanoadditives may not exceed 1 to 5% of the volume of the electrode.

To create an alloying electrode, which contains nanoparticles, the SHS method (self-propagating high-temperature synthesis) is widely used. However, obtaining such electrodes is technologically complex and economically expensive. Based on the foregoing, it should be noted that it is necessary to study the possibility of obtaining nanostructured coatings by the ESA method in a simpler way, namely, using special technological media created between the anode and cathode (part) and including nanocrystalline particles of a dosed amount.

The authors have shown that the introduction of the special processing media (SPM) into the inter-electrode space in the course of the ESA process makes it possible to obtain the coatings from non-conductive materials such as sulfur [20], boron [21], and nitrogen [22] and create the complex coatings [23]. Moreover, the use of the SPM provides an opportunity for significantly improving the quality indicators of the surfaces being treated, for combining such properties as running-in ability, corrosion resistance, heat resistance, wear resistance, etc., and also for increasing the reliability and durability of responsible machine parts.

To create the SPM containing nanoparticles, polymers are widely used as binding agents. In a polymer material containing nanofillers, the most important issue is the dispersion of the nanofillers in the polymer bulk. In practice, various methods for dispersing nanofillers in polymer bulk are used [24]. The most common technologies are ultrasonic dispersion (UD), high-speed mixing, and calendering with the use of a three-roll mixer. Carbon nanotubes are the most widely used non-metallic matrix nanofillers. There is a sufficient number of works devoted to the development of technology for the production and study of the structures of carbon nanomaterials.

The most widely used method for dispersing carbon nanotubes is the ultrasonic dispersion method, which allows uniformly distributing the carbon nanofillers both in various solvents and in low-viscosity epoxy oligomers, as well as in some polymers. The authors of work [25] showed that the process of mixing components by ultrasonic waves made it possible to distribute nanofillers sufficiently evenly in epoxy resin. The use of this method of the distribution of the nanotubes avoids the occurrence of defects associated with the violation of the integrity of their structure.

The analysis of the literary sources has shown that one of the most promising methods for improving the properties of the product surface layers is the ESA method. It has been also noted that when using the SPM, the metal surfaces of the parts can be alloyed by both the conductive materials and the dielectrics, which include, for example, the materials consisting of a combined matrix (epoxy resin with nanofiller (two-dimensional boron nitride and zero-dimensional silica)) [26].

Thus, the purpose of the work is to improve the quality of the surface layers of the responsible parts, which limit the life of the products. In this regard, the work is aimed at studying the peculiarities of the structure formation of the coatings obtained by the method of electrospark alloying the steel surfaces with the use of a special technological environment containing nanotubes.

3 Research Methods

For conducting the study, the samples were made of Armko iron. Such a material had been chosen, because when studying the coatings formed on technically pure iron, the factors of the alloying element influence on the structure formation during the ESA process could be excluded.

In forming the coatings by the ESA method, there were used the molybdenum electrodes in the form of a wire. Table 1 presents the SPM compositions being investigated.

To disperse the carbon nanotubes in the matrix, the ultrasonic dispersion method was used in compliance with the technology presented in [25]. The composition of the SPM had been chosen empirically. Such a situation was conditioned by the need to obtain a homogeneous distribution of nanoparticles in the non-metallic matrix.

The Armco iron samples were treated in three stages: at the first stage, the surface was alloyed with a molybdenum electrode at the discharge energy of 0.13 or 0.52 J. At the second stage, the SPM containing nanoparticles was applied onto the surface and, without waiting for drying thereof, the treatment was repeated using the ESA method by a molybdenum electrode at the same discharge energy as at the first stage.

To carry out the electrospark alloying process, the ESA unit of the Elitron-22A model was applied. For the research, the discharge energies of $Wp = 0.13$ and 0.52 J were used. For comparison, the samples made of Armco iron without the SPM were studied after alloying by a molybdenum electrode at the discharge energy of 0.13 or 0.52 J.

After completing the ESA process, the surface roughness was determined using the profilograph-profilometer device, mod. 201, of the Kalibr plant production. The measurement results were displayed through a special device onto the computer monitor in the form of graphs.

To identify the structural components by the metallographic methods, the sections in different cross-sections (transverse, longitudinal, oblique), mainly relative to the processing surface, were made from the samples after the ESA process. In order to

Table 1 SPM composition and alloying modes for obtaining functional coatings by the ESA method

Stages of applying the coatings	SPM composition	ESA modes, Wp, J
Stage I—ESA by Mo, Stage II—applying the SPM, Stage III—ESA by Mo	Multi-walled ARKEMA carbon nanotubes (0.2% by weight) in the epoxy resin of Epoxy 510 type without curing agent	0.13
		0.52
	Single-walled Tuball Ocsial carbon nanotubes (0.01% by weight) in the epoxy resin of Epoxy 510 type without curing agent	0.13
	Single-walled Tuball Ocsial carbon nanotubes (0.6% by weight) in polycarbonate	0.13

increase the optical contrast between different structural components and provide selective coloring, the process of chemical etching of the sections was carried out using special reagents. Manufacturing the thin sections, preparing the surfaces, and carrying out the metallographic studies were performed according to the known methods [27]. A 3% to 5% solution of nitric acid in ethyl alcohol was used as a reagent for the procedure of chemical etching of Armco iron.

The study of the microstructural parameters of the coatings was carried out on the microsections with the use of the scanning electron microscope of the SEO-SEM Inspect S50-B model equipped with the energy dispersive spectrometer of the AZtecOne type with the detector of X-MaxN20 model (manufactured by Oxford Instruments plc).

To analyze the Vickers hardness distribution in the surface layer and by the depth of the thin section from the surface, the NOVA 330/360 hardness tester (the manufacturer: INNOVATEST Europe BV, the Netherlands) was used in accordance with ISO 6507.

4 Research Results

Figure 1 represents the topographies of the surfaces of the Armco iron samples after the ESA processes at different alloying modes according to the Mo- SPM -Mo scheme. The SPM comprises 0.2% of the multi-walled ARKEMA carbon nanotubes. While analyzing the topography of the surfaces of the samples being studied, it can be concluded that with an increase in the discharge energy, the formation of a uniform and continuous coating is noted, however, with an increased roughness (Table 2). Figure 2 represents the surface roughness profilograms of the samples after the ESA process. During the ESA process with $Wp = 0.52$ J, the nano-sized inclusions have been formed in the coatings. Those are evenly distributed in the coating surface layers (Fig. 1d).

There is a need to consider the peculiarities of the coating structure formation in the course of processing Armco iron by the ESA method with a molybdenum electrode and using the SPM composed of the multi-walled ARKEMA carbon nanotubes (0.2%) in the epoxy resin of Epoxy 510 type without a curing agent. The ESA process was carried out at the discharge energies of 0.13 and 0.52 J. Figure 3 shows the microstructures of the coatings and the Vickers hardness distribution.

The microstructure after the ESA process consists of three zones, namely, the upper "white" layer, which is not etched in the reagent, the diffusion zone, and the base with a ferrite structure corresponding to Armco iron. While processing Armco iron, with an increase in the discharge energy, the thickness and continuity of the coating increase. In addition, there is an increase in hardness from 446 to 1300 HV. It is necessary to note the effect of the ARKEMA nanotubes on the hardness characteristics. At the ESA process with $Wp = 0.13$ J, the coatings without the SPM have a hardness of 446 HV, and those with the ARKEMA nanotubes (0.2%) of 608 HV. There is also a slight increase in the thickness of the "white" layer and the

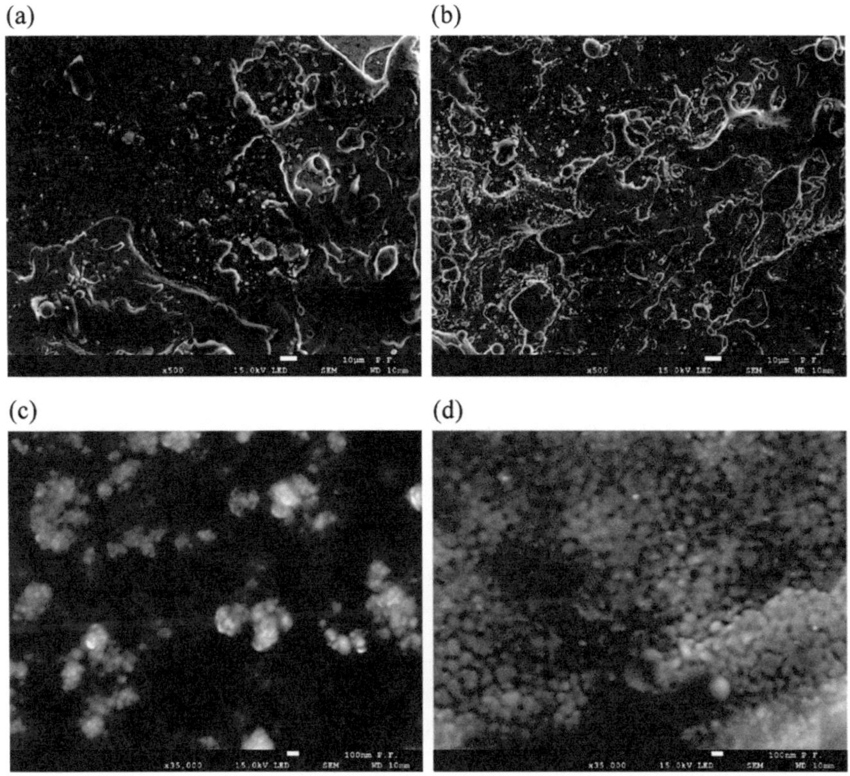

Fig. 1 The topographies of the surface areas of the Armco iron samples after the ESA process by molybdenum and the SPM components (multi-walled ARKEMA carbon nanotubes (0.2% by weight) in the epoxy resin of Epoxy 510 type without curing agent) at different magnifications: **a**, **c**—after the ESA process by the Mo- SPM—Mo at Wp = 0.13 J; **b**, **d**—after the ESA process by the Mo- SPM -Mo at Wp = 0.52 J

continuity of the coating due to the use of the SPM (Table 2). Obviously, the positive effect of the nanotubes on the quality parameters and the properties of the coatings is related to the formation of the nanostructures (Figs. 4 and 5). In the microstructures, the nanoscale phases of 40 to 60 nm size are found, and they are evenly distributed in the coatings.

Figures 6 and 7 present the maps of the distributions of the elements in the electrospark coatings obtained at Wp = 0.13 J and Wp = 0.52 J with the use of the SPM composed of the multi-walled ARKEMA carbon nanotubes (0.2%) in the epoxy resin of Epoxy 510 type without a curing agent. Because of the ESA process, the coatings with a uniform distribution of molybdenum are formed. Generally, during the ESA process, carbon, apparently in the form of the carbon nanotubes, has been concentrated on the surfaces of the samples being processed regardless of the discharge energy values.

Table 2 Summary data of the parameters for the obtained coatings

SPM composition	Wp, J	The thickness of the "white" layer, μm	Micro hardness of "white" layer, HV	Continuity of "white" layer,%	Surface roughness, Ra, μm
Without SPM	0.13	20–30	446	50	0.98
Multi-walled ARKEMA carbon nanotubes (0.2% by weight) in the epoxy resin of Epoxy 510 type without curing agent	0.13	30–40	608	50	1.26
	0.52	30–40	1300	70	2.67
Single-walled Tuball Ocsial carbon nanotubes (0.01% by weight) in the epoxy resin of Epoxy 510 type without curing agent	0.13	30–40	1438	80	1.72
Single-walled Tuball Ocsial carbon nanotubes (0.6% by weight) in polycarbonate	0.13	40–50	949	40	1.85

There is a need to consider the peculiarities of the coating structure formation in the course of processing Armco iron by the ESA method with a molybdenum electrode and using the SPM composed of the following:

1. The single-walled Tuball Ocsial carbon nanotubes (0.01% by weight) in the epoxy resin of Epoxy 510 type without a curing agent.
2. The single-walled Tuball Ocsial carbon nanotubes (0.6% by weight) in polycarbonate.

The ESA process was carried out at the discharge energies of 0.13 J. Like the classic electrospark coatings, the above ones each also consists of a "white" layer, a diffusion zone, and a ferrite structure of Armco iron (Fig. 8). The thickness and hardness values of the obtained coatings are affected only by the compositions of the SPMs. The discharge energy values and the initial electrode materials did not change for this study. It should be noted that the availability of the nanotubes in the course of the ESA process has a positive effect on the quality parameters of the coatings obtained. Thus, the content of the Tuball Ocsial nanotubes for 0.01% in the SPM contributes to increasing the continuity, thickness, and hardness from 446 to 1438 HV, i.e. 3 times. A further increase in the content of the nanoparticles up to 0.6% has no effect in ensuring the quality indicators of the coatings. In this case, sufficient continuity is not provided (Table 2).

The analysis of the microstructure of the coating, which had been obtained applying the ESA method by the molybdenum electrode without the use of the SPM, has shown that the dendritic structure has been being formed in the course and due to carrying out the process of accelerated cooling after processing (Fig. 9). The introduction of the SPM containing the Tuball Ocsial nanotubes (for 0.01% by weight)

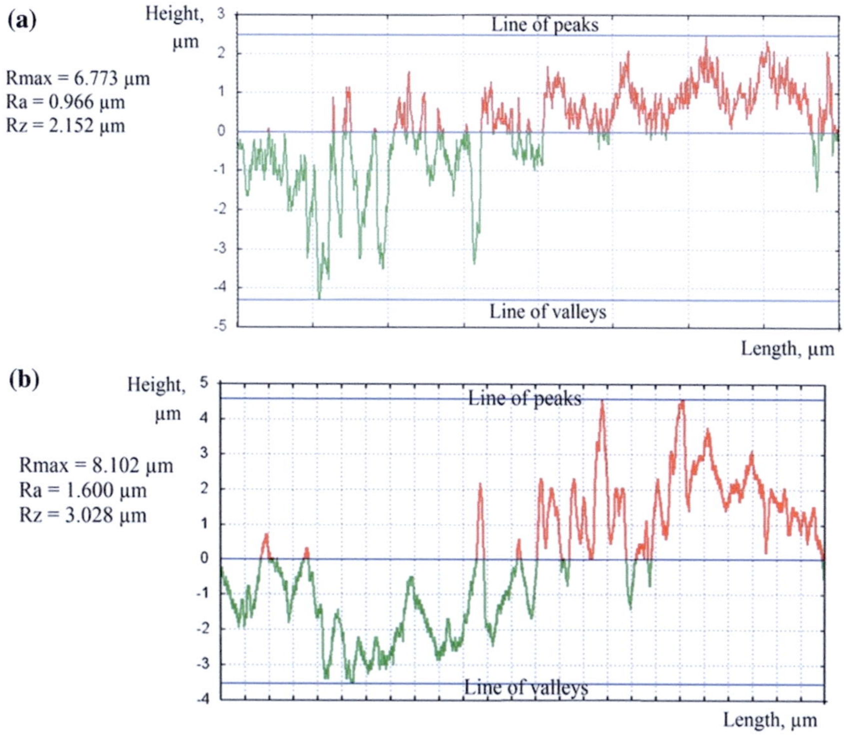

Fig. 2 Profilograms of the surface roughness of the samples after the ESA process: **a** without the SPM, **b** after the ESA process by Mo- SPM –Mo

significantly affects the microstructure of the coating. In the surface layer, there is a sharply defined zone of the crystals directed normally to the surface (Fig. 10). This testifies to the directed and accelerated heat removal during the crystallization process. Because of adding the nanotubes to the SPM, the process of grinding the crystallites is also noted in the coating.

The introduction of the Tuball Ocsial nanotubes for 0.6% into the SPM significantly changes the structure. Any dendritic structures are not detected (Fig. 11a). In the surface layer of the coating, the nanoscale inclusions of 90 to 130 nm (Fig. 11b) are recorded in the matrix, which is similar to the amorphous state, and these inclusions have different chemical compositions as compared to those of the coatings, as evidenced by the image of the microstructure in a Backscattered Electron Detector (BED) (Fig. 11c). It should be noted that the samples being studied were subjected to the ESA treatment by a molybdenum electrode at Wp = 0.13 J. Thus, the changes in the microstructure are associated with the influence of the nanotubes on the structure formation.

Figure 12 shows the results of the electron microscopic studies. Those are clear from the areas of the surface layers on the Armoco iron samples after the ESA process

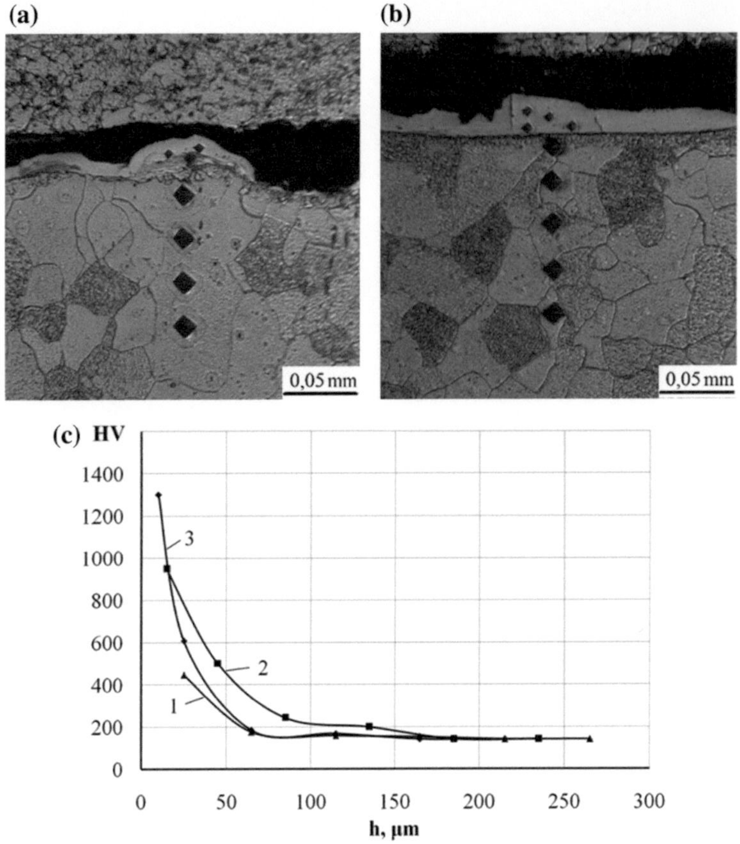

Fig. 3 Microstructures (a-b) and distribution of microhardness (c) in coatings on samples made of Armco iron after the ESA process by Mo- SPM -Mo, the SPM is composed of multi-walled ARKEMA carbon nanotubes (0.2%) in the epoxy resin of Epoxy 510 type without a curing agent: **a** coating obtained at Wp = 0.13 J; **b** coating obtained at Wp = 0.52 J; on the graph **c** 1—without the SPM, Wp = 0.13 J; 2—Wp = 0.13 J; 3—Wp = 0.52 J

at Wp = 0.13 J: without the SPM (Fig. 12a), and with the SPM, which contains the Tuball Ocsial carbon nanotubes in the amount of 0.01% by weight, (Fig. 12b). Also, it is shown the distribution of the carbon, iron, and molybdenum along the depth of the layer. The results of the analysis indicate that the carbon content remains unchanged in the both cases; the diffusion zone of molybdenum is of 12 to 14 microns. The addition of the nanotubes to the SPM contributes to obtaining the uniform chemical composition and the continuous coating.

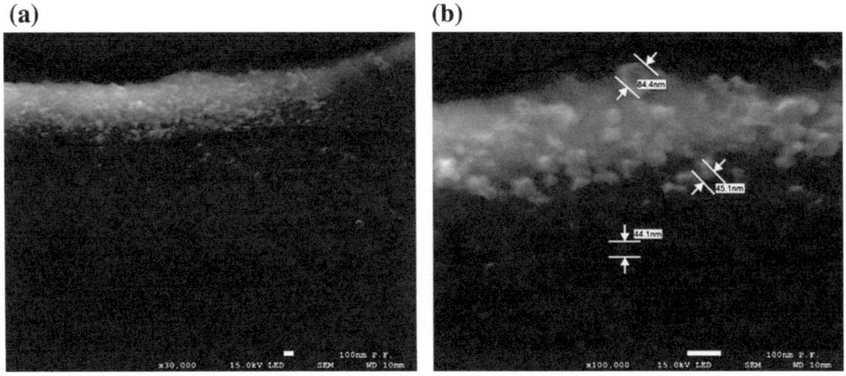

Fig. 4 The microstructures of coatings on the Armco iron samples after the ESA process, the Mo- SPM -Mo coating is obtained at Wp = 0.13 J, the SPM is the multi-walled ARKEMA carbon nanotubes (0.2%) in the epoxy resin of Epoxy 510 type without a curing agent: at the magnification: **a** 30,000 times, **b** 100,000 times

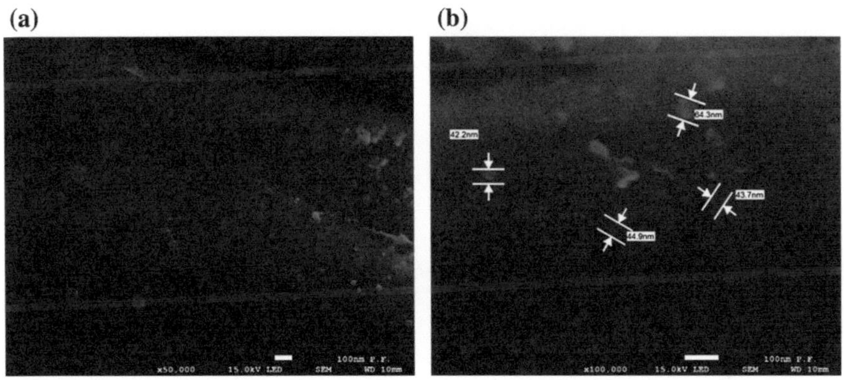

Fig. 5 The microstructures of the coatings on the Armco iron samples after the ESA process, the Mo- SPM -Mo coating obtained at Wp = 0.52 J, the—SPM is the multi-walled ARKEMA carbon nanotubes (0.2%) in the epoxy resin of Epoxy 510 type without a curing agent: at the magnification of: **a** 50,000 times, **b** 100,000 times

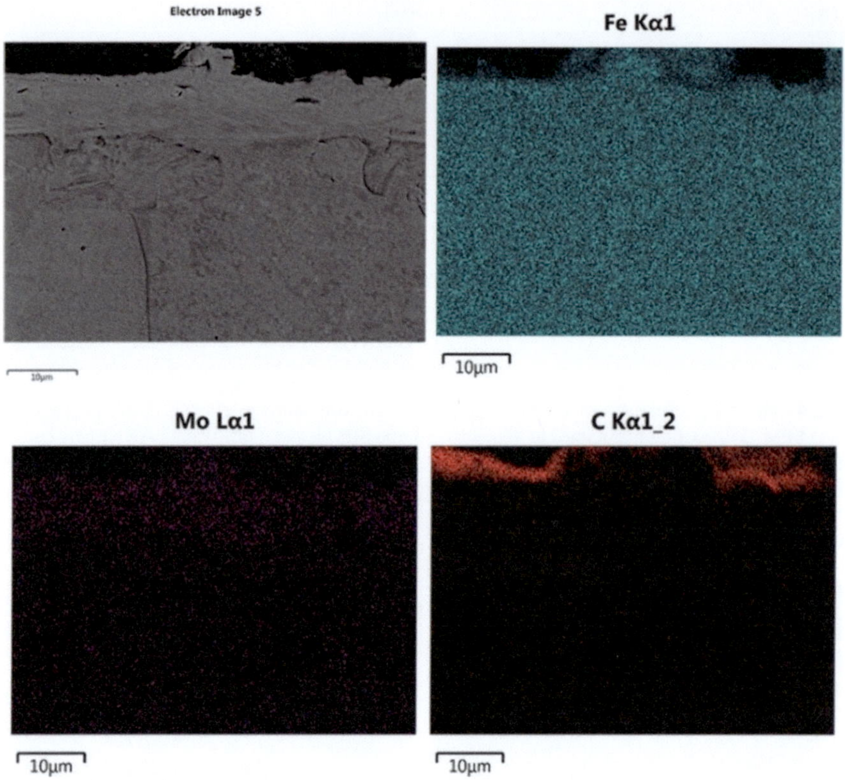

Fig. 6 The maps of the distribution of the elements in the electrospark coatings obtained at Wp = 0.13 J, the SPM is composed of the multi-walled ARKEMA carbon nanotubes (0.2%) in the epoxy resin of Epoxy 510 type without a curing agent

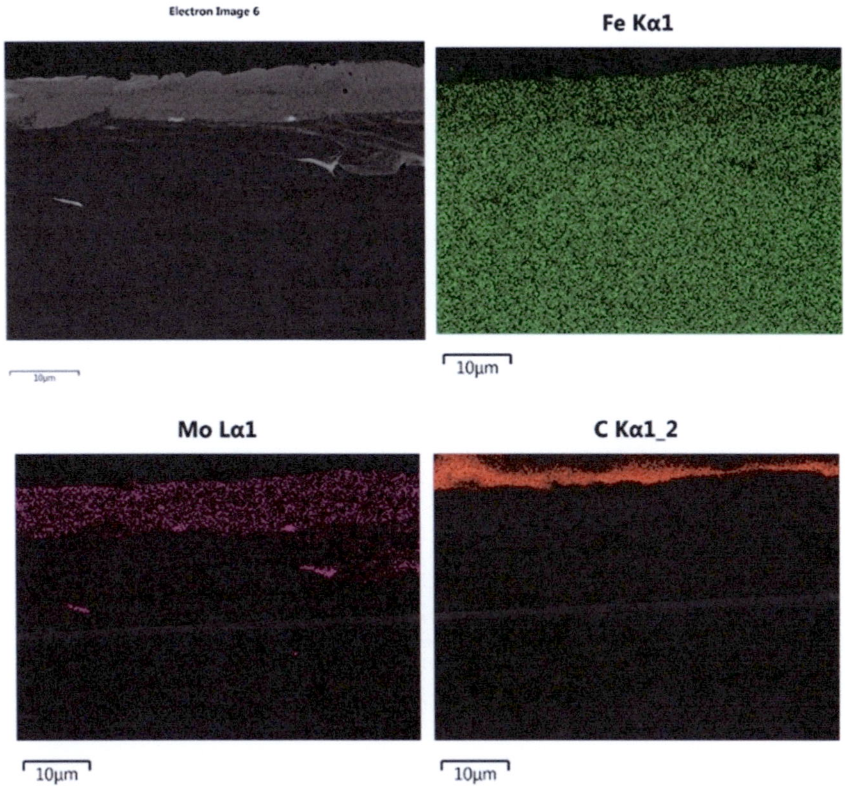

Fig. 7 The maps of the distribution of the elements in the electrospark coatings obtained at Wp = 0.52 J, the SPM is composed of the multi-walled ARKEMA carbon nanotubes (0.2%) in the epoxy resin of Epoxy 510 type without a curing agent

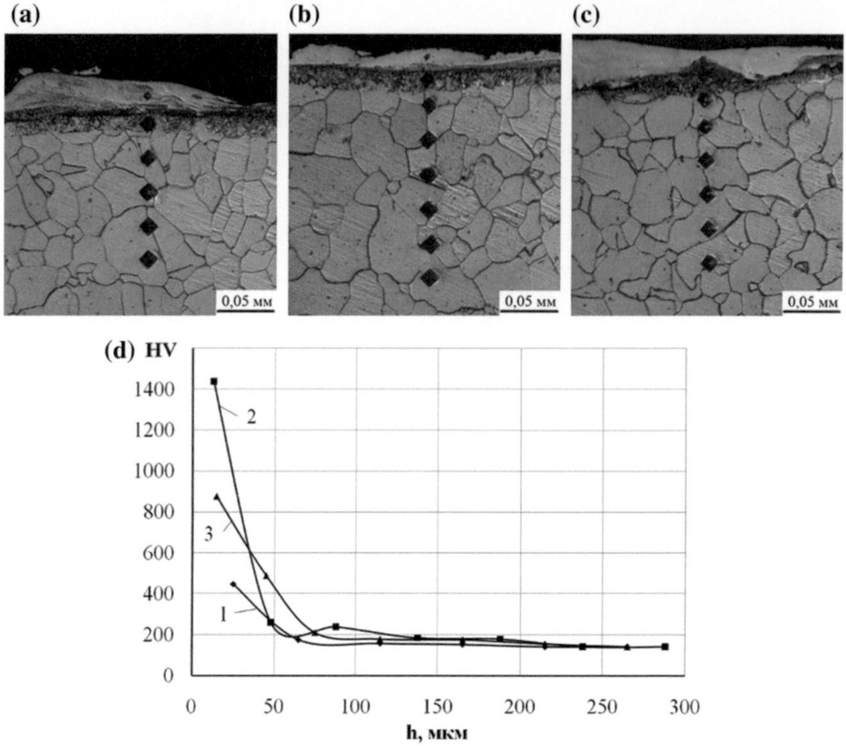

Fig. 8 The microstructures **a**–**c** and distributions of microhardness **d** of the Armco iron samples after the ESA process by the Mo-SPM-Mo: **a** without the SPM; **b** the Mo- SPM -Mo coating obtained at Wp = 0.13 J, the SPM—the single-walled Tuball Ocsial carbon nanotubes (0.01% by weight); **c** the Mo- SPM -Mo coating obtained at Wp = 0.13 J, the SPM—the single-walled Tuball Ocsial carbon nanotubes (0.6% by weight); on the graph **d** 1—without the SPM; 2—the Mo- SPM -Mo coating, the SPM contains the Tuball Ocsial nanotubes (0.01% by weight); 3—the Mo- SPM -Mo coating, the SPM contains Tuball Ocsial nanotubes (0.6% by weight)

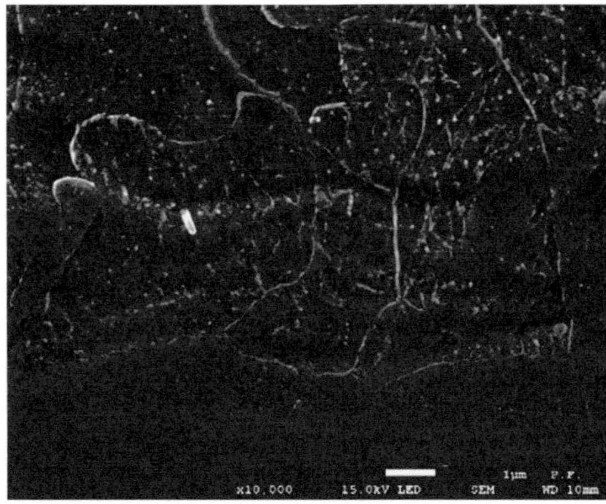

Fig. 9 The microstructures of the coatings on the Armco iron samples obtained after the ESA process performed by molybdenum without the SPM at Wp = 0.13 J

(a) (b) (c)

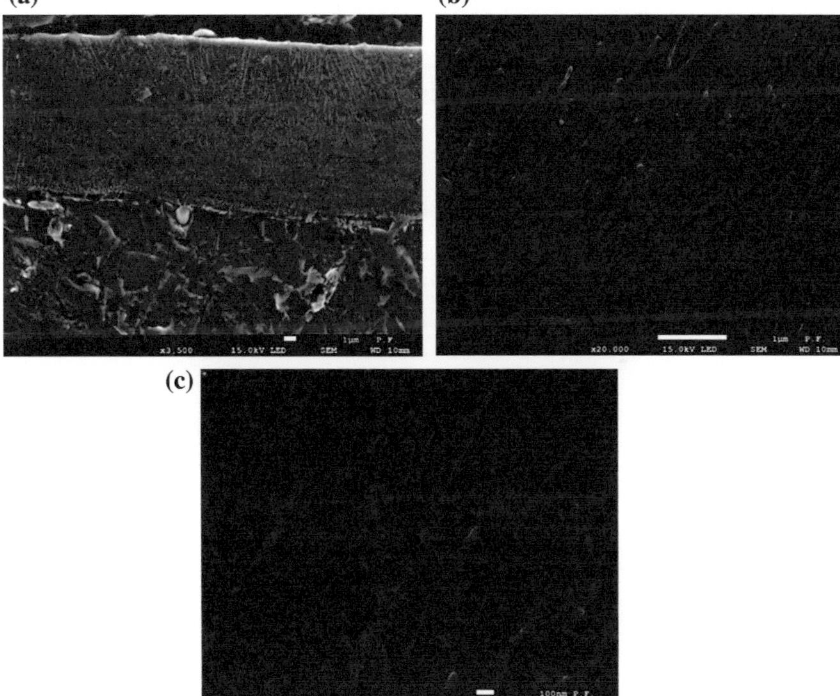

Fig. 10 The microstructures of the coatings on the Armco iron samples after the ESA process, the Mo- SPM -Mo coating obtained at Wp = 0.13 J, the SPM—the single-walled Tuball Ocsial carbon nanotubes (0.01% by weight) in the epoxy resin of Epoxy 510 type without a curing agent: at the magnification of: **a** 3,500 times, **b** 20,000 times, **c** 50,000 times

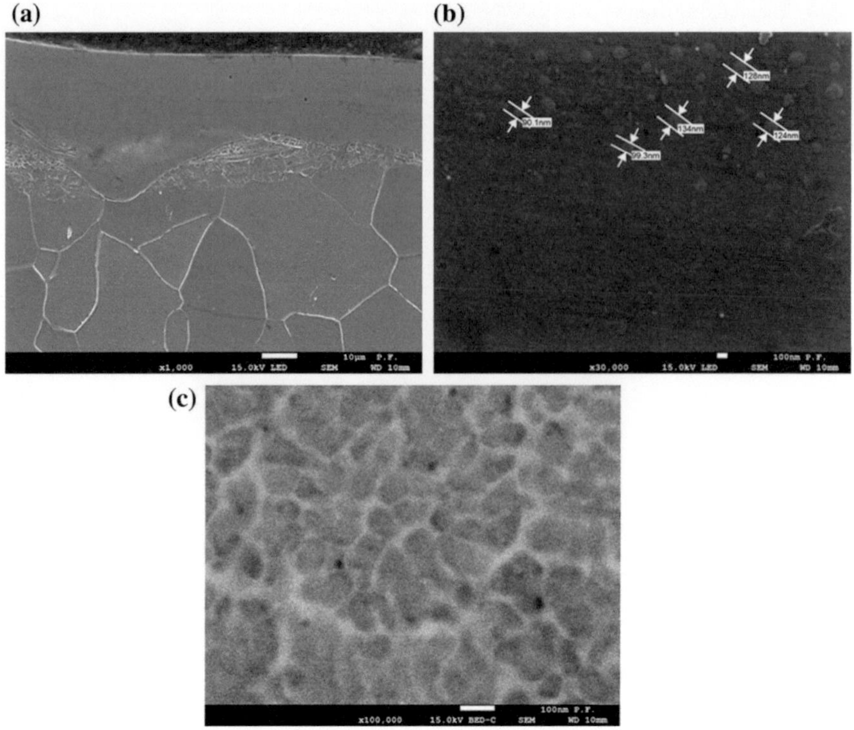

Fig. 11 The microstructures of the coatings on the Armco iron samples after the ESA process, the Mo- SPM -Mo coating is obtained at Wp = 0.13 J, the SPM—the single-walled Tuball Ocsial carbon nanotubes (0.6% by weight) in polycarbonate: the magnification: **a** 1,000 times, **b** 30,000 times, **c** 100,000 times

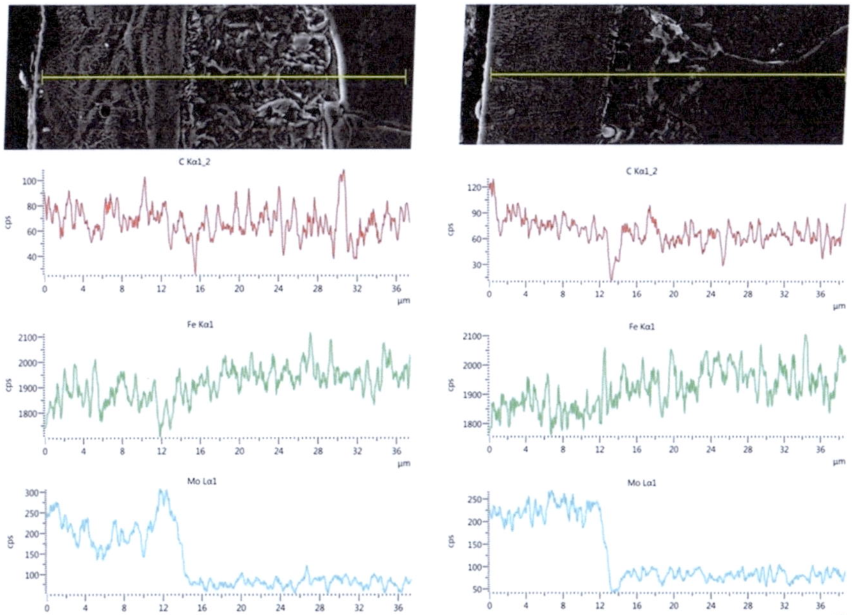

Fig. 12 The distribution of the elements in the electrospark coatings obtained at Wp = 0.13 J: **a** without the SPM, **b** with the SPM containing the Tuball Ocsial carbon nanotubes in the amount for 0.01% by weight

5 Conclusion

The analysis of the topography of the sample surfaces after the ESA process by molybdenum and with the use of the SPM has shown that with an increase in the discharge energy, the formation of a uniform and continuous coating is noted, however, with increased roughness. In addition, the coating contains the nano-sized inclusions, which are evenly distributed in the surface layer.

There have been considered the features of the coating structure formation in the course of the ESA processing the Armco iron by the molybdenum electrode with the use of the SPM composition of the multi-walled ARKEMA carbon nanotubes (0.2%) in the epoxy resin of Epoxy 510 type without a curing agent. The microstructures after the ESA process consist of three zones: the upper "white" layer, which is not etched in the reagent, the diffusion zone, and the base of a ferrite structure corresponding to the Armco iron. In the course of processing the Armco iron, with an increase in the discharge energy, the thickness and the continuity of the coating increase. Adding the ARKEMA nanotubes provides an increase in hardness from 446 to 608 HV at discharge energy Wp = 0.13 J and to 1300 HV at Wp = 0.52 J. The positive effect of the nanotubes on the quality parameters and hardness of the coatings is associated with the formation of the nanostructures. In the microstructures, the nanoscale phases of 40 to 60 nm are detected, and they are evenly distributed in the coatings. Because of the ESA process, the coatings with a uniform distribution of molybdenum are formed. Carbon, apparently in the form of the carbon nanotubes, is concentrated on the surfaces of the samples being processed, regardless of the discharge energy during the ESA process.

There have been considered the features of the coating structure formation in the course of the ESA processing the Aarmco iron by the molybdenum electrode with the use of the SPM compositions of the following:

1. The single-walled Tuball Ocsial carbon nanotubes (0.01% by weight) in the epoxy resin of Epoxy 510 type without a curing agent.
2. The single-walled Tuball Ocsial carbon nanotubes (0.6% by weight) in polycarbonate.

The presence of the nanotubes in the course of the ESA process has a positive effect on the quality parameters of the coatings. Thus, the content of the Tuball Ocsial nanotubes for 0.01% in the SPM helps to increase the continuity, thickness, and hardness from 446 to 1438 HV, i.e. 3 times. A further increase in the content of the nanoparticles up to 0.6% has no effect in ensuring the quality indicators of the coatings. The introduction of the SPM containing the Tuball Ocsial nanotubes (0.01% by weight) significantly affects the microstructure of the coating, the addition of the nanotubes to the SPM contributes to the grinding of crystallites in the coating and to the forming of the nanoscale inclusions of 90 to 130 nm in the matrix, which is similar to the amorphous state, and these inclusions have a different chemical composition as compared with that of a coating.

The introduction of the nanotubes into the SPM has a positive effect on the quality parameters of ESA coatings. The single-walled Tuball Ocsial carbon nanotubes for 0.01% are the most effective. Their introduction to the SPM helps to increase hardness and continuity. An increase in nanotubes worsens the effect of the positive influence.

Acknowledgements The results have been partially obtained within the research project "Development of new methods of surface nanostructuring steel surfaces with predicted quality parameters based on the electrospark alloying method" (Sumy State University) funded by the Ministry of Education and Science of Ukraine (State reg. no. 0122U000771) and Erasmus+ Jean Monnet Chair project 101085451 CircuMed.

References

1. Tarelnyk, V.B., Gaponova, O.P., Konoplianchenko, Ye.V., Martsynkovskyy, V.S., Tarelnyk, N.V., Vasylenko, O.O.: Metallofiz. Noveishie Tekhnol. **41**(2), 173 (2019)
2. Gaponova, O., Kundera, Cz., Kirik, G., Tarelnyk, V., Martsynkovskyy, V., Konoplianchenko, Ie., Dovzhyk, M., Belous, A., Vasilenko, O.: in Advances in Thin Films, Nanostructured Materials, and Coatings. Lecture Notes in Mechanical Engineering, p. 249. Springer, Singapore (2019)

3. Tarelnyk, V.B., Gaponova, O.P., Loboda, V.B., Konoplyanchenko, E.V., Martsinkovskii, V.S., Semirnenko, Yu.I., Tarelnyk, N.V., Mikulina, M.A., Sarzhanov, B.A.: Surf. Engin. Appl. Electrochem **57**, 173 (2021)
4. Pogrebnjak, A., Ivashchenko, V., Bondar, O., Beresnev, V., Sobol, O., Załęski, K., Jurgae, S., Coye, E., Konarski, P., Postolnyi, B.: Mater. Charact. **134**, 55 (2017)
5. Pogrebnjak, A.D., Ruzimov, S.M.: Phys. Lett. A **120**(5), 259 (1987)
6. Pogrebnyak, A.D., Kaverina, A.Sh., Kylyshkanov, M.K.: Protection of metals and physical chemistry of surfaces **50**(1), 72–87 (2014)
7. Kiilakoski, J., Langlade, C., Koivuluoto, H., Vuoristo, P.: Surf. Coat. Technol. **371**, 245 (2019)
8. Tarelnyk, V.B., Gaponova, O.P., Konoplianchenko, Ye.V., Martsynkovskyy, V.S., Tarelnyk, N.V., Vasylenko, O.O.: Metallofiz. Noveishie Tekhnol. **41**(3), 313 (2019)
9. Martsynkovskyy, V., Tarelnyk, V., Konoplianchenko, I., Gaponova, O., Dumanchuk, M.: in Advances in Design, Simulation and Manufacturing II. DSMIE 2019. Lecture Notes in Mechanical Engineering, p. 216. Springer Nature, Switzerland (2019)
10. Mashkov, Y.K., Korotaev, D.N., Baibaratskaya, M.Y., Alimbaeva, B.Sh.: Tech. Phys. **60**, 1489 (2015)
11. Sakhapov, R.L., Makhmutov, M.M., Sultanov, V.A.: J. Phys: Conf. Ser. **1588**, 012030 (2020)
12. Ivanov, V.I.: Lett. Mater. **10**(4), 427 (2020)
13. Tang, J.: Appl. Surf. Sci. **365**, 202 (2016)
14. Velichko, S.A., Kravchenko, I.N., Martynov, A.V., Chumakov, P.V.: Russ. Engin. Res. **41**, 661 (2021)
15. Yurchenko, V.I., Yurchenko, E.V., Dikusar, A.I.: Surf. Engin. Appl. Electrochem. **56**, 656 (2020)
16. Bazhin, P.M., Stolin, A.M., Zaripov, N.G., Chizhikov, A.P.: Surf. Engin. Appl. Electrochem. **52**, 217–224 (2016)
17. Wei, X., Chen, Z., Zhong, J., Xiang, Y.: Surf. Coat. Technol. **296**, 58 (2016)
18. Nikolenko, S.V., Igumnov, P.V., Gordienko, P.S.: Solid State Phenom. **316**, 745 (2021)
19. Aghajani, H., Hadavand, E., Peighambardoust, N.-S., Khamenehasl, S.: Surfaces Interfaces **18**, 100392 (2020)
20. Gaponova, O.P., Antoszewski, B., Tarelnyk, V.B., Kurp, P., Myslyvchenko, O.M., Tarelnyk, N.V.: Materials **14**, 6332 (2021)
21. Antoszewski, B., Gaponova, O.P., Tarelnyk, V.B., Myslyvchenko, O.M., Kurp, P., Zhylenko, T.I., Konoplianchenko, I.: Materials **14**, 739 (2021)
22. Myslyvchenko, O.M., Gaponova, O.P., Tarelnyk, V.B., Krapivka, M.O.: Powder Metall. Met. Ceram. **59**, 201 (2020)
23. Martsinkovsky, V., Yurko, V., Tarelnik, V., Filonenko, Yu.: Procedia Eng. **39**, 157 (2012)
24. Thakre, P.R., Bisrat, Y., Lagoudas, D.C.: J. Appl. Polym. Sci. **116**, 191 (2010)
25. Yasmin, A., Luo, J.-J., Daniel, I.M.: Compos. Sci. Technol. **66**, 1179 (2006)
26. Li, X., Wang, Q., Cui, X., Feng, X., Teng, F., Xu, M., Su, W., He, J.: Polymers **14**, 3618 (2022)
27. Colpaert, H.: Metallography of Steels: Interpretation of Structure and the Effects of Processing. ASM International, Ohio (2018)

Phase Composition and Structure of Nanocrystalline Films of NiCu, NiCo, and NiFe Alloys

V. B. Loboda, S. M. Khursenko, and V. O. Kravchenko

Abstract The work is devoted to the study of the crystal structure and phase composition of thin films of NiCu, NiCo, and NiFe alloys. Alloy films with thicknesses of 1–200 nm were obtained in vacuum at a residual gas pressure of ~10^{-4} Pa. The films were condensed at a rate of $\omega \geq 1$ nm/s. For electron microscopy studies, both NaCl chips and thin carbon layers were used as substrates. All films underwent stabilizing annealing in a vacuum to T = 700 K. In films based on NiCu and NiCo alloys with effective thicknesses of d < 10 nm, obtained on amorphous substrates at T = 300 K, an island structure was observed. In structurally continuous annealed films of NiCu and NiCo alloys at $C_{Co} < 70$ wt.%, an fcc structure was formed. For NiCu alloy films, depending on the concentration of the components, the crystal lattice parameter was a = 0.353–0.362 nm. Films of the NiCo alloy at $C_{Co} \geq 80$ wt.% had a two-phase composition (fcc + hcp)-solid solution NiCo. Films of the $Ni_{50}Fe_{50}$ alloy had a fcc lattice with a = 0.359–0.361 nm. Unannealed films exhibited a fine-grained structure; annealing led to an increase in grain size to 50 nm.

Keywords Thin films · NiCu alloy film · NiCo alloy film · NiFe alloy film · Crystal structure · Phase composition

The work is devoted to the study of the morphology, crystal structure, and phase composition of thin films of NiCu, NiCo, and NiFe alloys.

Films with thicknesses d = 1–200 nm were obtained in a vacuum at a residual gas pressure of ~10^{-4} Pa by the method of separate simultaneous evaporation of the components. For the evaporation of copper, a resistive heating method was used; for the evaporation of nickel, cobalt, and iron, an electron-beam method was used. The films condensed at a rate $\omega \geq 1$ nm/s. Both NaCl cleavages and thin layers of carbon were used as substrates for electron microscopic studies.

V. B. Loboda (✉) · S. M. Khursenko · V. O. Kravchenko
Sumy National Agrarian University, Sumy, Ukraine
e-mail: loboda-v@i.ua

A. D. Pogrebnjak et al. (eds.), *Nanocomposite and Nanocrystalline Materials and Coatings*, Advanced Structured Materials 214,
https://doi.org/10.1007/978-981-97-2667-7_8

The formation of the fcc NiCu alloy occurs directly during the condensation of the components in the entire thickness range; for structurally continuous annealed films, depending on the concentration of the components, the crystal lattice parameter is a = 0.353–0.362 nm in accordance with Vegard's rule. Unannealed and annealed films of the NiCo alloy at $C_{Co} < 70$ wt.% have an fcc structure; annealed films with $C_{Co} \geq 80$ wt.% have a two-phase composition (fcc + hcp)-solid solution NiCo. The change in the lattice parameter of fcc-NiCo films with increasing C_{Co} is described by Vegard's rule.

In films of NiCu and NiCo alloys with effective thicknesses d < 10 nm, obtained on amorphous substrates at T = 300 K, an island structure is observed.

$Ni_{50}Fe_{50}$ alloy films have an fcc lattice with a = 0.359–0.361 nm; unannealed films have a fine-grained structure; annealing leads to an increase in grain size up to 50 nm.

1 Introduction

The study of the physical properties of thin metal films is due both to obtaining results that contribute to the solution of a number of fundamental problems in solid-state physics and to the prospects for their practical application. The use of films has led to significant progress in microelectronics, high-frequency technology, optoelectronics, optics, and many other areas of modern science and technology.

A special place in the physics of thin films is occupied by films of magnetically ordered metals Ni, Fe, and Co. This is primarily due to the fact that their study allows solving a number of fundamental problems for «two-dimensional magnetism». In addition, in recent years, a number of new fundamental effects (giant magnetoresistance, spin-polarized tunneling, colossal magnetoresistance, etc.) have been discovered in film objects based on these metals, which form the basis for the development of miniature magnetoelectronic devices, new recording and storage methods information, and new types of highly sensitive sensors and transducers (nanoelectronics, spintronics).

Along with studies of films of pure metals (Ni, Cu, Co, Fe), it is promising to study films of metal alloys [1–6] and film multilayer structures [7–11], which include these metals. Alloy films (as well as bulk alloys) have a number of advantages over pure metal films: by changing the composition of the alloy, materials with other electrophysical, physicochemical, and operational properties can be obtained. Along with traditional film objects, in recent decades, a scientific direction has been formed related to the improvement of the service properties of massive metal samples by methods of creating layered structures in modified near-surface layers [12–14].

The properties of thin films, as a rule, differ significantly from the properties of bulk samples. Limiting the size of film objects in one of the directions leads to the appearance of so-called size effects, which are weakly expressed or not observed at all in the massive state. In the early stages of formation, thin films are not continuous. They are formed from small islands that may (or may not) be physically connected to each other depending on the thickness of the layer. The physical properties of island films differ significantly from the properties of both bulk metals and solid metal films. This paper presents the results of studying the morphology, crystal structure, and phase composition of NiCu, NiCo, and NiFe alloy films.

2 Technique and Methodology of Experiments

2.1 Preparation of Samples of Alloy Films

The method of obtaining metallic films by evaporation of the initial metal and its condensation in a vacuum (see, for example, [15]) has been widely used for many years in various fields of science and technology due to the universality of the technology, high productivity of the process, low energy consumption, absence of environmental pollution, etc. These and a number of other indicators determine the competitiveness of this method in comparison with other methods of film production (electrodeposition, cladding, metallization, sputtering, etc.). The quality of the formed films, the reproducibility of their functional properties, and the stability of characteristics under different operating conditions significantly depend on the patterns of evaporation and condensation.

To obtain thin alloy films, two main groups of methods are used: (1) the formation of multicomponent films directly on the substrate (separate evaporation of the components and subsequent diffusion annealing of multilayer structures, saturation in vapors); (2) the use of finished alloys as the initial metallizer (evaporation of the final portions of the alloy, stationary mode of evaporation, explosive evaporation).

Direct evaporation of compounds is quite widely used in the technology of obtaining multicomponent films and coatings due to its simplicity from the point of view of a constructive solution. At the same time, with this method of evaporation of alloys, the nature of the formation of films and their properties are determined by the fractionation of alloys during their evaporation, which leads to a discrepancy between the composition of the alloy film and the original alloy.

Some of the alloy films were obtained by us by the method of simultaneous condensation of the initial materials (Ni and Cu; Ni and Co) during their evaporation from two independent evaporators, which made it possible to directly obtain alloy films of arbitrary composition on the substrate.

To obtain part of NiCo film alloys, the method of evaporation of alloys of a given composition was also used. In doing so, the following considerations were taken into account. Data on «liquid–solid» phase equilibrium in bulk NiCo alloys [16, 17] show that the liquidus and solidus lines in this system practically coincide for any component concentration. As a result, no change in the chemical composition during melting can be expected. The difference between the vapor pressure of the Ni and Co metals at high temperatures is also insignificant, which also contributes to maintaining the composition of the alloy films in accordance with the initial composition.

The choice of methods for the evaporation of metal components of alloy films was also determined by such physical properties of materials as the melting temperature and saturated vapor pressure. Thus, during the condensation of copper (films of NiCu alloys), the method of thermal evaporation of Cu was used. In this case, 0.05 mm thick tungsten foil ribbons were used as evaporators. Films of Ni, Co, NiFe alloys and parts of NiCo alloys were obtained by electron-beam evaporation of the initial metals using an electron diode gun. It consists of anode and cathode units, mechanically connected by ceramic insulator rods. A high voltage of up to 3 kV was applied to the gun anode. A small piece of the original metal (nickel, cobalt, NiCo, or NiFe alloys) was attached to the tungsten anode by spot welding. The anode was heated to the required temperature by bombarding it with a defocused beam of electrons evaporating from a hot cathode (tungsten wire 0.3 mm in diameter). The purity of the evaporated metals was no less than 99.98%.

The initial material for some films of NiCo alloys was specially prepared by us with small weighed portions of NiCo alloys of the required composition. For the preparation of weighed portions of these alloys, pure (no worse than 99.9%) Ni and Co metals were used. The concentration of the components was varied by changing the mass ratios of metals (the relative error in determining the mass is no more than 5%). Samples were prepared by melting samples of starting materials in a ceramic crucible under high vacuum conditions with exposure for 1 h at a temperature close to the melting temperature for homogenization. The weight loss did not exceed 1–2%.

Massive weights of an alloy known as permalloy 50N (composition: 49.5–50.0% Ni, impurities (Si, Mn) < 1%, the rest Fe [18]) were the material for the manufacture of NiFe films.

A VUP-5 M industrial vacuum unit was used to condense thin films. The pressure of the gases of the residual atmosphere in the process of obtaining samples was 10^{-3}–10^{-4} Pa.

Condensation of alloy films carried out on a substrate holder developed by us is shown in Fig. 1. Its design made it possible to obtain four film samples of practically the same thickness but different compositions in one technological cycle of deposition, both for measuring the electrophysical and magnetoresistive properties (3) and for studying the structure (5). The geometrical dimensions of the studied samples (a, b) were set by special masks made with high precision from nichrome foil.

The films were condensed at room temperature (300 K) of the substrate. The rate of condensation was determined experimentally from the deposition time and final sample thickness. During the condensation process, it was maintained constant and amounted to 0.5–1.5 nm/s, depending on the evaporation modes. The alloy films were condensed in the absence of an external orienting magnetic field (if the Earth's magnetic field was ignored), which made it possible to obtain magnetically isotropic films.

Depending on the type of research, two types of substrates were used (Fig. 1). To study the electrophysical and magnetoresistive properties, polished optical glasses (6) were used as substrates. Previously, contact copper pads (2) with a chromium sublayer

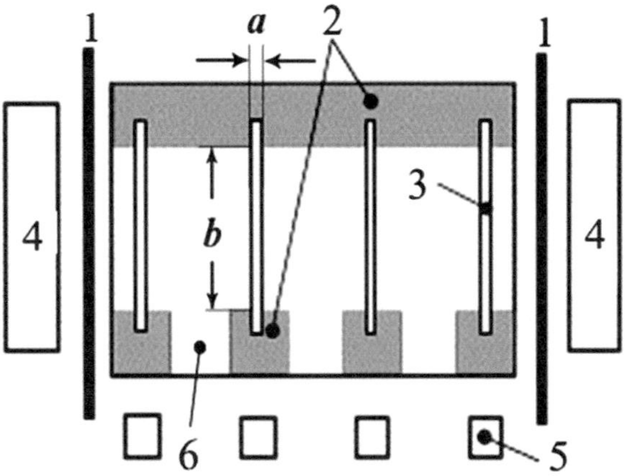

Fig. 1 Scheme of a substrate holder with substrates for condensation of film alloys: 1—screens; 2—contact pads; 3—film samples (a = 2 ± 0.05 mm, b = 10 ± 0.05 mm); 4—glasses «witnesses»; 5—single crystals of NaCl or carbon films; 6—glass substrate

were applied to them to improve the adhesion of copper to glass. For structural studies, single crystals of NaCl (orienting substrate) and carbon films (amorphous substrate) were used as substrates (5). In addition, glass plates «witnesses» (4) were fixed on the substrate holder to measure the thickness. The screens (1) ensured that the components of the alloys were hit separately on the «witness» plates.

The film thickness was measured using an MII-4 microinterferometer with a laser light source (miniature semiconductor laser, $\lambda = 647$ nm). The interference pattern was recorded using a digital camera with data transferred to a computer for subsequent recording and processing. With this method, it is possible to reduce the error in measuring the film thickness, especially in the thickness range d < 50 nm. The film thickness measurement error is 5–10% for thicknesses 50–200 nm and 10–15% for thicknesses d < 50 nm.

It should be noted that, in the case of ultrathin (d < 10 nm) films, the film thickness is commonly understood as the weight film thickness, which is determined from the deposition time and the known deposition rate. In this case, it is customary to speak of the effective film thickness (see, for example, [19]). The process of condensation of the «witness» films was 100–150 s, and their thickness was 100 nm or more. This made it possible to measure their thickness with an accuracy of 5–10%. Taking into account the error in measuring the condensation time t_c, the error in determining the effective thickness is 55–60% at $t_c = 1$ s (for ultrathin films) and decreases to 10–15% with an increase in t_c more than 10 s.

To calculate the concentration of the alloy film components during separate condensation of the components, a method was used, the essence of which is illustrated by the diagram, shown in Fig. 2.

Based on the geometry of the «substrate-evaporator» system, the thickness of individual components at point M can be calculated from the relations [15]:

$$d_1 = d_{01}\left[1 + \left(\frac{l_1}{h}\right)^2\right]^{-\frac{3}{2}} \text{ and } d_2 = d_{02}\left[1 + \left(\frac{l_2}{h}\right)^2\right]^{-\frac{3}{2}}, \tag{1}$$

Fig. 2 Geometry of the «substrate-evaporators» system for calculating the concentration of the alloy components: S1, S2—screens; E1, E2—evaporators; P1, P2—glass plates «witnesses» of thickness

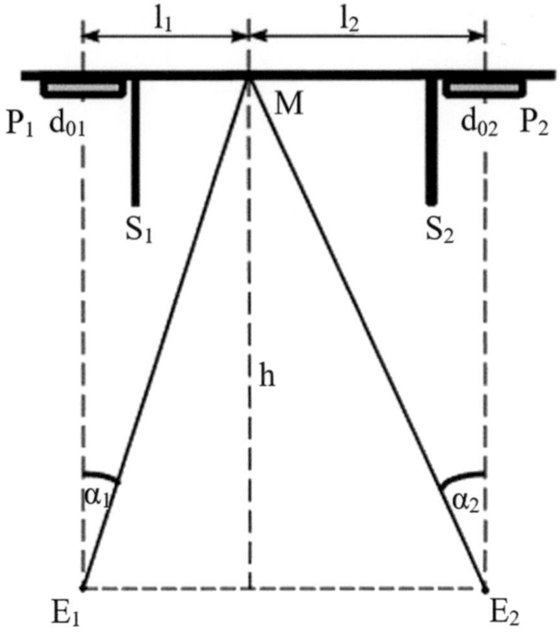

where d_0 is the film thickness at the point above the evaporator; d_{01}, d_{02} arethe thicknesses of the «witness» films; l_1, l_2 is the distance from the middle of the «witness» to the point M; h isthe distance from the plane of the evaporators to the plane of the substrate.

To find the concentration of the alloy film components at point M, one can use the obvious relations:

$$C_1 = \frac{D_1 d_1 \mu_1^{-1}}{D_1 d_1 \mu_1^{-1} + D_2 d_2 \mu_2^{-1}} \text{ and } C_2 = \frac{D_2 d_2 \mu_2^{-1}}{D_1 d_1 \mu_1^{-1} + D_2 d_2 \mu_2^{-1}}, \quad (2)$$

where D_1 and D_2 are the film densities of each of the materials; μ_1 and μ_2 are the molar masses of evaporated metals.

Taking into account that in bulk samples $D_{Ni} \approx D_{Cu} \approx D_{Co}$ and assuming that the same is true for films, relations (2) are transformed to the form:

$$C_1 = \frac{d_1 \mu_2}{d_1 \mu_2 + d_2 \mu_1} \text{ and } C_2 = \frac{d_2 \mu_1}{d_1 \mu_2 + d_2 \mu_1}. \quad (2a)$$

Since the concentration of the alloy film components was determined by the thicknesses d_1 and d_2, the error in determining the concentration of the components is the same as for the thickness.

2.2 *Study of the Structure and Phase Composition of Films*

To study the crystal structure and phase composition of film samples, electron microscopy (transmission electron microscopes EM-125, PEM-100-01) and electron diffraction (electronograph) studies were carried out. To carry out these studies, the films were condensed both on fresh (001) NaCl or KBr cleavages (single-crystal orienting substrates) and on carbon films (amorphous substrates). Carbon films simulated the conditions of deposition of film samples on amorphous glass.

To obtain information about the size and shape of crystallites, the data of microimages of the crystal structure of film samples were statistically processed, from which the size of individual Li grains was determined. The obtained data were used to construct a grouped frequency distribution and a histogram in the form of the dependence N = f(L). The average size of crystallites in the film plane was determined using the relation:

$$L = \frac{N_1L_1 + N_2L_2 + ... + N_nL_n}{N_1 + N_2 + ... + N_n}, \quad (3)$$

where N_n and L_n are, respectively, the number and average size of crystallites for the n-th interval.

To determine the sizes of crystallites along the normal to the substrate surface (crystallite heights), we used the method of analyzing the profiles of diffraction lines on electron diffraction patterns, similar to the method used in X-ray diffraction analysis to determine the sizes of coherent scattering regions during X-ray diffraction in bulk materials.

3 Physical Properties of Co, Ni, Cu, Fe and Their Alloys in the Bulk State [16– 22]

Bulk **nickel** has an fcc lattice (Fm3m) over the entire temperature range with the parameter a = 0.35243 nm.

Bulk **copper** has an fcc lattice (Fm3m) over the entire temperature range with a = 0.36148 nm.

Cobalt in the bulk state has two polymorphic modifications (low-temperature α-phase and high-temperature β-phase). The low-temperature α-phase has a hexagonal close-packed (hcp) lattice (structural type A3, space group $P6_3/mmc$) with parameters a = 0.2507 nm, c = 0.40686 nm (at 273 K). The high-temperature β-phase, which is stable up to the melting temperature of the metal, has a face-centered cubic (fcc) lattice (structural type A1, space group Fm3m, a = 0.3548 nm). The temperature at the beginning of the transition of α-Co to β-Co is 750 K; the temperature at the beginning of the reverse transition is 676 K.

Iron in a massive state crystallizes in four polymorphic modifications. The α-Fe modification has a body-centered cubic (bcc) lattice (structural type A2, space group Im3m) with parameter a = 0.286653 nm. At a temperature of 1042 K (Curie temperature), a transition occurs to the β-phase with the same structural type, but with the parameter a = 0.2895 nm. In the temperature range 1184–1665 K, γ-Fe is formed with an fcc lattice (a = 0.36468 nm); at higher temperatures, the δ-phase with the bcc structure and with the parameter a = 0.29322 nm is stable.

Nickel with Fe, Co, Cr, Cu, Mn, Mo, W, and other elements of the periodic system forms substitutional solid solutions with an fcc lattice in a wide range of concentrations. Such alloys are characterized by high ductility and strength, obtained as a result of a decrease in the energy of stacking faults, a decrease in the diffusion mobility of atoms, and a change in the electronic structure upon alloying. Nickel alloys are distinguished by high emission properties, increased mechanical strength and corrosion resistance, and can be processed in all types of machining. Nickel and copper in the entire range of component concentrations (from 0 to 100%) form a solid solution and have a diagram with unlimited solubility of the components in the solid and liquid states (Fig. 3). The solid phase of the massive NiCu alloy has an fcc lattice with a parameter from $a_0 = 0.35243$ nm to $a_0 = 0.36148$ nm depending on the concentration of the components, in accordance with Vegard's rule. The Curie temperature θ_{0C} of NiCu alloys is less than θ_{0C} for Ni and lies above room temperature at $C_{Cu} < 33\%$, approaching 70 K at $C_{Cu} = 50$–55%.

NiCo alloys (Fig. 4) form predominantly substitutional solid solutions with a single-phase composition based on lattices of pure metals. At room temperature, alloys with a Co content of up to 65–67 at.% have an fcc lattice over the entire temperature range of the solid state.

For $C_{Co} > 75$ at.%, a solid solution with an hcp lattice (low-temperature α-phase) is characteristic; with an increase in temperature, an fcc lattice (β-phase) is formed (the hcp → fcc transition temperature depends on the concentration of the components). In this case, due to the fact that both phases are close-packed, there is a crystallographic size-orientational relationship between them. In the intermediate range of concentrations, we have a two-phase composition.

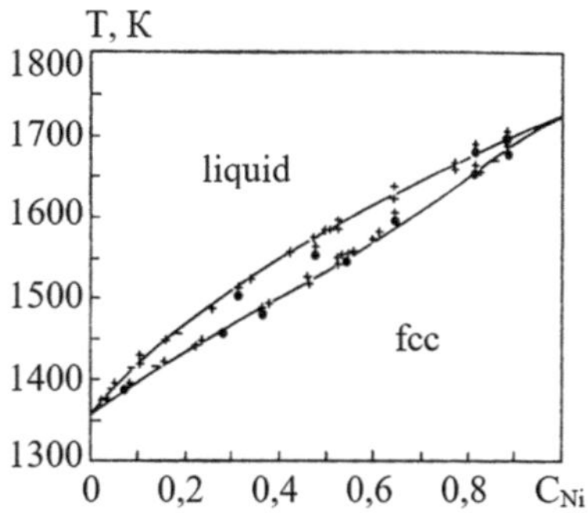

Fig. 3 Phase diagram of the state of the NiCu alloy

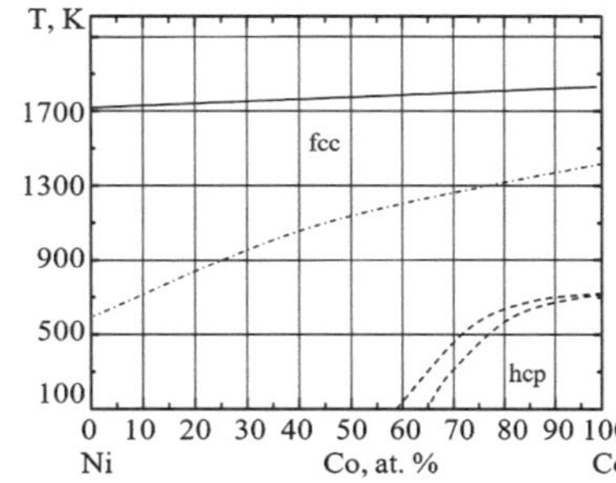

Fig. 4 Phase diagram of the state of the NiCo alloy

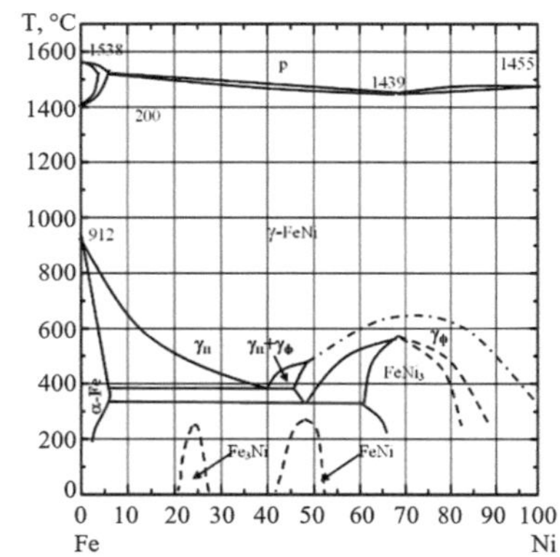

Fig. 5 Phase diagram of the state of the NiFe alloy

The phase composition of NiFe alloys (Fig. 5) is quite complex. Without going into a complete analysis of the phase diagram of this system, we note that we studied films obtained by evaporation of an alloy known as permalloy 50H (composition: 49.5–50.0% Ni, impurities (Si, Mn) < 1%, the rest Fe).

The alloy has high magnetic permeability, low coercive force (soft magnetic), almost zero magnetostriction, and significant (up to 5%) magnetoresistive effect. This alloy is characterized by an fcc lattice (Fm3m) with a = 0.359 nm.

4 Structure and Phase Composition of Ni, Co, Fe, and Cu Films

4.1 Growth Processes of Thin Metal Films

The physical properties of thin metal films largely depend on their structure, which, in turn, is determined by temperature, the state of the substrate, the vacuum conditions of condensation, the deposition rate, and so on. Summarizing the experimental results, we can distinguish four main stages of film growth [15]:

- formation of critical nuclei and islands;
- growth of islands due to coalescence;
- formation of a mesh structure with narrow channels;
- formation of a structurally integral film.

The first stage of film formation is the collision of a vapor atom with a substrate. As a result of such a collision, the atom transfers part of its energy to the substrate and can then either re-evaporate or remain on the substrate. If the substrate temperature is high enough, then in the process of surface diffusion, the adsorbed atom can join an already existing group (cluster) of other atoms. In turn, such complexes of atoms are thermodynamically unstable (due to high surface energy) and can decompose. At the same time, due to the addition of new atoms, such a cluster can grow and reach a critical size, above which it becomes thermodynamically stable. As a result, such a cluster can become the nucleus of an island of a metal film.

Theoretical calculations show that the size of the critical nucleus for metals with a higher boiling point is smaller than for metals with a lower boiling point. For metals with high boiling points (particularly Fe, Ni, and Co), even small nuclei are thermodynamically stable. In addition, the size of the critical nucleus is affected by the surface energy of the metal and the substrate, the substrate temperature, and the rate of condensation.

The substrate temperature T_s is one of the main factors determining the film condensation mechanism [23, 24]. At temperatures below a certain Q_1 value inherent in a given material, the «saturated vapor-crystal» (SV-C) mechanism is realized. At $T_s > Q_1$, condensation is carried out according to the scheme «saturated vapor–liquid-crystal» (SV-L-C). Numerous experimental studies show that the temperature Q_1 is approximately $2/3T_m$ (where T_m is the melting temperature of the film material). The existence of another mechanism «saturated vapor-supercooled liquid-amorphous phase» was also experimentally confirmed, according to which a film is formed at a temperature below $Q_2 = 1/3T_m$.

With further condensation of the film, the process of island coalescence begins, which leads both to an increase in their size and to an increase in the free surface on which the formation of secondary islands begins. The film is a mesh structure in the form of islands separated by channels. Continued condensation and the formation of secondary islands eventually lead to the filling of the channels and the formation of a continuous film.

On neutral substrates, molecular beam condensation occurs only under the condition $T_s < T_c$ (T_s is the substrate temperature, T_c is the critical condensation temperature). In this case, depending on the rate of condensation ω, the temperature T_c can be either lower than the melting temperature of bulk samples T_m or higher. In accordance with modern ideas about the mechanisms of formation of thin films in the literature (see, for example, [24–26]), it is customary to single out the following regimes of condensate growth:

(1) layer-by-layer (Frank-Van der Merwe regime) is realized if the atoms of the substance are connected with the atoms of the substrate (as a rule, single-crystal, with a close lattice parameter) stronger than one with the other; in this case, on the substrate, the construction of a monolayer of a substance is first completely completed, on which another monolayer, a third, etc., appears;
(2) island (Volmer-Weber mechanism) is realized in the case when the atoms of the substance are bound to each other more strongly than with the atoms of the substrate (usually neutral). In this case, small nuclei are formed on the surface of the substrate, which, as they grow, turn into large islands of the condensed phase. When the critical thickness is reached (determined by the kinetics of the condensation process), contacts appear between individual islands (a characteristic labyrinth structure with channels is formed). After the channels are filled, a continuous film is formed;
(3) intermediate (the Stranski–Krastanov regime), in which layer-by-layer condensation occurs first (1–2 layers are filled), after which the island growth regime begins. The main reason for changing the growth mechanisms is the change in the lattice parameter when the next layer is filled, which leads to a violation of the conditions for implementing the layer-by-layer growth mechanism and ensures the fulfillment of the island growth criterion.

A characteristic feature of the structure of thin films is the possibility of manifestation of the phase size effect in them [26]. In particular, in metal films, the appearance of new phases, which are not characteristic of a metal in the bulk state, and a decrease in the phase transition temperature in metal films, which have high-temperature phases, are possible. In island films, a change in the lattice parameter with thickness can also be observed; however, data on this effect are currently rather contradictory [24] (both an increase [26] and a decrease in the lattice parameter [19] were experimentally observed).

4.2 *Phase Composition and Crystal Structure of Ni, Co, Fe, and Cu Films*

A large number of experimental works have been devoted to the study of the crystal structure and phase composition of thin films of transition metals (for a detailed analysis, see, for example, [27]). Summarizing the experimental data, we can say that the structure and phase composition of transition metal films condensed in a vacuum quite definitely depend on a number of parameters: purity, microrelief and substrate temperature, substance deposition rate, the thickness of the condensed layer, subsequent thermal treatment of film samples, degree of vacuum, composition of the residual atmosphere, etc.

The fundamental influence on the phase composition and electrophysical properties of the films is the influence of such parameters as the pressure and composition of the gases of the residual atmosphere (P), the condensation rate (ω), and the substrate temperature (T_s). To describe the influence of these factors, we proposed an empirical condensation parameter α, which can be represented as follows:

$$\alpha = \frac{P}{\omega\sqrt{T_s}}$$

Analyzing numerous data as a whole, we can conclude that purer samples (without impurity phases) can be obtained by increasing the rate of their condensation (at P = const) or by decreasing the pressure of gases of the residual atmosphere P (at ω = const). Thus, at low values of α (low residual pressure or high rate of metal condensation), the best vacuum conditions will be realized. Then, in film samples, the phase corresponding to bulk samples should form with a higher probability. Conversely, at large values of α, the formation of impurity phases is possible.

Ni films [27]. It has been established that Ni does not have polymorphic modifications in films. In nickel films condensed under good vacuum conditions (for small values of the parameter α), only the fcc phase with the lattice parameter a = 0.351 ± 0.001 nm is observed.

The lattice parameters of pure Ni films almost coincide with the lattice parameters of bulk metal. The difference between the parameters is noted only in the case of ultrathin films. In this case, in films, depending on the conditions of production (primarily vacuum), the lattice parameter can either increase or decrease. When minimizing the influence of gaseous impurities, the lattice parameter only decreases with decreasing particle size. The reason for this decrease is usually associated with an increase in the concentration of vacancies in films compared to bulk samples.

Co films [27]. Studies of thin Co films have shown that, upon condensation on carbon substrates, either a low-temperature hcp phase or a high-temperature fcc phase is observed, depending on the particle size (film thickness) and substrate temperature. In the samples obtained on (001) NaCl, a two-phase fcc + hcp compound was observed both in island and continuous films. The phase composition in this case does not depend on the thickness and production conditions, but is completely determined by the size of the crystallites.

During the condensation of thin films in a high vacuum, a number of anomalous phases can form as a result of interaction with residual gases. Under ultrahigh vacuum conditions, anomalous phases are not observed. With a small value of the parameter α, and under high vacuum conditions, it is possible to obtain films with a crystalline structure of the corresponding bulk samples with a sufficiently high reproducibility.

Fe films [27]. Electron diffraction and X-ray diffraction studies of Fe films deposited on different substrates show that the films have a bcc lattice with a = 0.287 ± 0.001 nm. There is a large dispersion of the elements of the substructure and polyblockity in thickness even at thicknesses of 20–50 nm. This is due to the relatively high melting point of iron. As a consequence, condensation at room temperature is characterized by a significant degree of supercooling and a low diffuse mobility of atoms. It should be added to this that in films of refractory metals, a small size of critical nuclei is observed, as a result of which a large number of crystals simultaneously grow in the film.

A change in the substrate temperature leads to a noticeable increase in the size of the crystallites. In particular, as the temperature rises from 300 to 720 K, the grain size in iron films 40 nm thick increases from 5 to 50 nm. The annealing of the films in vacuum also leads to a similar increase.

Cu films [27]. According to the research results, Cu films have an fcc lattice with a = 0.359–0.367 nm. At low condensation rates, a two-phase composition of fcc-Cu + fcc-Cu_2O films is also observed.

The increase in the lattice parameter of both unannealed and annealed in a vacuum chamber at 700 K copper films is explained by the dissolution of gases of the residual atmosphere (primarily oxygen) in Cu films, which is confirmed by mass spectrometric studies.

4.3 Phase Composition and Crystal Structure of the Alloy Films

In most cases, the structure of alloy films does not fundamentally differ from the structure of bulk analogues. However, as an analysis of the literature data shows, there may be some differences associated with the features of the films, in particular, with the manifestations of the phase size effect (PSE).

The structure and phase composition of alloys based on Ni, Fe, Co, and Cu in the film state are studied in a large number of experimental works (see, for example, the list of references in [27]), which, as a rule, are narrowly focused.

It is noted that for NiFe alloys, the single-phase region of the α-phase based on the bcc iron lattice increases compared to bulk alloys (up to 50% Ni), the reason for this is explained on the basis of PSE concepts. On the whole, the nature of the $\alpha \rightarrow \gamma$ transition curve is similar to the diagram obtained for annealing nonequilibrium bulk samples. For the NiFe alloy, at the initial stages of condensation, the formation of an amorphous structure is observed, which, upon further deposition, crystallized into the α phase.

In films of the $Ni_{50}Fe_{50}$ alloy, as in the bulk state, the formation of an fcc phase is most probable.

Structural and phase studies of $Ni_{50}Fe_{50}$ film alloys obtained by electron-beam evaporation of the initial alloy in vacuum at room temperature were carried out in [28]. The samples condensed at a rate of 0.5–1 nm/s. It was found that the $Ni_{50}Fe_{50}$ films are finely dispersed even after annealing them up to 700 K (the grain sizes of the annealed samples do not exceed 25 nm). For films with thicknesses of 16–160 nm, an fcc phase was observed with a lattice parameter a = 0.359–0.361 nm, which slightly exceeds the parameter of the initial bulk material (a_0 = 0.3586 nm [29]).

Films of NiCo alloys are characterized by a multiphase composition (a mixture of bcc, fcc, and hcp phases). Depending on the composition, one can observe an fcc or hcp phase, while the region of existence of hcp-NiCo can expand due to the phase size effect. Some differences between the phase diagrams of bulk and film samples are associated with the nonequilibrium of vacuum condensates.

5 Research Results

5.1 Films of NiCu Alloys

The films of NiCu alloys obtained by us in the range of thicknesses d = 1–200 nm and concentrations $0 < C_{Cu} < 100$ at .% have an fcc lattice with a parameter that depends on the concentration of the components and varies from a = 0.352 nm to a = 0.362 nm (for bulk alloys $0.3524 < a_0 < 0.3615$ nm [20]). The formation of the fcc alloy occurs already at the stage of condensation, which is confirmed by electron diffraction (Fig. 6). For all unannealed NiCu films, the first line (111) is very broad, but not double (this would take place in the case of the eutectic structure of the fcc-Ni + fcc-Cu alloy), since the difference $\Delta d_{111} = d_{111}(Cu)–d_{111}(Ni) = 0.009$ nm is quite sufficient for separate electron diffraction observation of the (111) Ni and (111) Cu lines. Electron diffraction patterns from thermally stabilized structurally continuous samples (Fig. 6d) sometimes show additional extremely thin lines (from 2 to 5 lines) of Cu_2O oxide, which has an fcc lattice with a = 0.424–0.450 nm. The most intense lines (111) Cu_2O and (220) Cu_2O are observed in electron diffraction patterns and from freshly condensed films at their condensation rate $\omega < 0.5$ nm/s. Lines of unstable oxide NiO or stable nitride Ni_3N were not recorded, although a number of authors (see, for example, [30]) noted this phase formation in nickel films.

In films of NiCu alloys, the values of the lattice parameter are somewhat higher compared to bulk samples (Fig. 7), which can be partly explained by the penetration of atoms from the residual atmosphere into the crystal lattice of the alloy, and partly by the fact that the atoms of one of the alloy components during the condensation process can occupy positions, which do not correspond to an ordered alloy.

For ultrathin samples of the NiCu alloy with effective thicknesses d = 1–7 nm, the crystal lattice parameter is 0.002–0.003 nm smaller compared to bulk samples. In experimental works [19, 24, 25, 31, 32], devoted to the study of ultrathin films of fcc metals, it was shown that, depending on the conditions of preparation, the lattice parameter can be either smaller or larger compared to the lattice parameter of bulk metal, increasing or decreasing with increasing island size. However, under conditions where the influence of gas impurities is minimized [24], the lattice parameter always decreases with decreasing particle size. The reason for this is either the presence of an additional (Laplace) pressure arising due to the small radius of curvature of the nanoparticle [19] or an increased (compared with the macroscopic sample) concentration of surface and near-surface vacancies in nanoparticles [24].

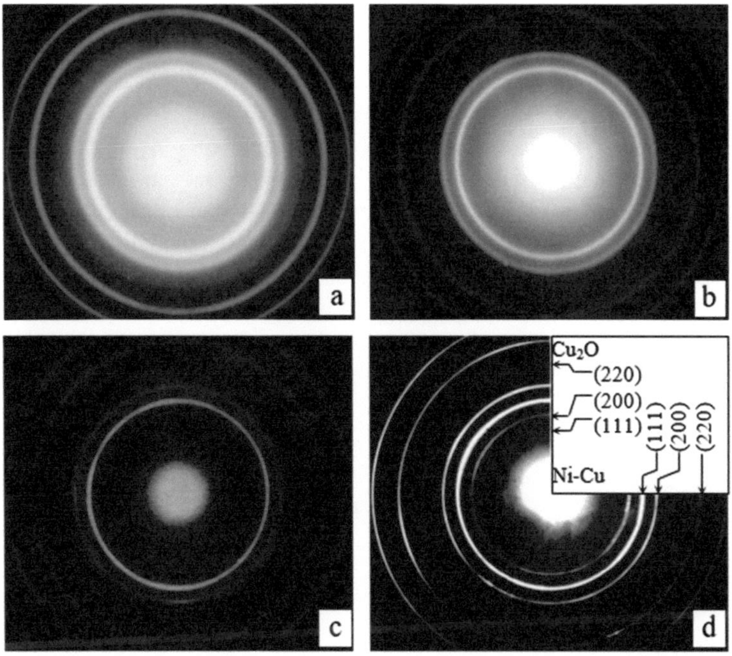

Fig. 6 Electron diffraction images of unannealed (**a**, **c**) and annealed to 700 K (**b**, **d**) NiCu alloy films: d ≈ 1 nm, C_{Cu} = 50.3 at.% (**a**, **b**); d = 40 nm, C_{Cu} = 38 at.% (**c**, **d**)

Fig. 7 Concentration dependence of the fcc lattice parameter of NiCu alloys (measurement temperature T = 300 K): ■—continuous films; ▲—island films; ——Vegard's rule

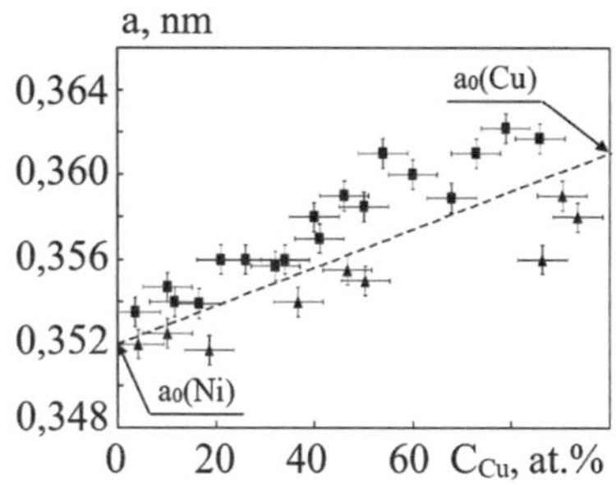

Ultrathin Films of NiCu Alloys

For ultrathin films of NiCu alloys (Fig. 8), an island growth mode is realized. In this case, in freshly condensed ultrathin films of NiCu alloys with an effective thickness d ≈ 1–2 nm, regardless of composition, almost the same structure was observed with an average size of individual islands 1.5–2 nm. When the effective thickness d ≈ 3 nm is reached, which, in all likelihood, is critical for NiCu film alloys, the so-called classical labyrinth structure [32] with irregularly shaped islands connected to each other and channels between the islands approximately the same width. A further increase in the effective thickness of the films to d ≈ 10 nm structurally manifested itself only in an increase in the density of islands and a decrease in the number of channels due to their gradual filling.

Since the condensation of film samples of the NiCu alloy was carried out on a neutral amorphous substrate (carbon film) at a temperature $T_s \approx 300$ K, less than $0.4T_m$ (T_m is the melting point of the alloy of a given concentration, 1400 K < T_m < 1700 K [16]), and diffusion the mobility of atoms in the islands was hindered, it can be assumed that the condensation of the alloys, in all probability, was carried out according to the «vapor → crystal» mechanism.

Subsequent annealing of the NiCu film alloys to T = 700 K leads to a significantly different morphological state of the samples, depending on their effective thickness. Thus, at d ≈ 1–3 nm, the islands are simply enlarged due to migration coalescence;

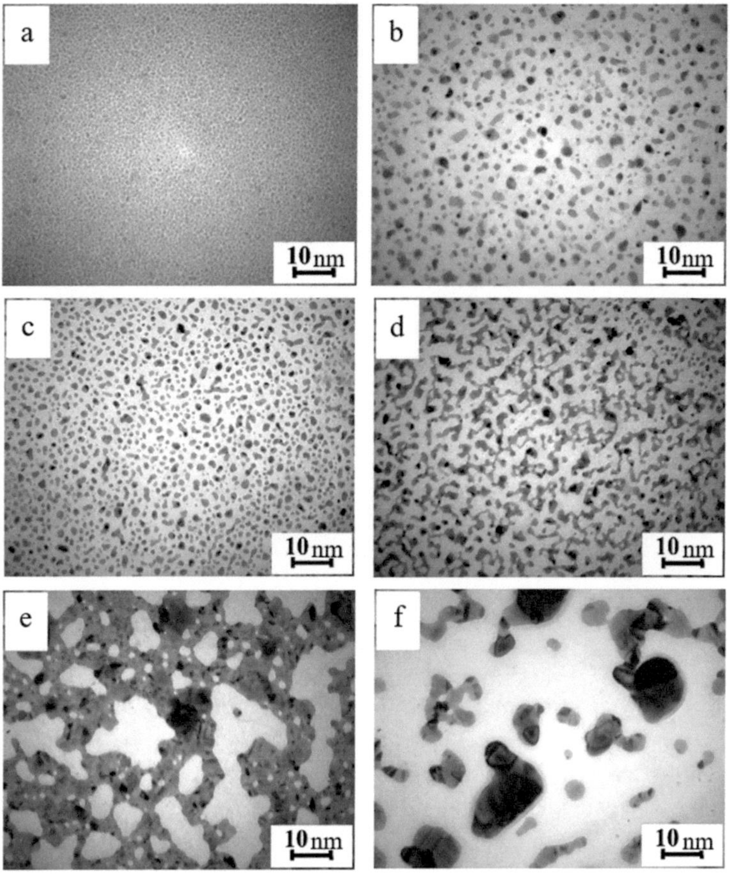

Fig. 8 Microstructure of unannealed (**a**) and annealed to 700 K (**b**–**f**) ultrathin NiCu alloy films of different thicknesses: d = 1 nm, C_{Cu} = 50.3 at.% (**a**, **b**); d = 2.5 nm, C_{Cu} = 36.7 at.% (**c**); d = 3.5 nm, C_{Cu} = 86.5 at.% (**d**); d = 6 nm, C_{Cu} = 90.5 at.% (**e**); d = 6.5 nm, C_{Cu} = 9.8 at.% (**f**)

along with this, there is also a significant decrease in their total number (Fig. 8b, c). At effective thicknesses d ≈ 3–4 nm, a characteristic «bridge» structure is formed in film alloys (Fig. 8d).

The most significant transformations during thermal stabilization annealing are experienced by films of approximately the same thickness d ≈ 6–6.5 nm, but with different component compositions. So, at high copper concentrations, the film becomes electrically continuous (see Fig. 8e), while at low copper concentrations (C_{Cu} = 9.8 at.%), the film remains island-like (Fig. 8f) with individual islands of complex shape, having sizes up to 20 nm. At the same time, crystals are visible inside the islands (even rather small ones), the shape, size, and number of which are not constant. Their appearance can be associated with the onset of recrystallization processes. The unification of the islands themselves into larger ones (Fig. 8f) obviously occurs according to the mechanism of «liquid-like» coalescence [32].

Features of the crystal structure of NiCu alloy films in a given range of thicknesses in a certain way affect the electrical properties (resistivity, TCR) of the samples under study.

With an increase in thickness (d > 10 nm), all the investigated alloy films in the freshly condensed state at T_s = 300 K become structurally continuous (Fig. 9a) and have an exclusively dispersed structure (average crystallite size L ~ 5 nm), which causes the angular expansion of the lines on electron diffraction patterns (see Fig. 6c). For NiCu films annealed to T = 700 K (Fig. 9b, c), the grain size increases to L ≈ kd, where k = 1–1.3.

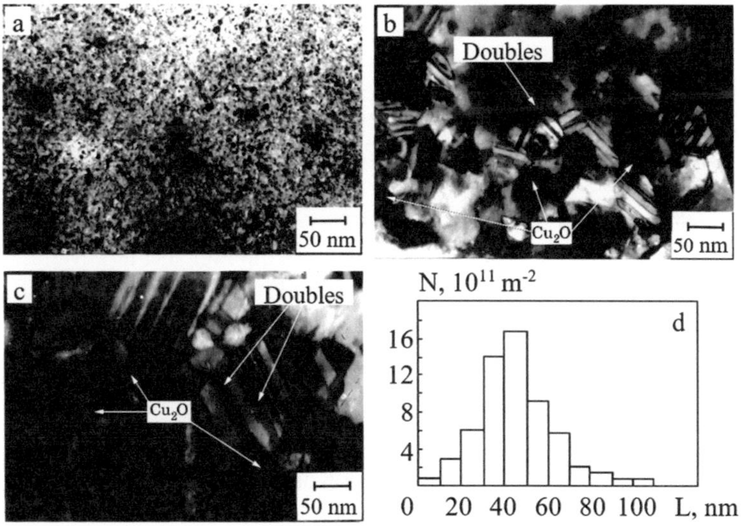

Fig. 9 Microstructure (**a–c**) and histogram (**d**) of unannealed (**a**) and annealed at 700 K (b-d) NiCu film alloys: d = 40 nm, C_{Cu} = 38 at.% (**a**, **b**, **d**); d = 80 nm, C_{Cu} = 60 at.% (**c**). Substrate-carbon film

As noted above, sometimes, at the first stages of annealing, electron microscopic images of NiCu film alloys show Cu_2O oxide crystallites in the form of translucent dark regions. Micrographs illustrating the processes of formation of Cu_2O and recrystallization of a film alloy with parallel polygonization (breaking of grains into individual subgrains by twins of thermal origin) are shown in Figs. 9. It should be noted that in the annealed samples, the average crystallite size before polygonization is 100–300 nm, which is much larger than in the original samples.

Influence of Temperature and Substrate on the Crystal Structure of NiCu Alloy Films

One of the main parameters that determine the process of crystallization of a film sample on a substrate is the substrate temperature T_s. Changing it leads to a change in a number of processes occurring simultaneously. Thus, with an increase in T_s, the mobility of condensed atoms migrating over the surface of the substrate increases, the desorption of adsorbed atoms of the deposited substance (as well as foreign particles and impurities) increases, the concentration of surface defects decreases, which leads to the formation of coarse-grained films with a more perfect structure, having an increased degree of texture (in the case of using oriented NaCl and KBr substrates).

To study the effect of substrate temperature on the crystal structure of NiCu alloy films, samples were deposited on (001) KBr substrates at various temperatures T_s at a rate of $\omega \approx 0.5$–1.5 nm/s. The nature of the influence of the substrate temperature on the crystal structure of the samples is shown in Fig. 10.

Films crystallized at room temperature are characterized by a fine-grained structure with an average crystallite size of ~5–10 nm (Fig. 10a). Such a crystal structure corresponds to a diffraction pattern in the form of slightly blurred rings (Fig. 10b), which indicates the small size of the crystals and their random orientation with respect to the substrate.

An increase in the substrate temperature to 373 K does not lead to a noticeable increase in the average grain size, as can be seen from Fig. 10c; however, the redistribution of the intensity of diffraction rings observed on the electron diffraction patterns (see Fig. 10d) may indicate the beginning of the oriented growth of some grains relative to the substrate.

A further increase in the substrate temperature during condensation to $T_s = 473$ K leads to the formation of a crystalline structure with grain sizes 50–70 nm (Fig. 10e). Simultaneously with the coarsening of crystallites, the number of oriented grains also increases, as evidenced by the contraction of the arches on the rings of electron diffraction patterns (see Fig. 10f).

In alloy films deposited on a substrate at a temperature of $T_s = 573$ K, a coarse-grained structure is observed with grain sizes reaching 70–100 nm (Fig. 10g). With a given crystal structure, an electron diffraction pattern is observed in the form of thin rings with point reflections of the corresponding (001) orientation placed on them (Fig. 10h). In addition, additional reflections are also observed on the electron

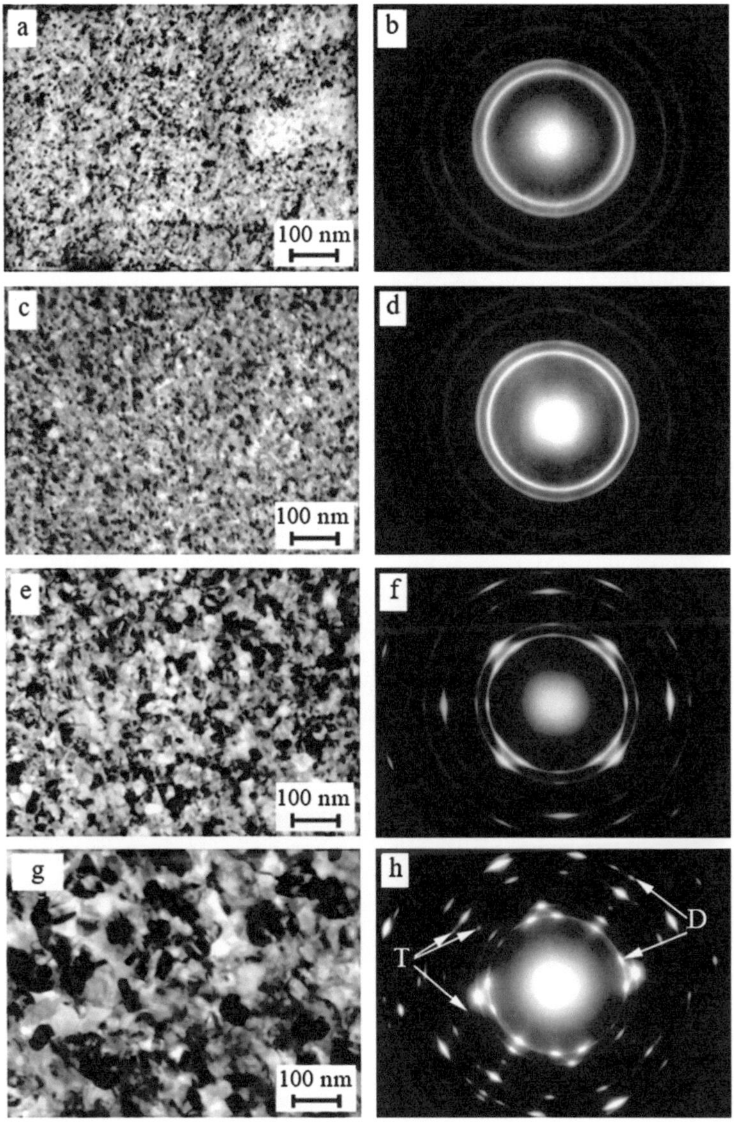

Fig. 10 Microstructure and corresponding electron diffraction patterns of NiCu alloy films (d = 40–60 nm, C_{Cu} = 22–28 at.%) deposited on (001) KBr cleavages at T_s = 293 K (**a**, **b**), T_s = 373 K (**c**, **d**), T_s = 473 K (**e**, **f**), T_s = 573 K (**g**, **h**)

diffraction patterns (marked D in Fig. 10h), which can be explained by the presence of doubles, the formation of which is a characteristic feature of thin films of many metals with an fcc lattice crystallized on alkali halides during thermal evaporation in vacuum (doubles are present on all {111} planes). The bands connecting the reflections (200) and (111) on the electron diffraction pattern (Fig. 10h) indicate the small size of the doubles.

Some dim additional reflections (marked T in Fig. 10h) can be explained by double diffraction (an electron beam diffracted from a matrix with orientation (001) hits the double and diffracts again, or vice versa). Their appearance occurs at some deviation from the exact orientation [100], i.e. with some film tilt. The oriented growth of the investigated thin films of NiCu alloys begins at the substrate temperature $T_s \approx 373$ K. NiCu alloy films obtained by deposition on KBr (001) cleavages at different substrate temperatures were subjected to thermal stabilization annealing in vacuum together with the substrate up to 673 K. The annealing of films on a substrate at higher temperatures is impractical due to the strong sublimation of KBr, which can lead to the destruction of the film sample.

An analysis of the results obtained shows that the annealing of alloy films obtained at room temperature and $T_s = 373$ K leads to a significant increase in the size of crystallites up to 100 nm or more (Fig. 11a, c). The formation of doubles and an increase in the number of grains oriented parallel to the plane of the substrate are also observed (textural maxima in Fig. 11d).

For films obtained at substrate temperatures of 473 K and 573 K, upon annealing to 673 K, an increase in the crystallite size up to 200 nm is observed. Along with this, the appearance of new crystallites with sizes of 5–10 nm is also observed (Fig. 11e, g) due to the polygonization process. Additional reflections in the corresponding electron diffraction patterns (Fig. 11f, h) are due to both twinning and the presence of stacking fault tetrahedra (SFT in Fig. 11g).

5.2 NiCo Alloy Films

Some NiCo alloy films were obtained using the finite sample evaporation method. A series of bulk NiCo alloys of known composition was prepared for evaporation.

According to the data of the phase diagram (Fig. 4), for NiCo alloys at temperatures close to the melting point and in the liquid state, complete mutual solubility of the components is observed in the entire concentration range. In this case, the liquidus and solidus lines coincide, which excludes the possibility of alloy fractionation during melting.

In general, during the evaporation of binary alloys, the composition of the film differs from the composition of the initial material due to the fact that one of the components may have a high evaporation rate. The evaporation rate of the material at temperature T is determined by the formula [33]:

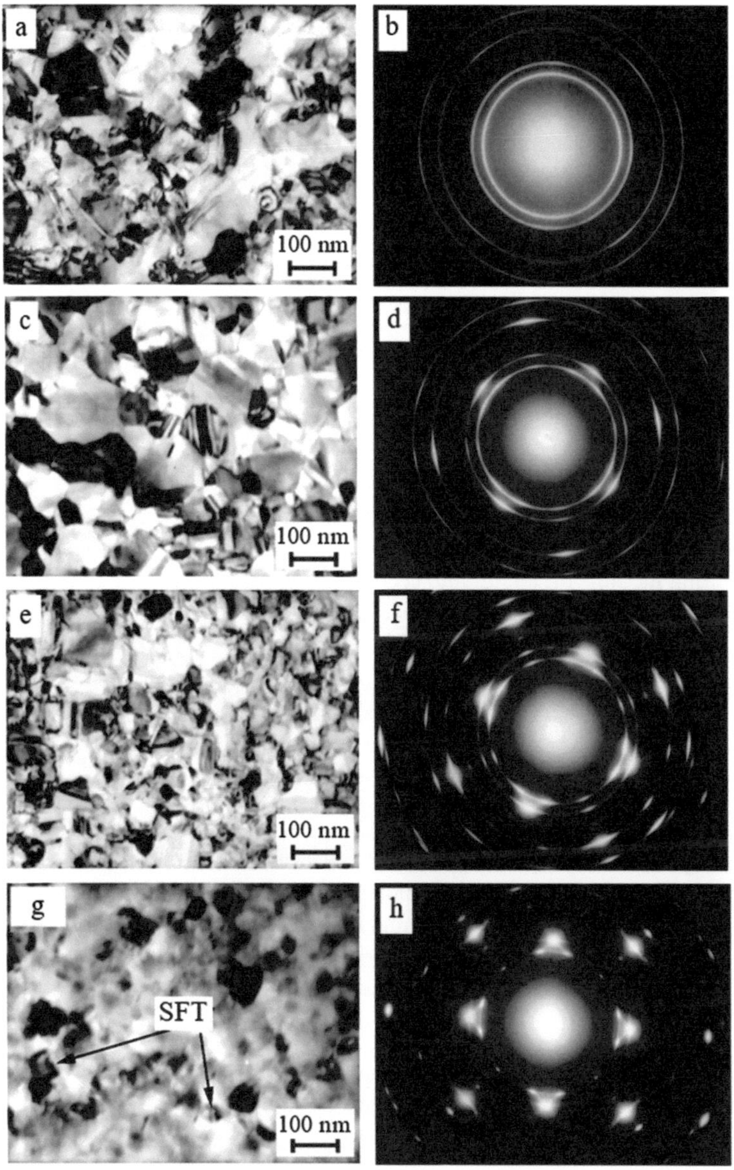

Fig. 11 Microstructure and corresponding electron diffraction patterns of NiCu alloy films annealed to 673 K (d = 40–60 nm, C_{Cu} = 22–28 at.%) deposited on (001) KBr cleavages at T_s = 293 K (**a**, **b**), T_s = 373 K (c, d), T_s = 473 K (**e**, **f**), T_s = 573 K (**g**, **h**)

$$v = kp\sqrt{\frac{M}{T}} = A_1 p, \tag{4}$$

where k is a constant; p is the vapor pressure; M is the molecular weight of the substance; T is the evaporation temperature.

When alloys are evaporated in a vacuum, the vapor pressure of each of the components above the melt is determined by the composition of the alloy and differs from the values for pure components. In the simplest case, the vapor pressure of component A (p_A) in the presence of component B (p_B) is determined by Raoult's law:

$$\frac{p_A - p_{AB}}{p_A} = C_B, \tag{5}$$

where C_B is the mole fraction of component B (the ratio of the number of moles of this component to the total number of moles of all components).

In [33], a method for calculating the fractionation of binary systems during the evaporation of final samples was proposed. Let at the initial moment of time in the evaporator there is a binary alloy with a mass content of the components m_{01} and m_{02}. During the time dt, taking into account the expression for the evaporation rate (4), the decrease in the mass of each of the components is

$$dm_A = -kp_{0A} f_A \sqrt{\frac{\mu_A}{T}} C_A(t)dt, \tag{6}$$

$$dm_B = -kp_{0B} f_B \sqrt{\frac{\mu B}{T}} C_B(t)dt, \tag{7}$$

where k is a constant depending on the choice of the system of units; p_{0A}, p_{0B} is saturation vapor pressure of pure components A and B of the alloy at temperature T; f_A, f_B areactivity coefficients of the components:

$$f_A = \frac{\alpha_A}{C_A}, \tag{8}$$

where $\alpha_A = \frac{p_{AB}}{p_A}$ is theactivity of component A (similarly for component B); $C_A(t)$, $C_B(t)$ are the atomic parts of the components; μ_A, μ_B are their molar masses. If Raoult's law is fulfilled (ideal melts), we have

$$1 - \frac{p_{AB}}{p_A} = C_B,$$

$$\alpha_A = \frac{p_{AB}}{p_A} = 1 - C_B = C_A, \tag{9}$$

whence $f_A = f_B = 1$.

According to the data of [15], the same saturation vapor pressures for Co and Ni are achieved at close temperatures (for example, $p_{Co} = 1.33$ Pa at T = 1790 K, p_{Ni} has the same value at T = 1800 K; a similar situation occurs for high pressures), so we can take $p_A = p_B$. In addition, the molar masses of Co and Ni are close ($\mu_{Co} = 58.9$ g/mol, $\mu_{Ni} = 58.7$ g/mol), which means that for the mass parts of Co and Ni we obtain the ratio:

$$\mathrm{C}_{Co} = \frac{\mathrm{C}_{Co}\mu_{co}}{\mathrm{C}_{Co}(\mu_{co} - \mu_{Ni}) + \mu_{Ni}} = \frac{58.9}{\mathrm{C}_{Co}(58.9 - 58.7) + 58.7}\mathrm{C}_{Co}$$
$$= \frac{58.9}{58.7 + 0.2C_{co}}\mathrm{C}_{Co} \approx \mathrm{C}_{Co}$$

Then

$$kp_{0A} f_A \sqrt{\frac{\mu A}{T}} \approx kp_{0B} f_B \sqrt{\frac{\mu B}{T}}.$$

As a result, the mass fraction of Co in the resulting alloy will be

$$c_{Co}\prime = \frac{dm_{Co}}{dm_{Co} + dm_{Ni}} = \frac{C_{Co}}{C_{Co} + C_{Ni}} = C_{Co} = c_{Co}. \tag{10}$$

Thus, under the condition that the Raoult law is satisfied, the concentration of the components in the film NiCo alloy should be the same as in the bulk sample, and the films should be homogeneous in chemical composition.

In accordance with the data of [16], bulk NiCo alloys with a Co content of up to 65–67 wt.% at room temperature form substitutional solid solutions with an fcc lattice over the entire temperature range of the solid state. For $C_{Co} > 75$ wt.% at 300 K, the solid solution has an hcp lattice based on α-Co; with an increase in temperature, an fcc lattice based on β-Co is formed (the temperatures of the direct and reverse hcp → fcc transitions depend on the concentration of the components). In the intermediate concentration range, a two-phase composition (fcc + hcp-tr.) is observed (Fig. 4).

The results of the analysis of electron diffraction patterns of films of the CoNi alloy with different concentrations of components (Fig. 12) show that patterns similar to massive alloys are observed for alloy films. For unannealed films with $C_{Co} \leq 70$ wt.% (Fig. 12a), the lines of only the fcc phase are recorded by electron diffraction. In this case, the diffraction maxima are quite wide, which indicates a small crystallite size. Annealing of the samples does not lead to a change in the phase composition of these films; only fcc-tr. NiCo (Fig. 12b–d). Thermal stabilization leads to a decrease in the width of the lines in the electron diffraction patterns and an increase in their intensity. The reason for this is the increase in grain size in the sample.

At a Co content of 75 wt.%, the films most likely have an fcc structure. The analysis of electron diffraction patterns in this case is difficult, since, due to the dispersion of the structure of the films, the lines in the electron diffraction patterns are quite wide, and the interplanar distances for the most intense lines of the fcc and hcp lattices are

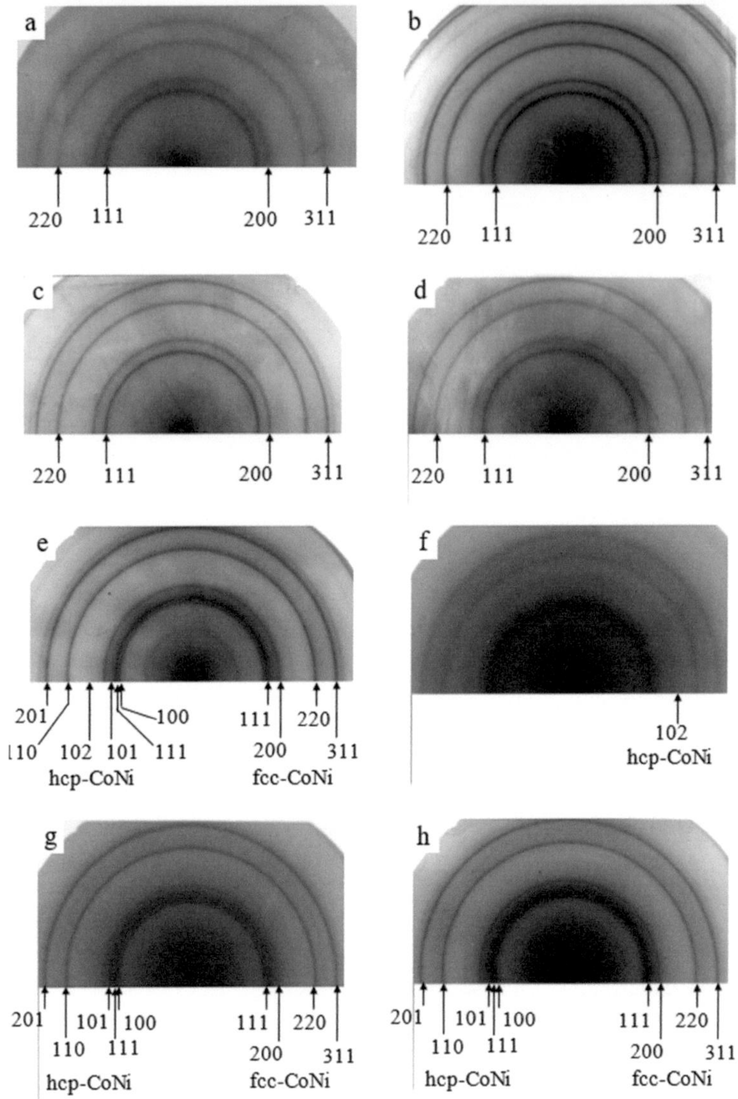

Fig. 12 Phase composition of unannealed (a, f) and annealed (b-e, g-h) NiCo films. Co content: **a**, **b** 10 wt.%; **c** 30 wt.%; **d** 70 wt.%; **e** 75 wt.%; **f**, **g** 80 wt.%; **h** 90 wt.%

almost the same. As a result, it is impossible to confidently separate the lines of each of the phases. Annealing to a temperature of 500–700 K leads to a decrease in the line width and the lines of both phases are fixed on the electron diffraction patterns (Fig. 12e), and the intensity of the lines of the fcc phase is higher. As is known [17], in a two-phase region, lines of both phases are observed in the diffraction pattern, when the composition of the phases is constant and corresponds to the boundaries

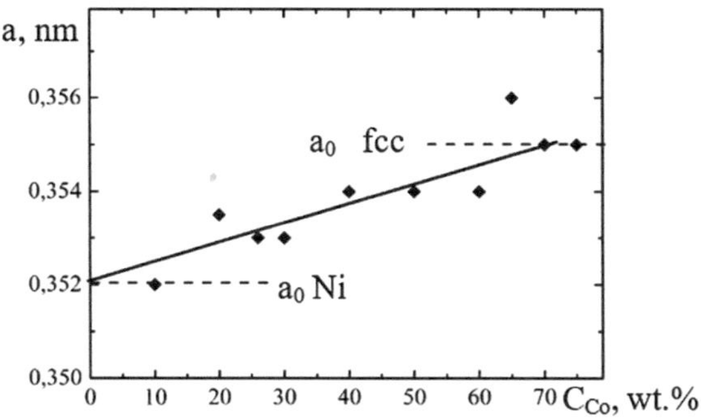

Fig. 13 Dependence of the fcc lattice parameter on the Co concentration for films of NiCo alloys

of the regions in the phase diagram. The lattice parameters of each of the phases also remain unchanged. A change in the concentration of the components causes only a change in the phase ratio, and, as a consequence, a change in the intensity of the lines. Thus, as the cobalt concentration increases, the intensity of the lines of the hcp lattice of the NiCo solid solution will gradually increase and, accordingly, the intensity of the lines of the fcc solid solution will decrease. Since the relative intensity of the phase lines is proportional to its volume (and, given the close values of the density of materials, and mass) fraction, the higher intensity of the fcc phase lines may indicate its higher content in the alloy of a given concentration. These reflections are confirmed by the presented electron diffraction patterns (Fig. 12).

Based on the results of studies of films of alloys with an fcc lattice, we built an experimental dependence of the lattice parameter on the concentration of Co (Fig. 13). It can be seen from the figure that an increase in the concentration of Co atoms leads to an increase in the fcc lattice parameter of the alloy according to a linear law, i.e. for films of fcc-NiCo alloys, Vegard's rule is observed.

For unannealed samples with CCo $\geq$ 80 wt.% (Fig. 12f), in full accordance with the state diagram of massive alloys, the electron diffraction lines of the hcp grating are observed. This can be confirmed by the presence of the (102) line, which is absent for the fcc lattice. As a result of annealing, a two-phase composition (fcc + hcp) is observed in these samples (Fig. 12g, h), while the intensity of the hcp lines is noticeably greater. Obviously, when the samples are heated to 700 K, a phase transition occurs in them with the formation of a high-temperature fcc phase, since the transition temperature is less than 700 K (with an increase in the Ni content in the alloys, the transition temperature decreases). However, despite the relatively slow cooling, the reverse transition does not completely pass, and a certain amount of the fcc phase remains in the sample. A possible reason for this is the size phase effect. According to the data of [34], the fcc phase can exist in small Co particles at temperatures below the end point of the reverse transition. Due to the rather significant size dispersion of crystallites in our samples (which is indicated by the data of electron microscopy studies), for that part of the crystallites, the sizes of which are much smaller than average, the fcc phase can remain stable, while in larger crystallites after cooling, an hcp phase is formed. According to electron diffraction data, the lattice parameters of these phases are: for the fcc phase a = 0.355–0.357 nm; for hcp a = 0.248–0.252 nm, c = 0.410–0.432 nm, which is in good agreement with similar data for bulk alloys.

As noted earlier, at the initial stages of film formation, features of the structural state are observed, in particular, at small thicknesses, the films are structurally and electrically discontinuous (island). In connection with this, we carried out a study of the structure of ultrathin (d < 20 nm) NiCo films, the results of which are shown in Fig. 14. Since the condensation was carried out on a neutral non-oriented substrate (carbon film) at room temperature ($T_s \approx 300$ K < $2T_m/3$), the diffusion mobility of atoms on the substrate was hindered. As a result, in unannealed films, regardless of composition and thickness (Fig. 14a), almost the same structure was observed: islands of irregular shape with sizes less than 3 nm, and the gaps between them look like channels of approximately the same width (classical «labyrinth» structure [35]).

Annealing such films up to 700 K leads to a significant morphological change in their structure. With a small thickness (d ≈ 5–10 nm), a simple enlargement of the islands occurs due to coalescence (Fig. 14b). With a greater thickness d ≈ 15–20 nm (Fig. 14c–d) «bridges» appear between the islands, and a further increase in thickness (Fig. 14f) leads to the formation of electrically and then structurally continuous films. When the islands are enlarged, the crystal structure is also improved, which is manifested in the formation of islands with a clear cut (Fig. 14d, f).

An analysis of electron microscopic images shows that for unannealed films of the entire concentration range, a strongly fine-grained structure with a crystallite size of 1–2 nm is observed (Fig. 14a), which explains the reason for the significant width of the diffraction maxima. Crystallites do not have a clear cut, so the statistical processing of data regarding their size is difficult. The micrographs indicate the formation of a large number of small crystallization centers during the deposition process and their simultaneous growth.

We have made an attempt to check the manifestation of the phase size effect in films of this alloy. For ultrathin films with a thickness of d = 9 nm with a cobalt content of 80 wt.% (Fig. 14b), lines of both phases are observed on the electron diffraction patterns; however, the relative intensity of the lines of the fcc lattice is noticeably greater than that of the hcp lattice, in contrast to thicker continuous films (Fig. 14g). Since the intensity of the electron diffraction lines is proportional to the volume content of this phase, it is obvious that in island films, compared to thicker continuous films, the fraction of the fcc phase is larger than the fraction of the hcp phase.

The reason for this is the small size of crystallites (phase size effect). Unfortunately, this statement is only qualitative, since the electron diffraction reflections from ultrathin film alloys are extremely weak.

The results of electron microscopic studies of the crystal structure of solid film samples (d = 50–60 nm) are shown in Figs. 15.

Unannealed films (Fig. 15a) of all studied component concentrations have approximately the same structure—small crystallites 5–10 nm in size without a clear cut. In the process of thermal stabilization, an increase in the size of crystallites is observed due to recrystallization processes (Fig. 15b–d).

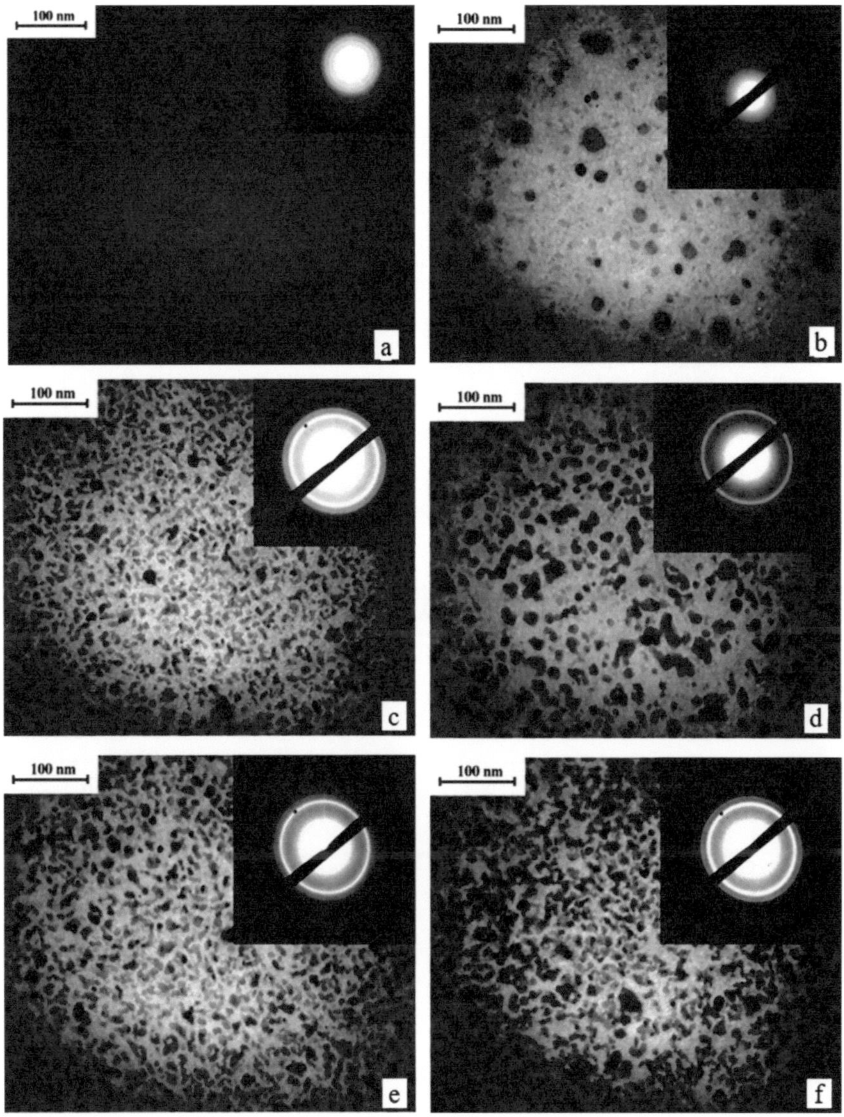

Fig. 14 Morphology of unannealed (**a**) and annealed to 700 K (**b**–**f**) ultrathin NiCo alloy films: **a** d = 20 nm, C_{Co} = 20%; **b** d = 9 nm, C_{Co} = 80%; **c** **d** = 9 nm, C_{Co} = 10%; **d** d = 10 nm, C_{Co} = 30%; **e** d = 15 nm, C_{Co} = 70%; **f** d = 20 nm, C_{Co} = 50%

For some of the crystallites, a clear faceting and the presence of stacking faults (most likely of growth origin) are observed. To estimate the size of crystallites in the film plane, statistical processing of micrographs was carried out with the construction of histograms (Fig. 16). The average grain sizes determined on their basis have a value of 110–200 nm and show a slight tendency to coarsen with increasing Co content.

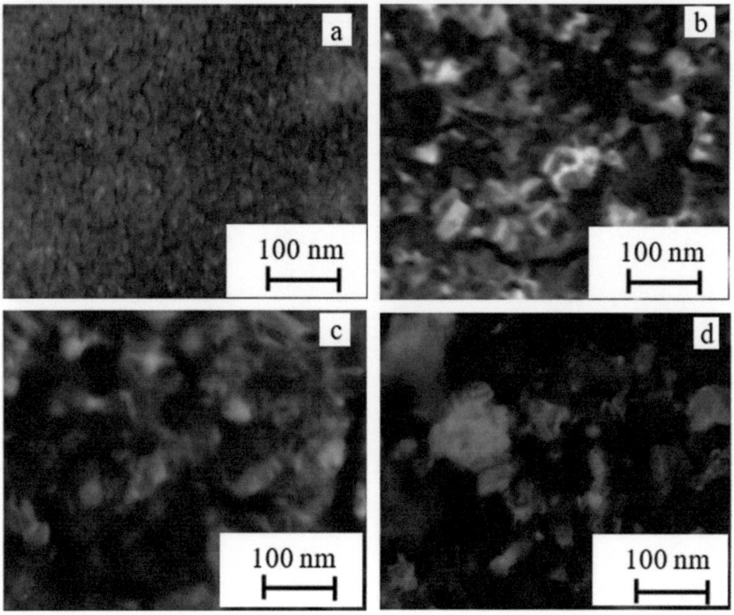

Fig. 15 Microstructure of NiCo film alloys with a thickness of d = 60 nm: **a** typical structure of an unannealed film (C_{Co} = 70 wt.%); **b–d** film structure after thermal stabilization (**b** C_{Co} = 10 wt.%; **c** C_{Co} = 50 wt.%; **d** C_{Co} = 90 wt.%)

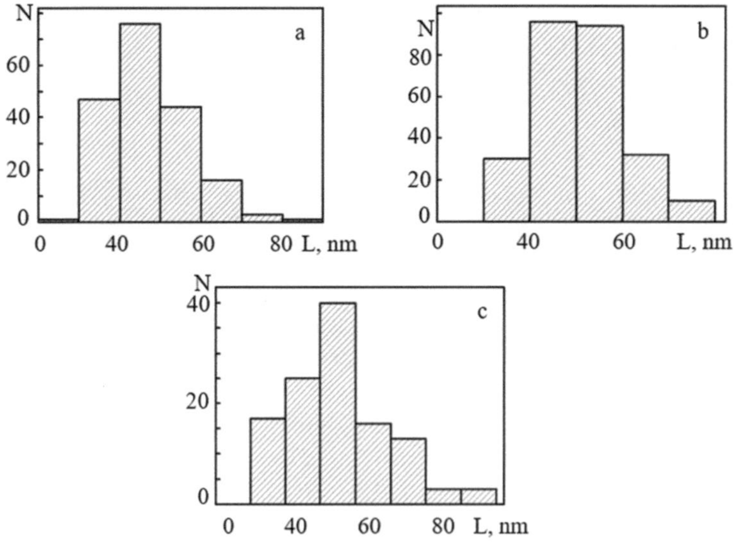

Fig. 16 Histograms for determining the average sizes of crystallites in the film plane: **a** C_{Co} = 50 wt.%; **b** C_{Co} = 70 wt.%; **c** C_{Co} = 90 wt.% (d = 60 nm)

Electron microscopic investigations make it possible to determine the sizes of crystallites in the film plane. We have made an attempt to estimate the sizes of crystallites in the direction of the normal to the film plane (to estimate the «height» of crystallites) by determining the sizes of coherent scattering regions (CSR). Determining the size of the CSR in the direction of the normal to the film is possible on the basis of an analysis of the broadening of the profiles of diffraction lines in the electron diffraction patterns.

The limited size of the CSR leads to an increase in the line width, and the line broadening is inversely proportional to the size of crystallites or blocks. The integral width β of the line is determined by the Selyakov-Scherer formula [36]

$$\beta = \frac{m\lambda}{L_{\perp}cos\theta}, \tag{11}$$

where m is a coefficient depending on the shape of crystallites, m = 0.98–1.38 (m ≈ 1 can be considered); λ is the de Broglie wavelength of electrons; $L_{\perp}$ is the average size of crystallites in the direction of the beam; θ is the Bragg angle.

In the electron diffraction analysis, the angles θ do not exceed 5°; therefore, cosθ ≈ 1 can be taken with sufficient accuracy. As a result, the crystallite size can be estimated from the formula:

$$\beta = \frac{m\lambda}{L_{\perp}\cos\theta}, \tag{12}$$

where $\beta = \frac{1}{f(0)}\int f(2\theta)d(2\theta)$.

The value of β represents the integral width of the peak, the shape of which is described by the function f(2θ). Since the integral in the numerator of this expression determines the area of the curve under the peak, and the denominator determines the maximum height of the peak, the value of β is the width of a rectangle of the same area and the same height as the peak.

It should be noted that the broadening of diffraction lines can also occur for other reasons. Among them, instrumental broadening, which is determined by the shooting conditions, plays a significant role. The instrumental broadening can be taken into account by shooting a reference sample, which is a coarse-grained film sample free from microstresses (we used the line broadening of the annealed Al reference sample).

Line broadening is also possible due to the existence of microstresses in the sample and a significant number of randomly located dislocations. For these cases, we have

$$\beta = 4\varepsilon tg\theta,$$

where $\varepsilon = \frac{\Delta d}{d_0}$ is therelative change in interplanar spacing (microstrain); Δd_m is thevolume-averaged maximum change in interplanar spacing; d_0 are interplanar distances in an ideal crystal. However, due to the small diffraction angles, the contribution of these factors to the total broadening is insignificant. Figure 17 shows the

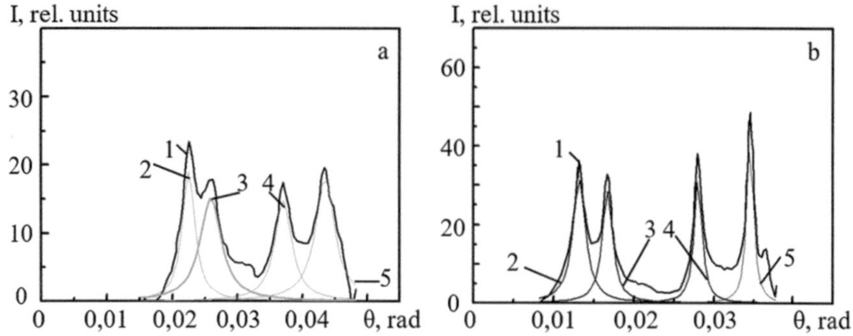

Fig. 17 Angular intensity distribution of diffraction maxima for NiCo alloy film d = 70 nm with C_{Co} = 10 wt.%: **a** unannealed; **b** annealed film. 1—experimental intensity distribution, 2-5—approximation of peaks

experimental (curves 1) used by us for the angular dependences of the intensity distribution of diffraction lines on the electron diffraction patterns and the approximation lines (curves 2–5) obtained using the Microcal Origin data processing package. When calculating the magnitude of the broadening, the approximation of the peak shape by the Cauchy function was used.

The computer selection of the function parameters made it possible to describe the peaks obtained after the background value was taken into account as accurately as possible. The half-width of the peak, expressed in angular units, was taken as the value of physical broadening. The values obtained for unannealed films (5–10 nm) coincide in order of magnitude with the data of electron microscopic studies, which indicates an approximately equiaxed shape of crystallites. In addition, these data show that the films are in a nanocrystalline rather than amorphous state. As a result of heat treatment, an increase in the size of crystallites is observed both in the film plane and in the direction normal to the sample. The values of grain sizes determined from line broadening have a value of the order of 20–25 nm. Thus, the annealed NiCo alloy films consist of crystallites whose dimensions in the substrate plane are 100–200 nm, and in the direction perpendicular to the substrate plane, they are 5–10 times smaller, i.e. crystallites have a significant unevenness.

We also note the features of the structural-phase state of NiCo film alloys obtained by simultaneous condensation of Co and Ni during evaporation from independent sources (electron-beam guns). In general terms, the structure of alloy films obtained by this method is identical to the structure of alloys obtained by evaporation of sample portions—unannealed films consist of small crystallites 5–10 nm in size; as a result of annealing, an increase in crystallite size up to 50–100 nm is observed, depending on the annealing temperature. However, the phase composition of such films has a number of significant differences.

In Figs. 18 and 19, for example, the microstructure of NiCo films with a thickness of d = 50 nm with a sharply different concentration of components—14 wt.% Co (Fig. 18) and 73 wt.% Co (Fig. 19) before and after annealing at T = 1300 K.

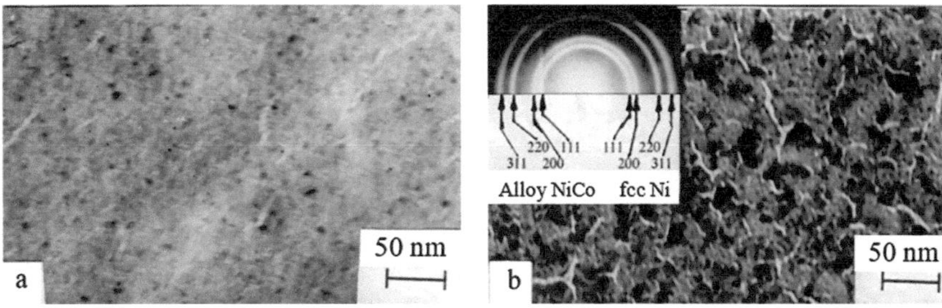

Fig. 18 Microstructure of NiCo alloy films (C_{Co} = 14 wt.%, d = 50 nm). Temperature: **a** 300 K; **b** 1300 K

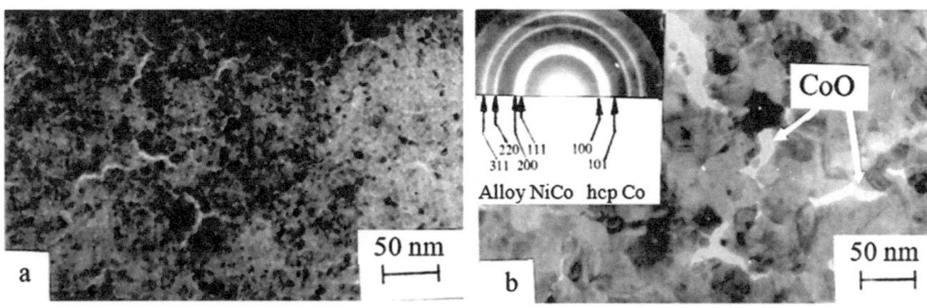

Fig. 19 Microstructure of NiCo alloy films (C_{Co} = 73 wt.%, d = 50 nm). Temperature: **a** 300 K; **b** 1300 K

The initial state (Figs. 18a, 19a) of both film samples is almost the same—small crystallites 5–10 nm in size, which does not allow reliable identification of their phase composition using electron diffraction patterns (significant width of diffraction maxima). After annealing the samples (Figs. 18b, 19b), an almost identical increase in the size of crystallites is observed as a result of recrystallization processes, while the phase composition (electron diffraction patterns in the insets) differs significantly.

In both cases (both at low and high cobalt concentrations), the films exhibit a two-phase composition: (fcc-Ni) + (fcc-NiCo) and (fcc-NiCo) + (hcp-Co), respectively, although, according to the state diagram of a massive alloy (Fig. 4) at low concentrations of Co, only the fcc-NiCo alloy is formed.

The reason for these differences, most likely, is that the formation of a NiCo solid solution with simultaneous condensation of Co and Ni on the substrates at $T_s = 300$ K does not occur to the end (nonequilibrium process), so that even high-temperature (T = 1300 K) annealing does not allow completely relax this disequilibrium.

It should also be noted that the micrographs of samples with $C_{Co} > 50$ wt.% show stacking faults (which appeared during annealing) and characteristic extinction contours (Fig. 19b). Here, the same feature occurs that is also observed in Co films: when the crystallite size is less than 10 nm, the formation of stacking faults in them is difficult, which is associated with a high energy barrier to the displacement of Shockley dislocations (it is known that the growth of a stacking fault begins on the

{111} fcc lattices as a result of the splitting of full dislocations into two partial ones). High-temperature recrystallization processes, leading to an increase in grains, cause only a slight decrease in the energy of partial dislocation nucleation.

When film alloys are annealed to 700 K, the formation of CoO and NiO oxides is practically not observed in them. In films that have been annealed to a temperature of T = 1300 K, in parallel with an increase in the size of crystallites, the formation of fine-grained oxide CoO is observed in samples with $C_{Co} > 15$ wt.% (Fig. 4b).

5.3 *NiFe Film Alloys*

To obtain films of NiFe alloys, we used the method of evaporating final portions of the finished NiFe alloy known as permalloy 50N.

The NiFe system obeys Raoult's law almost perfectly. It is also characterized by complete mutual solubility of the components in the solid state at high temperatures and in the liquid state (Fig. 5). In this case, for iron, the temperature at which its saturated vapor pressure becomes equal to the nickel saturated vapor pressure is approximately 50 K lower than the nickel temperature [15]. The evaporation rate of iron under such conditions becomes much higher than that of nickel.

Thus, during the evaporation of finished alloys, the fractionation of the composition of the films is possible (change in the composition along the thickness), and at the initial moments of evaporation, the films will be enriched in iron. Such conclusions are confirmed by the data of [37].

According to the literature data [38], in bulk samples of the NiFe alloy, a compound with an fcc lattice (taenite) is observed with a lattice parameter varying from 0.3524 nm (pure Ni) to 0.3596 nm ($C_{Ni} = 39\%$). In particular, the composition with $C_{Ni} = 49\%$ corresponds to the value of the lattice parameter a = 0.3586 nm. The study of the phase composition of the films by electron diffraction showed that in all films there is an fcc phase with a lattice parameter a = 0.359–0.361 nm (Fig. 20, inset).

Thus, in film samples (compared to bulk samples), some increase in the lattice parameter is observed. It should be noted that the electron diffraction patterns of the NiFe alloy films also show a number of lines that may belong to the NiFe intermetallic phase with the P4/mmm space group. Unannealed films have a fine-grained structure (Fig. 20a), their annealing leads to an increase in grain size; however, grain sizes do not exceed 50 nm (Fig. 20b).

6 Conclusions

- The formation of the fcc NiCu alloy occurs on the substrate directly in the process of simultaneous condensation of the components in the entire thickness range; for structurally continuous annealed films, depending on the concentration of the

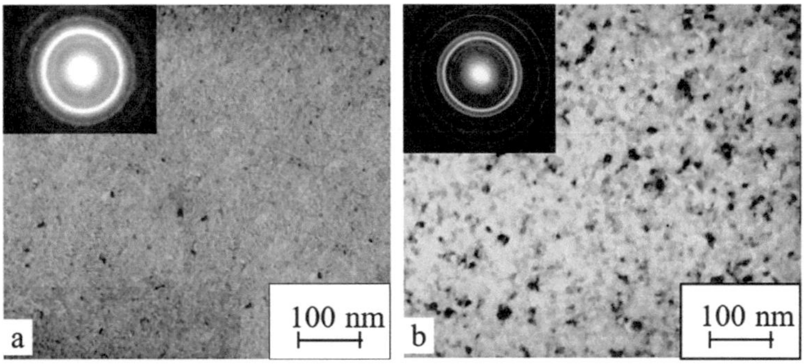

Fig. 20 Microstructure of NiFe alloy films (d = 36 nm) in the (**a**) unannealed and (**b**) annealed states. $T_s = 300$ K

components, the crystal lattice parameter is a = 0.353–0.362 nm in accordance with Vegard's rule;

- In films of NiCu alloys with effective thicknesses d < 10 nm, obtained on amorphous substrates at a temperature of T = 300 K, an island structure is observed with an average island size of 1.5–2 nm (unannealed samples) and up to 20 nm (annealed samples) depending on the effective thickness;
- Films of NiCu alloys with a thickness of d > 15–20 nm, obtained both on oriented ((001) KBr or NaCl) and non-oriented (carbon films) substrates at T = 300 K, have a fine-grained structure with an average crystallite size 5–10 nm; annealing to 700 K or condensation on a substrate heated to T_s = 400–600 K leads to an increase in the crystallite size to 50–100 nm; simultaneously with the coarsening of crystallites, the number of oriented grains also increases;
- In NiCo film alloys with a thickness of d < 15 nm, obtained on amorphous substrates at T = 300 K, an island structure is observed with an average island size of 1–2 nm in unannealed and 20–25 nm in annealed films, depending on the effective;
- Unannealed and annealed NiCo alloy films at $C_{Co} < 70$ wt.% have an fcc structure, annealed films with $C_{Co} \geq 80$ wt.% have a two-phase composition (fcc + hcp)-solid solution NiCo;
- The change in the lattice parameter of fcc-NiCo films with increasing C_{Co} is described by the Vegard rule;
- Unannealed film samples of NiCo alloys have a fine-grained structure with a crystallite size of 5–10 nm; as a result of annealing to T = 700 K, an increase in the average crystallite size in the film plane to 40–60 nm is observed;
- In unannealed films of NiCo alloys, crystallites have an equiaxed shape; after annealing in the direction normal to the film plane, their size is ~ 25 nm;
- For NiCo alloy films obtained by evaporation of Co and Ni from two independent sources, a two-phase composition is observed (fcc-Ni) + (fcc-NiCo) or (fcc-NiCo) + (hcp-Co);

– $Ni_{50}Fe_{50}$ alloy films have an fcc lattice with a = 0.359–0.361 nm; unannealed films have a fine-grained structure; annealing leads to an increase in grain size up to 50 nm.

References

1. Soni, V.K., Sanyal, S., Sinha, S.K.: Phase evolution and mechanical properties of novel $FeCoNiCuMo_x$ high entropy alloys. Vacuum **174**, 109173 (2020)
2. Hua, X., Li, J., Liu, H., Yang, Z., Liu, S.F.: Preparation of Cu_2Se thin films by vacuum evaporation and hot-pressing. Vacuum **185**, 109947 (2021)
3. Li, G., Song, G., Wang, N.: Influence of Cu content on the phase composition and thermoelectric properties of deposited Cu–Se films. Surfaces and Interfaces, p. 101651 (2022)
4. Pylypenko, O.V., Pazukha, I.M., Ovrutskyi, A.S., Odnodvorets, L.V.: Electrophysical and magnetoresistive properties of thin film alloy $Ni_{80}Fe_{20}$. J. Nano Electron. Phys. **8**(3), 03022 (2016)
5. Bereznyak, Y., Odnodvorets, L., Poduremne, D., Protsenko, I., Shabelnyk, Y.: High-entropy film alloys: electrophysical and magnetoresistive properties. Springer Proc. Phys. **210**, 17–24 (2018)
6. Shkurdoda, Y.O., Chomous, A.M., Shabelnyk, Y.M., Loboda, V.B., Khursenko, S.M.: Method of production and structural-phase state of medium-entropy Equiatomic FeNiCoCu film alloy. Proceedings of the IEEE 10th International Conference on Nanomaterials: Applications and Properties, p. 9309616 (2020)
7. Saltykov, D.I., Protsenko, S.I., Pazukha, I.M., Shkurdoda, Y.O.: Concentration and heat treatment effects on magnetoresistive properties of three-layer film systems based on Fe_xCo_{100-x} and Cu. Thin Solid Films **716**, 138422 (2020)
8. Bezdidko, O.V., Nepijko, S.A., Shkurdoda, Y.O., Shabelnyk, Y.M.: Structure and Magnetoresistive properties of three-layer films $Co_{(1-x)}Cr_x$/Cu/Co. J. Nano Electron. Phys. **13**(3), 1–4 (2021)
9. Loboda, V.B., Kolomiets, V.M., Khursenko, S.M., Shkurdoda, Y.O.: The electrical conductivity of the three-layer polycrystalline films Co/Ag(Cu)/Fe in the conditions of atoms interdiffusion. J. Nano Electron. Phys. **6**(1), 04032 (2014)
10. Loboda, V.B., Shkurdoda, Yu.O., Kravchenko, V.O., Khursenko, S.M., Kolomiets, V.M.: Structure and magnetoresistive properties of polycrystalline Co/Cu/Co films. Metallofiz. Noveishie Tekhnol. **33**(2), 161–169 (2011)
11. Loboda, V.B., Kolomiets, V.M., Shkurdoda, Yu.O., Kravchenko, V.O., Dekhtyaruk, L.V.: Structure and magnetoresistive properties of nanocrystalline film systems based on Co, Fe, Ag, and Cu. Metallofiz. Noveishie Tekhnol. **34**(8), 1043–1055 (2012)
12. Pogrebnjak, A.D., Lebed, A.G., Ivanov, Yu.F.: Modification of single crystal stainless steel structure (Fe–Cr–Ni–Mn) by high-power ion beam. Vacuum **63**(4), 483–486 (2001)
13. Pogrebnjak, A.D., Shablya, V.T., Sviridenko, N.V., Valyaev, A.N., Plotnikov, S.V., Kylyshkanov, M.K.: Study of deformation states in metals exposed to intense-pulsed-ion beams (IPIB). Surf. Coat. Technol. **111**(1), 46–50 (1999)
14. Pogrebnjak, A.D., Yakushchenko, I.V., Abadias, G., Chartier, P., Bondar, O.V., Beresnev, V.M. Takeda, Y., Sobol, O.V., Oyoshi, K., Andreyev, A.A., Mukushev, B.A.: The effect of the deposition parameters of nitrides of high-entropy alloys (TiZrHfVNb)N on their structure, composition, mechanical and tribological properties. J. Superhard Mater. **35**(6), 356–368 (2013)
15. Meissel, L., Gleng, R.: Thin Film Technology: Handbook. Moscow: Sov. Radio, Vol. 2, 768 p. (in Russian) (1977)

16. Ganina, N.I., Zakharov, A.M., Olenichev, V.G.: Diagrams of the State of Metallic Systems. Moscow: VINITI, Vol. 32, 626 p. (in Russian) (1988)
17. Lyakishev, N.P.: State Diagrams of Binary Metallic Systems: Handbook. Moscow: Машиностроение, 1996, Vol. 2, 1024 p. (in Russian) (1997)
18. Kikoin, I.K.: Tables of Physical Quantities. Handbook. Moscow: Atomizdat, 1008 p. (in Russian) (1976)
19. Nepiyko, S.A.: Physical Properties of Small Metal Particles. Kyiv: Naukova dumka, (in Russian) (1985)
20. Bar'yakhtar, V.G.: Solid State Physics: Encyclopedic Dictionary. Kyiv: Naukova dumka, 656 p. (in Russian) (1996)
21. Pushin, V.G., Blinov, S.G., Yurchenko, L.I.: Investigation of the microstructure of cobalt-nickel alloys metastable with respect to fcc-hcp martensitic transformation. Fizika Metallov i Metallovedeniye **75**(3), 96–109. (in Russian) (1993)
22. Langvagen, S.E., Lyubashevsky, O.P., Chizhov, P.E.: Magnetic properties of small particles of Fe–Ni alloys. Fizika Metallov i Metallovedeniye **9**, 69–73 (1991). (in Russian)
23. Komnik, Yu.F.: Physics of Metal Films. Dimensional and Structural Effects. Moscow: Atomizdat, 264 p. (in Russian) (1979)
24. Gladkikh, N.T.: Surface Phenomena and Phase Transformations in Condensed Films. Khar'kov: KHNU im. V.N. Karazina, 276 p. (in Russian) (2004)
25. Kukushkin, S.A., Osipov, A.V.: Processes of condensation of thin films. Uspekhi fizicheskikh nauk **168**(10), 1083–1116. (in Russian) (1998)
26. Protsenko, I.Yu., Saenko, V.A.: Thin Metal Slabs (technology and power). Sumi: SumDU, 187 p. (in Ukrainian) (2002)
27. Loboda, V.B., Saltykova, A.I., Khursenko, S.N.: Physical properties of thin nanocrystalline metal films: Book 1. Crystal structure, phase and elemental composition of films. Saarbrucken: LAP LAMBERT Academic Publishing, 113 p. (in Russian) (2014)
28. Loboda, V.B., Kravchenko, V.O., Shkurdoda, Yu.O. et al.: "Structure, electrical conductivity and galvanomagnetic power of thin melts of $Fe_{0.5}$–$Ni_{0.5}$ alloy", Visnyk SumDU. Seriya: Fizyka, matematyka, mekhanika **8**(67), 115–123. (in Ukrainian) (2004)
29. Goldshmidt, H.J.: Interstitial Alloys. Moscow: Mir, 424 p. (in Russian) (1971)
30. Semiletov, S.A., Zavyalova, A.A., Imamov, R.M.: About anomalous crystal structures in thin films of transition and rare-earth metals. Izvestiya AN SSSR **41**(11), 2230–2237. (in Russian) (1972)
31. Morokhov, I.D., Trusov, L.I., Chizhik, S.P.: Ultrafine Metallic Medium. Moscow: Atomizdat, 264 p. (in Russian) (1977)
32. Trusov, L.I., Kholmyansky, V.A.: Island Metal Films. Moscow: Metallurgiya, 320 p. (in Russian) (1973)
33. Kravchenko, V.A., Loboda, V.B., Protsenko, I.Yu. et al.: Structure and electrophysical properties of Ni–Co alloy films within 300–1300 K temperature range. Funct. Mater. **6**(5), 892–895 (1999)
34. Kuzmenko, A.I., Prokofiev, T.A., Protsenko, I.E. et al.: Phase diagram of the fcc-hp transition in small particles of cobalt. Izvestiya AN SSSR, №8, pp. 1590–1592. (in Russian) (1986)

35. Palatnik, L.S., Fuchs, M.Ya., Kosevich, V.M.: Formation Mechanism and Substructure of Condensed Films. Moscow: Nauka, 318 p. (in Russian) (1972)
36. Umansky, Ya.S., Skakov, Yu.A., Ivanov, A.N. et al.: Crystallography, X-ray Diffraction and Electron Microscopy. Moscow: Metallurgiya, 632 p. (in Russian) (1982)
37. Kostrzhitsky, A.I., Karpov, V.F., Kabanchenko, M.P., Solovieva, O.N.: Operator's Handbook for Vacuum Coating Plants. Moscow: Mashinostroyeniye, 176 p. (in Russian) (1991)
38. Maurice, F., Meny, L., Tixier, R.: Microanalysis and Scanning Electron Microscopy. Moscow: Metallurgiya, 392 p. (in Russian) (1985)